Technological Determinism and Social Change

Communication, Globalization, and Cultural Identity

Series Editor: Jan Servaes, City University of Hong Kong

The *Communication, Globalization, and Cultural Identity* series explores and complicates the interlinked notions of "local" and "global" by integrating global dependency thinking; world-system theory; local, grassroots, interpretative, and participatory theory; and research on social change.

In the current world state, globalization and localization are seen as interlinked processes, and this marks a radical change in thinking about change and development. It also marks the arising of a new range of problems. One of the central problems is that the link between the global and the local is not always made clear.

The debates in the field of international and intercultural communication have shifted and broadened. They have shifted in the sense that they are now focusing on issues related to "global culture," "local culture," "(post)modernity," and "multiculturalism," instead of their previous concern with "modernization," "synchronization," and "cultural imperialism." With these new discussions, the debates have also shifted from an emphasis on homogeneity towards an emphasis on differences. With this shift towards differences and localities, there is also an increased interest in the link between the global and the local and in how the global is perceived in the local.

Titles in Series

Technological Determinism and Social Change: Communication in a Tech-Mad World, edited by Jan Servaes.

Technological Determinism and Social Change

Communication in a Tech-Mad World

Edited by
Jan Servaes

LEXINGTON BOOKS
Lanham • Boulder • New York • London

Published by Lexington Books
An imprint of The Rowman & Littlefield Publishing Group, Inc.
4501 Forbes Boulevard, Suite 200, Lanham, Maryland 20706
www.rowman.com

Unit A, Whitacre Mews, 26–34 Stannary Street, London SE11 4AB

The cover image was taken by Rico Lie on a Samsung smartphone on the MTR in Hong Kong.

British Library Cataloguing in Publication Information Available

Library of Congress Cataloging-in-Publication Data Available

ISBN 978-0-7391-9124-8 (cloth : alk. paper)
ISBN 978-0-7391-9125-5 (electronic)

Printed in the United States of America

Contents

List of Tables and Figures

TABLES

FIGURES

List of Acronyms

AAUFF	All About Us Film Factory
AIDS	Acquired Immune Deficiency Syndrome
BBS	Bulletin Board System
BCC	Behavior Change Communication
BRAC	Bangladesh Rural Advancement Committee
CCTV	China Central Television
CfD or C4D	Communication for Development
CDSC	Communication for Development and Social Change
CIU	Compulsive Internet Use
CNNIC	China Internet Network Information Center
CSC	Communication for Social Change
CSO	Civil Society Organization
CSSC	Communication for Sustainable Social Change
CSSC&D	Communication for Sustainable Social Change and Development
DVD	Digital Versatile/Video Disc
ECA	Ethnographic Content Analysis
EFTA	European Fair Trade Association
FAO	Food and Agriculture Organization
FGD	Focus Group Discussions
GAID	Global Alliance for ICT and Development
GDP	Gross Domestic Product
GED	Gender Development Index
GEM	Gender Equity Measure
GFATM	Global Fund to Fight AIDS, Tuberculosis and Malaria
GLTB	Gay, Lesbian, Transgender and Bisexual
GNH	Gross National Happiness
GNP	Gross National Product

GSE	General Self-Efficacy
HDI	Human Development Index
HIV	Human Immunodeficiency Virus
HIV/AIDS	Human Immunodeficiency Virus/Acquired Immune Deficiency Syndrome
HPI	Human Poverty Index
ICT	Information and Communication Technologies
ICTD	Information and Communication Technologies for Development
IM	Instant Messenger
IMF	International Monetary Fund
INEXSK	Infrastructure, Experience, Skills, Knowledge
IT	Information Technology
ITU	International Telecommunications Union
KAP	Knowledge, Attitude, and Behavior
MDGs	Millennium Development Goals
M&E	Monitoring and Evaluation
NEWS	Network of European Worldshops
NGO	Nongovernmental Agency
NICT	New Information and Communication Technologies
NWICO	New World Information and Communication Order
OLPC	One Laptop Per Child
PAR	Participatory Action Research
PBS	Public Broadcasting Service
PEPFAR	United States of America President's Emergency Plan for AIDS Relief
PPP	Public–Private Partnership
PSA	Public Service Advertisement
PV	Participatory Video
PVR	Participatory Video Research
R&D	Research and Development
RME	Research, Monitoring and Evaluation
RTC	Right to Communicate
SDGs	Sustainable Development Goals
SCOT	Social Construction of Technology
SPOT	Stichting Promotie Televisiereclame/Foundation for the Promotion of TV-advertising
STI	Sexually Transmitted Infections
TB	Tuberculosis
UDHR	Universal Declaration of Human Rights
UN	United Nations
UNAIDS	Joint United Nations Programme on HIV/AIDS

UNCDP	United Nations Capital Development Programme
UNCSTD	United Nations Commission on Science and Technology for Development
UNCTAD	United Nations Conference on Trade and Development
UNDP	United Nations Development Programme
UNESCO	United Nations Educational, Scientific and Cultural Organization
UNFCCC	United Nations Framework Convention on Climate Change
UNFPA	United Nations Population Fund
UNICEF	United Nations Children's Emergency Fund
UNIFEM	United Nations Development Fund for Women
USAID	United States Agency for International Development
WB	World Bank
WCC	World Council of Churches
WFTO	World Fair Trade Organization
WHO	World Health Organization
WTO	World Trade Organization
WWF	World Wildlife Fund

Introduction to the 3 A's

Technology is Great!?

Jan Servaes

"Technology is Great" was the first thing I saw when arriving at a *Forum on the Arts and the Cultural Sector in a Digital Age*, organized by the British Council, in collaboration with the Arts & Humanities Research Council, the National Endowment for Science, Technology and the Arts, and the School of Creative Media at the City University of Hong Kong in March 2013. Approaching the poster a bit closer it turned out that there was "Britain" added in small print. "Technology is Great Britain." This smart or slickly devised advertisement, sponsored by the UK Trade & Investment Department, claimed that "90% of the world's mobile phones are powered by British technology from ARM," a leader in microprocessor Intellectual Property. Since, I have seen the slogan displayed at other conferences and trade shows in the world.

Is technology great? A recent Pew Research survey (Smith, 2014) on *US Views of Technology and the Future* finds that most Americans (59%) believe that the technological developments of the coming half-century will have a net positive impact on society. However, 39% think these changes will lead to a future in which people are worse off than they are today. "In the long run, Americans are optimistic about the impact that scientific developments will have on their lives and the lives of their children—but they definitely expect to encounter some bumps along the way," writes Aaron Smith (2014, pp. 5–6), the author of the report. "They are especially concerned about developments that have the potential to upend long-standing social norms around things like personal privacy, surveillance, and the nature of social relationships."

Similar concerns lead to the organization of a panel discussion on *Technological determinism and Communication for Sustainable Social Change* during the London conference of the International Communication Association in June 2013. The aim of the panel was to shed new light on a theoretically and practically significant issue in communication studies: technological determinism and communication for sustainable social change. The objective was to provide insight into the role of technology and culture in social change, an issue that has been increasingly central to communication for social change theories and practices around the world over the past decades.

The technological deterministic approach played a significant role in the so-called modernization perspectives (Servaes, 1999). It saw technology as a value-free and politically neutral asset that can be used in every society and historical context as a driving force of social change.

At the risk of oversimplification, we could take Karl-Erik Rosengren's typology (1981) as our starting point. Rosengren returns to the old but still basic question of the relationship between culture and social structure. He distinguishes between two extremes: "social structure influences culture" versus "culture influences social structure." For our purpose, we read social structure as social change, and culture as communication (including technology). Next Rosengren puts these basic propositions in a cross-tabulation in order to classify the four main types of the relationship: autonomy, technology creates social change (idealism), social structure creates communication technology (materialism), and the interdependence or mutual influence between both (cf. Table 0.1).

So far the discussion on the relationship between social change and communication technology concerned mainly the axis materialism–idealism. But it may well be, Rosengren argued, that in the coming years the focus of attention will be that of interdependence–autonomy, and that it will be moving from a holistic, overall perspective to a more differentiated one. And, indeed, newer theories and practices embrace a more contextualist and multiplicity approach in which the relation between particular technologies and social change should be examined and understood in specific cultural, economic, and political contexts (Servaes, 1999, 2013). Most of the contributions in this book further elaborate, expand, and redefine these tensions between information and communication technologies and social change.

Though the above may be an assessment of the discussion at an academic and theoretical level, the more applied and professional world continues to

Table 0.1 Types of Relations between Communication Technology (ICT) and Society (Social Change)

		Social change influences communication technology (ICT)	
		YES	*NO*
Technology Influences	YES	**Interdependence** (two-way influence)	**Idealism** (strong ICT influence)
Social change	NO	**Materialism** (ICT are dependent)	**Autonomy** (no causal connection)

embrace the technological deterministic perspective. Christine Ogan's chapter in this book and the timely article she coauthored with some of her graduate students in 2009 reminds us of these ongoing strong dominant beliefs in "technology is great!": "We believe that the more recent attention to ICTs has to do with the constant search for the magic solution to bringing information to people to transform their lives, allowing them to improve their economic condition, educate their children, increase literacy and the levels of education and spread democracy in their countries. Despite years of research that tells us that information is necessary but insufficient to bring about this change, ICTs have become the most recent iteration of the holy grail for development. And even if communication scholars know better because critical scholarship written over the last 30 years has told them so, newcomers to this field from other information-based disciplines may not have such close acquaintance with that literature. Furthermore, because of the appeal of the modernization paradigm, there is a tendency to forget that it cannot work" (Ogan et al., 2009, pp. 667–8).

The London gathering stimulated interesting discussions from both theoretical and applied perspectives. All participants agreed that, given the variety and depth of challenges in sustainable social change, both researchers and practitioners need to espouse a broad and contextualized understanding of the role of technology in social change that transcends conventional technological deterministic approaches. In order to reach a wider audience and invite more contributions and perspectives, a book project was considered the logical next step. The result is the book you are currently reading.

The book is structured around three A's. The first section provides some "Appetizers" and general introductions; the second section develops the more substantial "Arguments"; while the third section attempts to assess and broaden these in "Applications" and case studies. The conclusion provides a brief summary with recommendations for further study.

APPETIZERS

Christine Ogan opens this section with a quote from the late Rob Kling, a visionary in his writing on social informatics. In 1989 Kling wrote, "One of the fascinating and important sociological questions surrounding computerization is the extent to which the use of computer-based systems really transforms any part of the social order—and if so, how?" These words were part of his plenary address to the *Society for Social Studies of Science*. In a later article he further articulated his views and research on the utopian versus dystopian views of the role of technology in society. Yet more than 20 years later, Ogan observes, scholars and practitioners are still arguing about technology's role in social change in this deterministic way. In the 2009 meta-analysis of writing on the role of communication in development over a 10-year period, she found that technology (e.g., access to the internet) still appeared to be driving the discourse on social change. Though this research addressed mostly poor people in developing countries, digital inequality still exists in the more developed parts of the world as well. In recent work coming out of the EU Kids Online project, Ogan found that one key to closing that divide for ethnic minorities and even majorities in one or more countries in the study was self-efficacy, not higher socioeconomic status or other variables. The large number of participants in the study provide some solid evidence for understanding what is most important to improving children's skill levels in using ICTs.

Next Rich Ling examines the idea of ubiquitous social expectations based on Weiser's notion of ubicomp. He positions ubicomp in the context of networked mediation technologies. A central function of these technologies is to allow people to perpetually be in contact with one another. As these technologies gain a critical mass in society we increasingly expect them of one another. Thus, these mediation artifacts are not simple devices that facilitate our individual lives, they are the basis of reciprocal expectations. Therefore, Ling argues, as this process continues, we increasingly do not have the option of opting out. Rather we are, to one degree or another, compelled to have them.

The chapter by David Morley surveys perspectives on technological change and globalization which might be seen as "media-centric" in respect of their assumptions about the extent to which media technologies themselves are necessarily the driving force of cultural and political change. It argues for the need to situate contemporary forms of technological "newness" in historical perspective, if we are to avoid a narrowly foreshortened form of "presentism." It also offers an alternative, "contextualist" view of how we might analyze the ways in which particular technologies are mobilized in different cultural contexts and considers how, rather than focussing on "universalized" models of technologically driven change, we might develop comparative forms of analysis informed by anthropological and comparative perspectives on technology use.

Patchanee Malikhao starts from a slightly different position. Her chapter concerns the critical sociological studies of mediated culture, intercultural communication for development, technology and Buddhism. Critical sociology aims at studying how the meaning, values, attitude, norms, belief or disbelief systems, and worldview of the audience in a given postmodern culture change due to the messages perceived from media institutions. She complements this from a Buddhist phenomenological perspective in order to find strategies to alter the unfavorable changes embedded in media technology and media institutions to a more sustainable orientation. Phra Dhammapidok (P.A. Payutto), a famous Thai monk, argues that the root cause of unsustainable development lies in the lack of ethics. Sustainable development and social change is impossible without a change in the way of thinking that mankind can steward nature by using technology. Technology has become a tool of the cult of success, implying the dwelling on growth and progress in line with the modernization ideology. To achieve desirable social change for sustainability, according to Buddhists, is to develop "*evolvability*," or the potential to live in harmony with other human beings, wildlife, and nature. This needs a redefinition of "Western ethics." Intercultural communication for a sustainable social change and the competencies associated with an intercultural sensibility are presented as possible ways out of the stalemate.

ARGUMENTS

Colin Sparks presents the first of the more developed arguments on technological innovation and social change. The enthusiasm generated by the application of new communication and information technologies to issues of social change is considerable. These technologies, through empowering individual social communication on a wider scale and through the greatly reduced entry barriers to their implementation, most certainly contain the promise of more equitable and dialogic communication. While there remain serious problems of access, and even more of access quality, it is nevertheless clearly the case that mobile technologies have made at least some forms of connectivity, a reality even in situations that were apparently unpromising.

There are, however, good grounds for expressing some skepticism toward some of the claims commonly made in this context. The issues of equity and empowerment, and thus ultimately those of "development" or social change in anything other than the crude and unilinear definition of raising economic productivity and embedding Western modernity, are not primarily technological. As many people have noted, the key issues at stake are those of usage, and thus of human agency rather than technical affordance. It needs to be remembered, however, that human agency and the practices that it entails, are not the result of individual choices but are deeply inflected by social structures. In this respect, the implementation of new communication and information technologies in a developmental context implies a confrontation with those intractable issues that have in the past proved effective barriers to the liberatory potential of earlier technologies. There is, indeed, even some evidence that the conditions upon which the widespread utilization of the potential of these new technologies—notably a relatively educated, self-confident and socially equal user base—are ones that run directly counter to the needs of more traditional drivers of development like industrial development, infrastructural construction and economically driven mass migration. Therefore, Colin Sparks takes the view that while technologies might provide new means of self-organization (alongside enhanced means of surveillance) the *issues of development and human liberation ultimately remain questions of politics and economics*.

Marko Skoric, Yong Jin Park and Yalong Jiang empirically examine the relationship between dominant cultural values of a society and the level of ICT development, with an aim of uncovering the process of cultural shaping of digital networks. They also analyze the links between ICT development and use, and the establishment of robust democratic institutions and practices. Their analyses show that neither cultural values nor technologies could be considered neutral or inconsequential with regard to sustainable political development; instead, they have specific and predictable implications for democracy. They find that ICT development mediates the relationship between certain cultural values and democratic development, suggesting that *culture acts as a catalyst for technological progress*, which in turns promotes the development of democratic institutions and practices.

The broad question Jo Tacchi addresses in her chapter is "what makes mobility, particularly related to mobile phones, meaningful?" She argues that, while mobile phones can be seen as active agents and facilitators of change or development, they must be understood within specific uses and settings in order to understand the part they play in social and economic meaning making. She wants to explore the concept of *meaningful mobilities* further in order to challenge two types of determinism in relation to technology and development.

First is technological determinism and the need to balance what she calls *contextualized affordances*. The second is a form of *cognitive determinism* that shapes the field of communication for development and social change and the types of research questions we ask.

In relation to the first determinism, she argues that there are at least two important things that need to be taken into account when considering technological determinism within a communication for social change agenda: the affordances, or constraining and enabling material possibilities of media technologies; and the contexts. By way of example Tacchi discusses the implications of mobile phones and disadvantaged women in India.

In the second determinism she explores how disciplinary and other fields can be seen to frame and in some cases over-determine what questions are asked when we think about technology and development. Here she enters the roam of framing; how development and other powerful structures and systems can approach mobile phones to frame perceptions of groups of people in society and therefore highlight and prioritize particular characteristics and differences over others.

In the concluding part Tacchi discusses what these examples tell us about technological determinism and media and communication for development and social change, and how the concepts of contextualized affordances and meaningful mobilities (when appropriately framed) can help us to move forward.

The core aim of Emily Polk's research is to explore the communication processes of community-led responses to the current global financial and climate crisis via an analysis of *The Transition Network*, an online network comprised of over 1000 initiatives in 34 countries that have used the Transition model to start a variety of projects aimed at achieving greater sustainability. This research seeks to better understand how global digital networks are being used to facilitate local communication processes around sustainable social change. Transition builds upon and has developed alongside similar social movements that have been mobilized in part due to digital technology and social media, like the Occupy Wall Street movement and the Arab Spring, both of which used digital media to communicate and coordinate their actions. The emerging social movements share primary characteristics in relation to their communication online: (1) They have been and are mobilized and coordinated online, yet remain decentralized and participatory, (2) They suggest that they are not just trying to achieve new goals but rather shift the processes by which the goals are reached, (3) They have cultivated numerous actor networks that recognize the necessary relationship between the physical public sphere and the virtual public sphere with regard to mobilizing action.

Polk first questions the ways in which the Transition movement used digital technology, specifically its Transition Network, the online hub of the movement and other social media such as Facebook and Twitter. Second, she tries to understand when, and how and in what ways Transition Amherst used the online Transition Network to support its own activities.

Her data suggest that the Transition movement supports *glocalization processes* by strategically using digital technology, specifically the Internet, for two primary reasons: (1) To transmit information on multiple levels to aid local Transition town initiatives, and (2) To maintain control of its branding and sustain its legitimacy and authority as a global movement toward community resiliency.

At a more critical level she suggests that there is an important connection to be made between the digital facilitation of social movements like Occupy, the Arab Spring and Transition and the *leaderless* ethos that inform an important part of their identities. She argues that strong leadership development is crucial to sustaining a long-term movement, and the valorizing of the leaderlessness of these initiatives is ultimately to its detriment.

John Hartley focuses on the methodology of *digital storytelling.* Digital storytelling is an international movement for self-representation and advocacy, especially in educational, arts, and therapeutic communities. It has begun to attract a significant body of scholarship, with Australia an important player. Digital storytelling is a well-established technique for combining expert facilitation with bottom-up creative expression, often with social change objectives in mind for minority or vulnerable identities.

Hartley explores some of the issues that have emerged for activists and scholars, including the problem of how to "scale up" from self-expression to communication (i.e., self-marketing), and the question of the role that stories play in constituting "we"-communities.

New communication technologies don't determine their own uses; they are accompanied by informal social-learning efforts to enable ordinary people's participation. But how does individual self-expression articulate with large-scale technological systems? And how do existing institutional forms—such as community arts and community media—accommodate to new technologies?

The chapter pursues the relationship between storytelling and political narrative over the extreme long term, using well-known and lesser-known connections between Australia and Turkey to tell the tale. It considers how digital self-representation intersects with that political process, and what activists need to know in order to intervene more effectively.

APPLICATIONS

These theoretical and methodological considerations are further explored and complemented with more findings and cases in the third section of this book.

Karin Wilkins starts with the by now classic case of the *Grameen Bank* and its micro-financing programs for poor Bangladeshi women. Ideally it has always been claimed that microenterprise programs designated toward improving women's economic conditions intend to create opportunities that will live on after the project itself has ceased its funding, by providing loans to collectives and supporting skills and education to individuals. In this chapter, Wilkins considers the extent to which those programs that focus on communication technologies, such as cell phones and computers, as part of microenterprise projects, have demonstrated sustainability over time.

She contends that participatory processes, while critical in terms of promoting ethical imperatives for engagement, are *not* enough. Strategic intervention must now recognize the importance of creating sustainable social change. While projects that offer education and loans to women have that potential, broader issues, such as patriarchy within households and economic crises within communities, nations, and global conditions, might constrain their ability for positive consequences. She considers the conditions that suggest more sustainable paths in this sector, based on an empirical review of the available literature, to arrive at identifying the limitations of a neoliberal development agenda.

The contribution by Ellen Hommel, Rico Lie and Anneke Smelik explores the *impact of visual technology on youth participation*. They look for benefits of the use of visual technology in education and the implications for contemporary adult educators. While developing their arguments they empirically look at two specific projects, which use participatory video as an intervention method with youth in order to strengthen participation.

First they sketch the characteristics of multimedia with the fast development of new media technologies and the vast impact of media in contemporary culture. Then they discuss the impact of technological convergence on youth culture. They argue that youth participation, defined as active citizens' involvement, is a *prerequisite* for learning and for sustainable change. In the third part they discuss in more detail participatory practices that are embedded and implemented in everyday life. They select two cases from the Netherlands, exploring the conditions of participatory video for strengthening youth participation, and focusing on the added value of visual media. In the final part of their chapter they discuss the role that adult professionals—as filmmakers, facilitators and mediators—play in these projects, using the theoretical work of John Dewey and Paolo Freire.

Melissa Brough's study investigates the relationship between *digital communication, participation and social change* in Medellin, Colombia, a city that has been recognized across Latin America for its digital initiatives. Through two distinct case studies—a bottom-up, community media project, and a top-down, government initiative—this chapter explores the social implications of using digital technologies to promote civic/political engagement and development, and interrogates the conflation of digital citizenship and participation. The ideologies, practices, and technology choices of these two initiatives reflect different conceptions of digital citizenship and participation, resulting in distinct communicative architectures and practices, all of which have fundamental implications for how we understand social change in the digital age. Contrasting them highlights four key dimensions for analysis to avoid technological determinism: horizontality, dialogical communication, openness, and autonomy.

Brough concludes that while digital communication technologies are widely seen as affording more citizen participation than traditional media, this is clearly not guaranteed; it depends greatly on the surrounding cultures and practices of use (including *digital literacies*), as well as on the structure and management of the communication platforms.

Valentina Baú positions her analysis of a participatory video project against the aftermath of the 2007/2008 postelection violence in Kenya. She explains how, through the application of participatory approaches to video-making, a process of appropriation of video technologies occurs within communities who want to tell their own story. The possibility of using video to facilitate communication between enemy groups, victims and perpetrators, can open the path towards social change and sustainable peace.

She argues that, in contrast with theories of technological determinism, rather than lessening human agency, technology here empowers people by offering a channel for dialogue. A participatory approach also allows for flexibility in the use of the medium; this prevents the product from being defined by its own technology's structure. Within this context, technology is neither to be viewed as an external element that is forced upon a group, nor as a value-free asset as it is still being regarded in certain practices, but as a new tool that allows communities affected by conflict to take ownership of their own development through communication.

Last, Song Shi, in the case of *Tiger Gate*, explains how ICTs empowered activists were engaged in social change actions to increase government accountability and transparency in China. The Tiger Gate event was a nationwide grassroots ICT for social change action that questioned the authenticity of a group of wild tiger photos published by government agencies in China.

The chapter explores how collective bottom-up actions launched by grassroots activists connected and empowered by ICTs could increase government accountability; and how the collective intelligence of the networked activists played a corrective role for the government policy-making. Shi argues that rather than a powerful driver of social change (a technology deterministic understanding of the relation between ICT and social change), in this case, ICTs are tools that may provide a new potential for combining the information embedded in an ICT system with the knowledge embedded in people.

By way of conclusion we will add some more concerns to the opening question in the book: Is technology great? Or, rather, do people matter more?

ACKNOWLEDGMENTS

Most of the chapters in this volume were first presented at the ICA Conference of June 2013 in London. In order to further complete the different ways of approaching this "wicked" problem of technological determinism and social change, three more contributions were invited. Two articles have been originally published in the Elsevier Journal *Telematics and Informatics. An interdisciplinary journal on the social impacts of new technologies*, of which Jan Servaes is the Editor-in-Chief:

Rich Ling, From ubicomp to ubiex(pectations), *Telematics and Informatics*, Volume 31, Issue 2, May 2014, Pages 173–183, ISSN 0736-5853, http://dx.doi.org/10.1016/j.tele.2013.09.001.

David Morley, On living in a techno-globalised world: Questions of history and geography, *Telematics and Informatics*, Volume 30, Issue 2, May 2013, Pages 61–65, ISSN 0736-5853, http://dx.doi.org/10.1016/j.tele.2012.08.001.

Rich Ling and David Morley's articles are here reproduced with the formal permission of Ms. Laura Pritchard, Permissions Supervisor at Elsevier's Oxford office.

Chapter 9, "A Trojan Horse in the City of Stories? Storytelling and the Creation of the Polity," by John Hartley, is modified from Cultural Science. © John Hartley and Jason Potts, 2014, Cultural Science: A Natural History of Stories, Demes, Knowledge and Innovation (pp. 34–76), published by Bloomsbury Academic, an imprint of Bloomsbury Publishing Plc. Reproduced under license.

Marko Skoric and his colleagues preferred to revisit their original article and have written a new piece for this book.

We also wish to acknowledge and thank Ms. Julia E. Kirsch, Vice President/Publisher, and Ms. Sabah Ghulamali and Mr. Brian Hill, Assistant Editors, at Lexington Books, A division of the Rowman & Littlefield Publishing Group, for their support and assistance.

REFERENCES

Ogan, C., M. Bashir, L. Camaj, Y. Luo, R. Pennington, S. Rana, and M. Salih (2009). "Development Communication: The State of Research in an era of ICTs and Globalization," *International Communication Gazette*. 71(8), 655–670.

Rosengren K.E. (1981), "Mass media and social change: some current approaches," Katz E. & Szecsko T. (Eds.), *Mass media and social change*. Beverly Hills, CA: Sage.

Servaes, J. (1999). *Communication for Development. One world, multiple cultures*. Cresskill, NJ: Hampton Press.

Servaes, J. (Ed.) (2013). *Sustainability, Participation and Culture in Communication. Theory and Praxis*. Bristol-Chicago: Intellect-University of Chicago Press.

Smith, A. (2014). *U.S. Views of Technology and the Future*. Washington DC: Pew Research Center. http://www.pewinternet.org/2014/04/17/us-views-of-technology-and-the-future/

Part I

APPETIZERS

Chapter 1

The Same Old Theory, the Same Old Quick Fix

But Please Add Efficacy to the Mix

Christine Ogan

My friend Rob Kling, who died too soon at the age of 58, was a visionary and a brilliant analyst on the place of information technology in society. He was fond of locating examples of technological determinism in popular discourses in the mass media, referring to them as utopian and dystopian discourses.

> Technological utopianism does not refer to a set of technologies. It refers to analyses in which the use of specific technologies plays a key role in shaping a utopian social vision, in which their use easily makes life enchanting and liberating for nearly everyone. In contrast, technological anti-utopianism examines how certain broad families of technology facilitate a social order that is relentlessly harsh, destructive, and miserable. (1996, p. 42)

He characterized writing about technology in utopian or dystopian terms as a type of limiting genre that prevented the authors from discussing important social relationships related to the positioning of technology within a society (1996). Kling's sociological interest in technology discourses focused on computerization, which he thought of as a social movement in which groups formed around a computer technology with shared ideological beliefs (Kling, 1991, p. 354). In research that he conducted with Iacono (see also, Iacono and Kling, 1996), Kling identified these beliefs:

1. Computer-based technologies are central for a reformed world
2. The improvement of computer-based technologies will help reform society
3. No one loses from computerization.
4. More computing is better than less, and there are no conceptual limits to the scope of appropriate computerization.
5. Perverse or undisciplined people are the main barriers to social reform through computing (1991, p. 355).

Kling was focused mostly on computing and technology use in developed countries, not in less economically well-off parts of the world. And his work in social informatics grew out of his background in computer science, not communications. So it is striking how much of his analysis of computing in society applies equally to the parts of the world that are striving to become networked and to apply every kind of computing application to the efforts to improve their economies, their educational systems, and other areas of their development.

LEAPFROGGING AS DETERMINISM

One way in which technology was supposed to address communication problems was in its ability to leapfrog over existing technologies that were more difficult to disseminate to remote places and to disadvantaged potential users. Leapfrogging was the perfect technologically deterministic idea in which the new technology would eliminate the need for infrastructure and likely not require great public investment or new public policy to implement. In a 2000 article by Davison, Vogel and Harris, the authors discuss the use of cellular phones as leapfrogging devices being used in China. They note that:

> while leapfrogging may appear as an attractive option for the late adopters, it may not provide the intended results in all circumstances. The greatest danger is that a new "cargo cult" may arise, where the developing economies observe the benefits which later and succeeding generations of IT bring to the industrialized nations. Hurrying to acquire the same technology, developing countries rely on the blind belief that similar benefits will quickly accrue to themselves. (p. 3)

One of the earliest leapfrogging communication technologies was the transistor radio. With no need for electricity or electrical outlets to power it, messages from the capitals and other large cities, where radio stations were located, could be disseminated to anyone in nearly any location. However, the need for batteries, which were expensive and not widely available meant that the radios might have limited use. The small size allowed them to be transported anywhere, but the sound quality was not always so good. Of course the radios could not solve the problems of development, even though listening required no literacy skill, because of the one-way nature of the communication and the inability of the sender to know how the message was being received and what barriers might be involved in putting the message to use.

Other communication technologies followed on the transistor—and many were powered by the transistor technology (portable CD players, personal audio players and boom boxes). The introduction of each one was heralded with a deterministic message about the magic bullet solution to development problems. CD-ROMs would solve the problem of getting vast amounts of information to the people cheaply. Where libraries in tropical environments were impractical or expensive, the CD-ROM would not carry the high cost and storage issues of books. Low orbiting satellites (or LEOs), another *deus ex machina*, could bring digitally delivered information to remote areas without electricity and allow people to send packets of information back to major urban centers on a twice-per-day basis. Audio cassette recorders were documented to have brought down repressive governments in the Soviet Union and Iran (see for instance A. S. Mohammadi and A. Mohammadi, 1994) and also used to bring personalized development lessons to women who came to community wells to get water or gathered in someone's home for a social occasion.

One of the largest scale leapfrogging projects was designed by Nicholas Negroponte from the MIT lab. A simple computer called the XO and billed as the "one laptop per child" solution to educating children in developing countries has been described by its originators as the "size of a small textbook" with "built-in wireless and a unique screen that is readable under direct sunlight for children who go to school outdoors," and "extremely durable, brilliantly functional, energy-efficient, and fun." Also described as a "real world laptop for real world change," it is said to be the result of a design by "experts in academia and industry," "a product of the very best thinking about technology and learning" (laptop.org, 2014).

Unlike many of the other technologies that only require individuals to make decisions about adoption, and which can catch on with others by word of mouth, the XO computer adoption requires huge numbers of people to buy in to make it a success.

> The scale of the operation is immense, simply because the XO is meant to enroll for—profit manufacturers in the quest for a new goal—low prices and a stable set of features. But because of the need for scale, governments, and more specifically education bureaucracies, also become essential members of the XO network as they are only institutions capable of efficiently distributing the machine on such a scale through the already existing network of schools that they control. It is here, however, that the OLPC [One Laptop Per Child] Project appears to be running into problems as their own constructivist educational philosophy is difficult to fully reconcile with the need to enroll teachers in the project as well as children. (Luyt, 2008)

Accompanying the scale problem, social forces, such as the changing nature of global capitalism, the desire for technological salvation, and the "prospect of ruthless competition from for-profit firms anxious to establish their dominance in an industry that would view laptops as essentially mass commodities" could prevent the leapfrogging effect Negroponte envisioned (Luyt, 2008).

In all the cited examples, technological determinism was a key part of the narrative that attempted to convince people who are eager, and even desperate, for a simple fix for solving the problems of world poverty and economic development. In her analysis of the discourses involving leapfrogging technologies, Wilson (2002) writes:

> in many countries the factors needed for such an approach to be successful are not present. Only when understandings of development as catch-up to western ideals are questioned are alternative approaches with alternative outcomes, which are likely to be more relevant in a developing world context, able to emerge. Further . . . the availability of technology cannot be equated with access and use since both are intricately woven within the local sociotechnical system including the political and economic constraints facing developing countries. (p. 90)

It seems that even when we are aware that sociotechnical systems are essential for supporting the adoption and widespread use of technological applications in development, the idea that this time will be different and this technology will bring dramatic change is seductive. Several studies I next describe are illustrative.

TECHNOLOGY'S ROLE IN DEVELOPMENT HISTORY

Though the early work in communication and development by Everett Rogers (1962, 1971, 1976), Wilbur Schramm (1964). Daniel Lerner (1958) and others claimed a causal role to technology in development, Rogers later rejected that argument in his 1976 article on the passing of the dominant paradigm. On the role of capital-intensive technology as a factor propelling development, he said:

> More developed nations possessed such technology. Less developed nations had less of it. So the implication seemed plain: Introduce technology to the less developed countries and they would become more developed too. It was assumed that appropriate social technology would appear to accompany the externally introduced material technology. When the needed social structures did not always materialize in less developed countries, the fault was accorded to "traditional" ways of thinking, beliefs, and social values. (p. 215)

Given the great stature of Rogers in the academic community, this article, and indeed the whole issue of *Communication Research* where the critical article appeared might have influenced the direction of academic research in communication's role in development. And indeed it did—at least for a short while. Around this time, communication researchers were beginning to look at alternative models for development—models that focused on the structural problems in developing countries and on dependency relationships with the West or with the East, as well as the shortcomings of the diffusion of innovations approach. These critical perspectives generally had a more limited view of a positive role for communication technologies in development efforts (see James, 2005, or Servaes, 1999). At best, they discouraged technology transfer and looked to appropriate (usually smaller, less expensive and less complicated) technologies that better suited the needs in less developed environments (see B. Eres, 1981; Hudson, 1979 for discussions of this movement).

The passing years have brought with them a changed world now focused on policies and practices of globalization, and most of the research of communication's role in development is now couched in terms of globalization. Meta research of the trends found in the literature on communication and development have revealed patterns the scholars' topical focus, use of particular theories, and perspectives on technology's role in "lifting countries out of poverty" in the twenty-first century, as the current saying goes. Three meta-analyses of the major work in communication and national development have been conducted that span the years from 1958 to 2007 that document the trends in this research over that period (Fair, 1989; Fair and Shah, 1997; Ogan et al., 2009).

When we examine that research for evidence of technological determinism, we find that scholars are still looking for a magic wand to be waved and communication technologies to bring positive results. In the first study conducted by Fair, 207 journal articles, books, book chapters and dissertations were examined that spanned the years from 1958–1986 (1989). Given that the criticisms of the modernization (or dominant) paradigm were being leveled beginning in the 1970s, and particularly beginning with the previously mentioned issue of *Communication Research* in 1976, Fair broke down her analysis into two periods (1958–1975 and 1976–1986). The first thing to be considered is that mass media technologies were used as variables irrespective of their content, meaning that the very existence of a media technology was considered enough to cause development. Fewer than 18% of the all the studies (in both periods) examined content (p. 142) and even fewer (about 5%) analyzed the quality or relevance of that content.

Of the articles in the entire time period that were theoretically based, 30.8% tested Lerner's modernization model—a model that attributed powerful effects to the mass media (p. 137). In this model, media were said to have a direct and positive impact on development. Though Fair found that the number of articles using Lerner's model had declined in the second period over the first, she concluded that "the field has continued on the tradition of the modernization paradigm" (p. 144). This paradigm could be said to provide a positive and technologically deterministic role for the mass media (primarily television and radio) in development as Fair pointed out.

> The media, of course, were key technologies, facilitating the bridging of traditional and modern sectors. By adjusting the attitudes and behaviors of individuals in developing societies to those seen as consistent with individuals in the developed world, researchers were making and continue to make normative judgments. (1989, p. 146)

More evidence of a technologically deterministic perspective was found in the examination of the independent variables in the studies that were based on quantitative empirical methods. Overall 62.8% of such studies used mass media technologies as independent variables (p. 139). Fair expected to see a major decline in such studies in the second time period, following the critical assessments appearing in the mid-1970s. Though some decline was found (from about 54% to 46%), media technologies continued to drive the design of the studies as playing the primary role in development outcomes. She also expected a corresponding increase in mass media as intervening variables in response to the literature critical of the modernization paradigm, but there was little change in that percentage—in both time periods it held to around half of all the studies with mass media as any type of variable (p. 140).

The second study by Fair and Shah (1997) continued the meta-analysis of communication and development research, examining 140 items of published work from 1987–1996, to determine the state of media effects research during that time period. During this time Fair and Shah document the attacks on the modernization paradigm arising from neo-modernization revisionist theory, dependency theory, world systems theory, and the postmodern critique that led to an approach called "another development." In other words, a general discrediting of modernization theory occurred over this time period, and the authors asked whether this would be reflected in the research on media's role in development.

Interestingly, the modernization paradigm disappeared from the literature during the 10-year period analyzed, while participatory theory (where those in developing countries are involved in the decision making related to solutions to their own development problems)—a particularly nondeterministic approach—took its place at the top (39.2% of the 45 studies that included theory). Fewer studies in this time period claimed that media have a direct role in individual modernization (39.2% vs. 17.6%) or in national development (39.2% vs. 10.6%) than previously. Surprisingly, the percentage of studies using media as independent variables increased from 62.8% to 74.3%

over time, with fewer studies (5%) placing media in an intervening variable location. Fair and Shah explain the latter finding by saying that "though researchers in the second period focused more exclusively on media as independent variables, at the same time they recognized that the media and development linkage is more complex, as revealed by the larger number of mediating variables used to explain the media effects relationship" (p. 14). They comment that except for the interpersonal and family relationships, that relationship was less often personal and more often structural" (p. 14), leaving us to conclude that the technologically deterministic thinking had largely passed from consideration by the researchers conducting these studies. When the consequences of media's role in development was discussed in the analyzed studies, the conclusion that "media activate modernity" had declined from 36.4% in the Fair (1989) study to 27.4% in the 1976 study, also leaving us hopeful about the conceptualization of media's role in the process in future studies.

However, the 1998–2007 focus on similar questions in another meta-analysis of the development and media literature (Ogan et al., 2009) had less optimistic findings in this regard. Modernization popped up again as THE major theme or paradigm used to frame the study (37.4% of the total). The reason for the resurgence of a deterministic approach, the authors believe, was the shift away from mass media effects to ICT (information and communication technology) effects in the studies. Of the 211 articles (no books or dissertations included), only 24 were published in the first 5 years of the study as mass media and development studies appeared to have gone out of fashion (p. 660). But interest in the Internet, mobile technologies, telecenters, etc.—all ICT-based—rose sufficiently over these 10 years to attract new research into the relationship between media technologies (now individual, one-to-one or one-to-many technologies) and development.

For the most part, the work on ICT4D (or Information and Communication Technology for Development) was not based on the several-year critique of mass media technologies in a deterministic role. Lacking this historical context, the new research viewed ICTs like the magic bullet once again. This can be seen in the subset of articles on ICT4D in the study as an even higher percentage—44.7%—focused their work within the modernization paradigm (p. 661). Interestingly this trend has been occurring in development and communication research, while other U.S. and Western-based research has moved way beyond any consideration of a hypodermic needle or magic bullet role for media.

Other indications that technology loomed largest in these studies is the amount of research that characterized media (either mass media or ICTs) as the "prime mover" in the development process (77.7%), while only 12.8% considered media as a complement to development and the rest assigned it a minimal or no role at all (2009, p. 664). As with mass media in previous decades, ICTs were frequently viewed as "having the power to leapfrog over other societal problems to achieve development goals—including the development of women, the obtainment of measurable increases in economic development and the adoption of more democratic policies" (p. 666).

DETERMINISM IN THE POLICY ARENA

As Christensen (2011) has illustrated, technologically deterministic assessments of the role of social media in the uprisings in North Africa and the Middle East in recent years have led some policymakers to believe that investments in ICTs and particularly social media can bring positive outcomes for democratic development to other countries (2011). Christensen's analysis of policy documents and publicly released statements by the Swedish government regarding the decision to allocate $50 million for Internet projects "in support of democratization and freedom of speech" in developing countries (p. 234) presents a convincing case for the ways in which stakeholders in Sweden assumed a causal relation between use of ICTs and social change (2011).

In Bangladesh's digital policy that aimed to employ ICTs to improve social equity, education, health care, and economic growth, Asaduzzaman (2013) found instead that it only benefitted those who were already privileged and that the marginalized populations were not able to attain any of the benefits. The analysis of the actual outcomes vs. the technologically deterministic goals of the policy brought greater inequality to the majority of the people (2013).

These examples support the now rather old study of Nulens and Van Audenhove (1999) that examined the ICT discourses resulting in policies of three major international development organizations—the International Telecommunications Union (ITU), the World Bank, and the United Nations Economic Commission for Africa (ECA). Using the terms technophilia and technophobia instead of utopian and dystopian, the authors detail the approaches of these NGOs. In the case of the World Bank "there is still a strong belief that ICTs can change the society" and in its general reports "technology is assumed to overcome all economic problems" (1999, p. 466). The authors' analysis of both ECA and the World Bank documents found also a focus on extra-technological issues, but "the impact of ICTs (as such) could have on Africa are believed to be tremendous" (p. 466). Though not uncritical of the use of ICTs, all three organizations were found to hold "a strong belief in the market-driven approach" where the "discourse is associated with privatizations, liberalizations and deregulations" (p. 466–67).

WHAT IS NEEDED?

In 2010 a group of researchers headed by Sonia Livingstone and Leslie Haddon conducted a 25-country study of children and adolescents' use of computers and the Internet, including social media across Europe. Turkey, a European developing country, was included in this study. Two of us on the research team (Leen d'Haenens and I) who have been interested in the communication patterns and technology uses of Turkish ethnic minorities in Europe decided to examine the subgroup of Turkish minority respondents in Europe and compare their computer-based competencies with the majority populations in their countries and analyze them alongside the Turkish respondents who live in Turkey (2013).

It was once thought that access to computers was the only requirement for closing what has been called the digital divide. That idea was, of course, borne of technological determinism. As Warschauer (2003) said, "it is clear that the underlying framework of a 'digital divide' is technological determinism—in other words, the view that the mere presence or absence of a technology has a determining affect (sic) on behavior and social development" (p. 297). Most of those writing about the solution claimed that technology would drive the change that allowed everyone to become computer literate, thus closing the divide. These days, however, most scholars refer to digital inequality, not a divide, and despite widespread access to computing and the Internet, inequality exists all over the world (cf. van Dijk, 2005; Hargittai, 2010; Hargittai and Hinnant, 2008; Warschauer, 2003a, 2003b; Zillien and Hargittai, 2009).

In our study we reasoned that such inequality would exist despite access to the Internet by all of our respondents. If a utopian (and deterministic) world existed we should find few, if any, differences in the skills of these children (aged 11–16), despite their geographic home or their ethnic minority status. Ethnic minority children were identified through a question on the survey asking what languages were spoken in the home. Only five countries had sufficient Turkish-speaking families among the respondents to be included in our study. Those were Denmark, the Netherlands, Germany, Austria and Bulgaria. Turkish-speaking children (n = 167) in the study were compared with the majority children in those countries (n = 4, 949) and Turkish children living in Turkey (1018).

Online competencies were measured through self-report of eight questions in the study.[1] We also asked questions about the frequency, location, and types of Internet use as well as several demographic questions. The findings showed that access was insufficient for achieving equal levels of reported competency when using the Internet.

> Majority European children in the five countries concerned have a markedly higher score than their peers in the Turkish diaspora and a very much higher one than children in Turkey. It is worth noting that only on the issue of finding information on how to use the internet safely do Turkish children believe they are better informed than their peers in the Turkish diaspora. On average, majority European children report that they have mastered 4.7 of the 8 online competencies, as against 3.2 for Turkish diaspora children ($t=6.8$; $df=3{,}818$; $p=0.000$) and 2.7 for Turkish children living in Turkey ($t=20.2$; $df=4{,}411$; $p=0.000$). Note the significant gap between the reported competences of Turkish diaspora children on the one hand and of Turkish children in Turkey on the other ($t=2.1$; $df=850$; $p=0.04$). Across the full set of reported online skills (the least frequent skill being the changing of filter settings (31%) and the most frequent (69%), the blocking of undesirable messages) the subjects' socio-economic background correlates with the reported number of skills; that is, the lower the social status of the child's family, the less varied the child's range of online skills. (d'Haenens and Ogan, 2013, p. 49)

Despite the self-report issues that always arise, the differences between the groups were large enough that we thought it worth pursuing in a regression analysis. In addition to the demographic variables, the frequency and amount of time spent online, the location of the computer in the home, the number of online activities for which the respondents reported participation, and the child's parents' frequency of Internet use, we also measured the child's self-efficacy through a scale adapted from Schwarzer and Jerusalem (1995).[2] Self-efficacy is defined by the concept's originator as "the belief in one's capabilities to organize and execute the courses of action to manage prospective situations" (Bandura, 1995, p. 2).

We believed that self-efficacy was an important variable differentiating the groups, and that proved to be the case. Though the correlational analysis found that socioeconomic status was significantly related to the number of skills reported, in the regression analysis, self-efficacy and to a lesser extent, age were the only variables that significantly predicted higher skill levels in all groups. Socioeconomic status was only predictive for the majority European children. Self-efficacy was much more predictive for the children in Turkey and the Turkish children in the diaspora (thought to be less privileged overall) than the majority children in Europe (d'Haenens and Ogan, 2013, p. 56).

This finding shows the power of self-efficacy, not access to technology or even high socioeconomic status, to increase perceived computing skill levels.

It provides some evidence to negate the utopian or dystopian deterministic perspectives we so often buy into when technology is introduced to a social system.

The children living in Turkey and in the diaspora in Europe perceived their skill levels as much lower than those of the European majority children. Bandura's research illustrates that these groups might face obstacles, such as discrimination or low expectations for performance, that undermine their ability to perform at higher levels.

> The mere presence of a highly confident individual undermines effective use of routine skills. Attending to what is strange in new tasks, rather than what is familiar and clearly within one's range of capability, may similarly hinder performance. And when people are cast in subordinate roles or are assigned inferior labels, implying limited competence, they perform activities at which they are skilled less well than when they do not bear the negative labels or the subordinate role designations. (Bandura, 1982, p. 142)

Self-efficacy as a construct has been tested cross-nationally to determine whether it has validity in cultures outside the West. The one analysis that included both developed and developing countries supported the "assumption that general perceived self-efficacy is a unidimensional and universal construct. The evidence of its universality originates from the assessment of GSE [General Self Efficacy] in 25 countries" (Scholz et al., 2002, p. 249). The authors' research replicated similar results by Schwarzer and Born (1997) in 13 nations.

While self-efficacy is an individual level variable, Bandura (1982), who developed and conducted extensive research of the concept, also viewed it as a group variable. In an article which received the Distinguished Scientific Contribution award from The American Psychological Association he wrote:

> People do not live their lives as social isolates. Many of the challenges and difficulties they face reflect group problems requiring sustained collective effort to produce any significant change. The strength of groups, organizations, and even nations lies partly in people's sense of collective efficacy that they can solve their problems and improve their lives through concerted effort. Perceived collective efficacy will influence what people choose to do as a group, how much effort they put into it, and their staying power when group efforts fail to produce results. (p. 143)

Bandura linked personal efficacy to that of the group in demonstrating that collective efficacy is "rooted in self-efficacy. Inveterate self-doubters are not easily forged into a collectively efficacious force." But when the members of the group have high levels of self-efficacy, they are capable of bringing about social change in a group or society (Bandura, p. 143).

Recent research on computer efficacy in fourth and fifth-grade children was conducted by Shank and Cotton (2014), who broke down the general self-efficacy construct into domains (following on Bandura and others). In addition to General Self-Efficacy, the authors measured technological, mathematics/science, and academic self-efficacy. They first found relationships between second-level digital divide scales (measuring multimedia and content creation, communication, and social networking) and self-efficacy (p. 189). When they focused on specific computer and Internet activities that might drive the relationships, regression analyses showed that frequency of emailing and frequency of communication were the only variables that related positively to all the self-efficacy domains (p. 189), while frequency of social media use was negatively related to general efficacy and academic self-efficacy.

Shank and Cotton summarize:

> Our conclusions from this research also suggest that education interventions, youth development programs and technical change in elementary schools might empower students through technology use. First, of all domains of self-efficacy, technology influences technological efficacy the least. Technology use most often related to general and academic self-efficacy, suggesting how technology serves as a medium to be utilized by educators. While Bandura points out that domains of self efficacy lead to behavior in that domain (1997), here we find evidence that engagement with technology affects many domains of empowerment, not just the most proximate. (2014, p. 191)

In the end, it isn't just access, but it also isn't just access plus education that will reduce digital inequality in individuals or societies as is illustrated by Shank and Cotton's findings. Not all uses of the Internet or computers are equally empowering as the authors demonstrated when they found that social network use may work against self-efficacy development. So even the idea that self-efficacy is an important variable when social change is the goal is insufficient. Our research and our deeper thinking about the issues needs to be ongoing and not allow us to fall into the easy deterministic traps.

AVOIDING EASY ANSWERS

This discussion of research findings related to digital inequality should provide lessons for policymakers and researchers alike regarding the role of technology in development. I am not suggesting that providing technology, developing high-speed networks, and training citizens to make the best use of computers and the Internet should not be part of sound development goals. Whatever we think of globalization, it is important to note that every country, rich or poor, is part of the global society we all live in. And when some societies are not networked to global systems, and their citizens are not technology literate, they will be disadvantaged. But I am suggesting that using a technologically deterministic discourse, and deciding that throwing technology at a problem will bring about the solution is not the appropriate way to address technology's role in development. Because we have all uttered words that reflect the utopian or dystopian narratives related to technology and its misconceived autonomous role in social change, we need to think harder about technology's embeddedness in social systems before we speak or write or even contemplate ideas about its power.

NOTES

1. Those competencies included the ability to block messages from persons you do not want to have contact with, the ability to find information on how to use the Internet safely, to bookmark a website, to change the privacy settings on a social network profile, to assess websites as to correctness of the information, to delete the history of visited websites, to block advertisements or junk mail and spam, and to change filter settings.

2. Self-efficacy was measured by five questions in the survey. Respondents were ask to rate their relative agreement to the following statements: (1) It is easy for me to stick to my aims and achieve my goals; (2) I am confident that I can deal with unexpected problems; (3) I can generally work out how to handle new situations; (4) If I am in trouble I can usually think of something to do; and (5) I can generally work out how to handle new situations

REFERENCES

Asaduzzaman, A.S.M. "Digital Bangladesh: Information and Communication Technology for Empowerment." Doctoral Thesis, Goldsmiths, University of London, U.K.

Bandura, A. (1995). *Self-efficacy in Changing Societies*. Cambridge University Press.

Bandura, A. (1982). "Self-Efficacy Mechanism in Human Agency." *American Psychologist*. 37(2), 122–147.

Christensen, C. (2011). "Discourses of Technology and Liberation: State Aid to Net Activists in an Era of 'Twitter Revolutions.'" *Communication Review*. 14, 233–253.

D'Haenens, L. and C. Ogan. (2013). "Internet-using Children and Digital Inequality: A Comparison between Majority and Minority Europeans." *Communications: The European Journal of Communication Research*. 38(1), 41–60.

Davison, R., D. Vogel and R. Harris. (2000). "Technology Leapfrogging in Developing Countries: An Inevitable Luxury." *The Electronic Journal of Information Systems in Developing Countries.* 1(5), 1–10.

Eres, B.K. (1981). "Transfer of Information Technology to Less Developed Countries: A Systems Approach." *Journal of the American Society for Information Science.* 32(2), 97–102.

Fair, J.E. (1989). "29 Years of Theory and Research on Media and Development: The Dominant Paradigm Effect." *International Communication Gazette.* 44, 129–150.

Fair, J.E. and H. Shah (1997). "Continuities and Discontinuities in Communication and Development Research Since 1958." *Journal of International Communication.* 4(2), 3–23.

Fisher, E. (2010). "Contemporary Technology Discourse and the Legitimation of Capitalism." *European Journal of Social Theory.* 13(2), 229–252.

Hargittai, E. (2010). "Digital Na(t)ives?Variation in Internet Skills and Uses among the 'Net Generation.'" *Sociological Inquiry.* 80(1), 92–113.

Hargittai, E. and A. Hinnant. (2008). "Digital Inequality: Differences in Young Adults' use of the Internet." *Communication Research.* 35(5), 602–621.

Hudson, H.E. (1979). "Implications for Development Communications." *Journal of Communication.* 29(1), 179–186.

Iacono, S. and R. Kling. (1996). "Computerization and Tales of Technological Utopianism." *Computerization and Controversy.* Second Edition. San Diego: Academic Press, pp. 85–104.

James, J. (2005). "Technological Blending in the Age of the Internet: A Developing Country Perspective." *Telecommunications Policy.* 29, 285–296.

Kling, R. (1991). "Computerization and Social Transformations." *Science, Technology, & Human Values.* 16(3), 342–367.

Kling, R. (1996). "Hopes and Horrors: Technological Utopianism and Anti-Utopianism in Narratives of Computerization." In R. Kling, (Ed.), *Computerization and Controversy,* Second Edition. San Diego: Academic Press, pp. 40–58.

Lerner, D. (1958). *The Passing of Traditional Society: Modernizing the Middle East.* Second Printing. Glencoe, IL: The Free Press.

Luyt, B. (2008). "The One Laptop Per Child Program and the Negotiation of Technological Meaning." *First Monday.* 13(6). http://firstmonday.org/ojs/index.php/fm/article/viewArticle/2144/1971. Accessed March 7, 2014.

Mohammadi, A.S. and A. Mohammadi. (1994). *Small Media, Big Revolution: Communication, Culture, and the Iranian Revolution.* Minneapolis: University of Minnesota Press.

Neulens, G. and L. Van Audenhove. (1999). "An Information Society in Africa? An Analysis of the Information Society Policy of the World Bank, ITU and ECA." *International Journal Gazette.* 61(6), 451–471.

Ogan, C., M. Bashir, L. Camaj, Y. Luo, R. Pennington, S. Rana, and M. Salih. (2009). "Development Communication: The State of Research in an era of ICTs and Globalization." *International Communication Gazette.* 71(8), 655–670.

"One Laptop per Child." Laptop.org. http://laptop.org/en/laptop/hardware/index.shtml. Accessed March 7, 2014.

Rogers, E. (1976). "Communication and Development: The Passing of the Dominant Paradigm." *Communication Research.* 3(2), 213–240.

Rogers, E. (1962). *Diffusion of Innovations.* New York: The Free Press.

Rogers, E. and F.F. Shoemaker. (1971). *Communication of Innovations.* New York: Free Press.

Schramm, W. (1964). *Mass Media and National Development: The Role of Information in Developing Countries.* Stanford: Stanford University Press.

Scholz, U., B. Dona, S. Sud, and R. Schwarzer. (2002). "Is General Self-Efficacy a Universal Construct?" *European Journal of Psychological Assessment.* 18(3), 242–251.

Schwarzer, R. and A. Born. (1997). "Optimistic Self Beliefs: Assessment of General Perceived Self-Efficacy in Thirteen Countries." *World Psychology.* 3, 177–190.

Schwarzer, R. and M. Jerusalem. (1995). "Generalized Self-efficacy Scale." In J. Weinman, S. Wright and M. Johnston (Eds.), *Measures in Health Psychology: A User's Portfolio. Causal and Control* Beliefs (pp. 35–37). Windsor, U.K.: NFER_NELSON.

Servaes, J. (1999). *Communication for Development. One World, Multiple Cultures.* Cresskill, NJ: Hampton Press.

Shank, D. B. and S. R. Cotton. (2014). "Does Technology Empower Urban Youth? The Relationship beween Technology Use and Self-Efficacy." *Computers in Education.* 70, 184–193.

Warschauer, M. (2003a). "Dissecting the 'Digital Divide': A Case Study in Egypt." *The Information Society.* 19, 297–304.

Warschauer, M. (2003b). *Technology and Social Inclusion.* Cambridge, MA: MIT Press.

Van Dijk, J. (2009). *The Deepening Divide: Inequality In the Information Society.* Thousand Oaks et al.: Sage.

Wilson, M. (2002). "Understanding the International ICT and Development Discourse: Assumptions and Implications." *African Journal of Information and Communication* 3, 80–93.

Zillien, N. and E. Hargittai. (2009). Digital Distinction: Status-specific Internet Uses *Social Science Quarterly.* 90(2), 274–291.

Chapter 2

From Ubicomp to Ubiex(pectations)

Rich Ling

SOCIAL MEDIATION TECHNOLOGIES AND UBIQUITOUS EXPECTATIONS OF USE

Ubicomp has provided us with profound and useful insights into the trajectory of information technology in everyday life. Based on the work of Mark Weiser (1991, 1993, 1996), this approach to understanding the diffusion of technology has led us to see the integration of technology into a wide variety of devices. Indeed, we are often not aware of the sophistication of the devices and systems that we are using. Ubicomp along with the kindred areas of pervasive computing, ambient intelligence and "the internet of things" undeniably traces the arc of contemporary society. We have seen information technology being embedded into a wide variety of machines and devices. The ease with which digital devices can be made means that this type of technology has flourished. In this context, we have seen the rapid adoption of smart phones and tablets and the embedding of their use in our daily lives. Most recently with the development of app stores, we are also seeing the growth of new distribution channels that go outside the traditional path dominated by telecomm operators.

While the flux of technological development is important, it is only half the story. The other half is how we make sense of these developments, how we choose to integrate them into our daily lives (Silverstone et al., 1992) and how they are becoming a part of the structure of society (Ling, 2012). There are many social practices that we have had to recalibrate to accommodate the developments in the technology world. These are also different for different parts of the phone-using public. There are those who seem to take to the new digital world like fish to water. Others are less graceful in their interaction with the technology. Our accommodations have, in turn, been routinized and have also structured our social interaction.

Our use of these technologies is often seen as an individual decision. I choose to buy a PC or a game consol. I use the latest digital camera, GPS, digital watch and so on. In some cases the choice is truly my own. The consumption is largely an individual decision (perhaps pushed on by trying to keep up with the neighbors or the cajoling of ad agencies). In this chapter, however, I ask about the effect of our social context in the use of a particular set of technologies, namely those technologies that mediate social interaction. I also suggest that these technologies become a part of the structure of society that coerces us to use them.

When we think about the technologies that we use to communicate/coordinate with others then we can see that, as Weiser suggests, they are becoming more ubiquitous. We also see that we are not necessarily independent actors with regard to their consumption. Indeed there is a type of social compulsion in our use of these technologies. That is, we use them because of the needs and wishes of other people, not just our own needs. This may operate at the local level or at the broad social level. Regardless, the choice to use a particular set of technologies is not completely our own nor is it the volition of particular individuals to demand their use. Rather, these technologies are a taken for granted part of the structure of society (Ling, 2012).

Clocks and timekeeping illustrate this. The clock (in its various forms) is widely available. We can have a functional device or one that is quite elaborate. We might only have a clock on our mobile phone or we may keep time by using publically available devices. Clocks help us to coordinate meetings, the comings and goings of trains, busses and planes. We use time as a way to make and keep appointments and as a way to gauge how much longer we can have a casual talk with the colleague in the neighboring cubicle before we have to leave and pick up the kids at day-care. That is, timekeeping gives us (in the collective sense) a structure so as to facilitate interaction. In all of this, it is generally expected that we respect timekeeping when meeting and interacting with others. We do not have the leave to ignore timekeeping. There is a web of mutual expectations associated with punctuality. There are ubiquitous expectations associated with mastering and using timekeeping that coordinate our collective interactions. As Simmel notes: "If all the watches in Berlin suddenly went wrong in different ways even only as much as an hour, its entire economic and commercial life would be derailed for some time" (Simmel, 1903, p. 308). Thus, time provides a guide for arranging our collective lives. Further, its use is not optional. To be seen as a competent member of society we necessarily have to surrender ourselves to the dictates of time and timekeeping. This is a part of the baggage we accept to be a part of society. Other systems of social mediation might include automobile-based transportation for people living in the suburbs, ownership and use of a fax machine by hotels in the 1990s and, to an increasing degree mobile communication (Ling, 2012).

Moving to the more purely digital world, there is also the development of ubiquitous expectations associated with many ICTs. Networked personal devices allow individual addressability. They provide us the ability to communicate as needed. However, as they gain a critical mass, we are increasingly expected by others to use and/or own one. Thus, on the one hand there is freedom, but on the other there is tied coercion. This social pressure is indeed a significant element in the development and maintenance of the new distribution systems that are arising with the use of smart phones. In this chapter, I will look at the history of ubicomp and its relationship to ubiex and then I will examine how ubiex develops.

COMPUTER DIFFUSION AND THE FOCUS OF UBICOMP

Technology in Service to the Individual

In the original work on ubiquitous computing, Weiser's basic notion was that with time, digital processing would diffuse into many different devices. Living on the cusp of the digital revolution and in the spirit of Moore's Law, Weiser's insight was fundamental. Rather than thinking of computers as huge mainframe devices, they were being scaled down from the macro, through the meso into the realm of the micro. Along the way, they were also diffusing into a wide variety of other devices.

In his well-cited Scientific American article, Weiser talks about how computers will "vanish into the background" (1991, p. 94). Computing devices, according to this vision, would become smaller, less expensive and more robust. The additional ability to connect the devices into networks further facilitated the drift toward embedding digital functionality into the social world. This is the vision of ubiquitous computing or ubicomp. It presented designers with the idea of a shift in the way we interact with devices. So long as the computer was a device with a command line or a GUI-based form of interaction, there were limitations with regard to how and where we would work with digital devices. Weiser suggests that there was an alternative way to use them. The command line or the keyboard/ mouse form of using the PC provided the ability to carry out certain tasks. These approaches were, however, not always applicable to more diffused forms of computing.

According to Weiser, computation would recede into the background. Computing devices would no longer require our active attention but, like Bertie Wooster's man Jeeves, they would be quietly available in the background as needed. Weiser writes "Today's multimedia machine makes the computer screen into a demanding focus of attention rather than allowing it to fade into the background" (1991, p. 21). The reduction in size and power requirements would also lead to the development of new computing devices. These included the famous tabs, pads and boards described by Weiser (1991). This trajectory can be seen in the development of other digital devices such as the digital cameras, GPS devices, laptop PCs, game consoles, MP3 players stand-alone video disk players and mobile phones. These illustrate the embedding of computing technology into ever-smaller devices.[1]

Calm Technology

There is often an assumption in the original ubicomp work that technology was there to support the needs of the individual. For example, Weiser and Brown (1996) write that ubicomp "will bring information technology beyond the big problems like corporate finance and school homework, to the little annoyances like, Where are the car-keys, Can I get a parking place, and 'Is that shirt I saw last week at Macy's still on the rack' " (1996, p. 6)? The notion of technology in the service of the individual is beguiling. In the conclusion of his Scientific American article, Weiser compared ubicomp to taking a walk in the woods.

Most important, ubiquitous computers will help overcome the problem of information overload. There is more information available at our fingertips during a walk in the woods than in any computer system, yet people find a walk among trees relaxing and computers frustrating. Machines that fit the human environment, instead of forcing humans to enter theirs, will make using a computer as refreshing as taking a walk in the woods (1991, p. 104).

This is a wonderfully appealing vision. Technology will become the servant of humanity. We will need to simply conceive of a need and then set the digital world into motion to solve the problem. The individual is in control. Weiser and Brown take this tact when developing the notion of "calm computing" (1996). Computing is supposed to give the individual control over his or her life. Technology will gently allow us to deal with the exigencies of life in a more deliberate way that does not disrupt the flow of our preferred activities. While the ringing of a phone—or to use the example of Weiser and Brown the ringing of a pager—may destroy calm, other technologies can restore serenity. In their vision, ubicomp technologies will do this by existing at the periphery of our attention until there is some reason for the individual to take control of the situation. In that case, one can focus on the issue being presented by the technology and then place it back at the periphery of his/her consciousness.

A calm technology will move easily from the periphery of our attention, to the center, and back. This is fundamentally encalming for two reasons.

First, by placing things in the periphery we are able to attune to many more things than we could if everything had to be at the center. Things in the periphery are attuned to by the large portion of our brains devoted to peripheral (sensory) processing. Thus the periphery is informing without overburdening.

Second, by recentering something formerly in the periphery we take control of it. Peripherally we may become aware that something is not quite right, as when awkward sentences leave a reader tired and discomforted without knowing why (1996, p. 9).

It is the ease of the transition that is the key. A calming technology is one that moves easily between the center and the periphery of our attention and that gives the user the appropriate level of detail as needed. It is a technology that will not surprise the user unnecessarily. They describe how the user should have a sense of being located and "at home" with the technology and the information that it provides.

The Routinization of Technology

Moving somewhat more into the social sphere of analysis, Tolmie et al. (2002) looked at the use of ubicomp in the home and its interaction with established routines. They correctly noted that routines are a type of glue. They write, "There is a sense in which routines are the very glue of everyday life, encompassing innumerable things we take for granted such that each ordinary enterprise can be undertaken unhesitatingly" (2002, p. 399). Routines are social devices that are often taken for granted, but that provide an essential consistency. Always having tomato soup on Fridays or the first person who wakes up that makes the coffee give us a sense of what Silverstone and Haddon called ontological security (Silverstone and Haddon, 1996).

Tolmie et al. give us an historical insight into the organization of society previous to the ubiquity of the mobile phone. They provided the illustration of a person, Michael, who needed to coordinate a task with a friend named John. In the vignette, Michael uses the landline to call John's landline, but is not able to reach him. Michael knows, however, that the John had the habit of visiting the local coffee shop before he opened his own store in the morning. "[Michael] goes to phone John. However, John doesn't answer. He notes that the time is about a quarter to ten—He says he thinks he will go to the coffee shop [a small coffee house just around the corner] where he thinks he'll catch John because John usually goes in there for a coffee before opening up the Farm Shop at ten o'clock—When he gets to the coffee shop he sees John waiting at the counter" (2002, p. 304). Tolmie et al. discuss how there was no requirement between these people that they knew of one another's routines. However, it was a part of their tacit stock of knowledge.

There is no matter-of-course requirement upon Michael that he should specifically coordinate his routine with John's and he has no particular accountability placed upon him that he should attend to John's routine at all. In direct contradistinction to our previous observations Michael quite specifically marks out what he knows of John's routine for comment—he knows that John goes to the coffee shop every morning before he goes to open up the shop. Here John's activities have been made a matter of note for Michael in a way that John himself might not ordinarily take note of them. (Tolmie et al., 2002, p. 304).

This tacit knowledge suggests that Michael goes to the coffee shop to talk with John. The stock of tacit knowledge could vary between individuals based on their history and their social closeness. The vignette also suggests that in the pre-mobile world, we legitimately could be unreachable. There was no social requirement that we be available to one another at all times. There were core periods of availability when, for example, we were at work. Beyond that, there was no expectation that we could talk to one another at the drop of a hat.

A half decade after the work of Tolmie, the vision of Weiser and his colleagues was reexamined by Bell and Dourish (2006). According to Bell and Dourish the vision of ubicomp was by then a long way toward being fulfilled. They note that ubicomp was defined by its visions of a technical future and how we experience this. Specifically, they noted that the proliferation of mobile telephony was, in some ways, the fulfillment of the ubicomp vision. At the time of their writing, they found that nearly all the people—presumably over a certain age—owned a mobile phone and that the phones were progressively being interwoven into many aspects of everyday life. Bell and Dourish conclude by saying that there is a need for a better understanding of the nexus between society and the new information and communication technologies. They write that "we need a deeper understanding of how social and cultural practice is carried out in and around emerging information technologies" (2006, p. 142). This assertion suggests that we move beyond the ego-based notion of ubicomp. That is, we look at the structural consequences of the technology on society and not simply the convenience that it might provide.

Mobile Telephony and the Loss of Unreachability

Picking up the thread from Bell and Dourish, the adoption of mobile telephony means that we are no longer unreachable. This illustrates how one sense of ubicomp has run its course. The scenario between Michael and

John would probably have played out another way if it took place today. In addition to drawing on tacit knowledge that John often visits the coffee shop, Michael could have just called John on his mobile phone, an option that was not possible in the landline-based world. Indeed, it is often the case that our eventual interlocutors are not as easily available. In the vignette, it is convenient that the coffee shop is "just around the corner" but in many cases for people living in today's suburban society, meetings such as this require driving (Elliott and Urry, 2010).

The widespread adoption of the mobile phone means that we are accountable to one another outside of the core times of the previous regime. This does not say that there is no cultivation of tacit knowledge. Indeed, the insight we have into one another's lives may be finer grained. Following from Licoppe's notion of connected presence (2004), we can be continually in contact with our social sphere. This paradigm shift that arose following the diffusion of mobile communication, texting and more recently mobile social networking, means that we have access to one another's situation at a far different level. Thus rather than knowing the broad outline of one another's routines, we may know that, for example, at this time of the day our friend usually goes to the store, but when I texted him 10 minutes ago he was at the barber shop. This will remind us that we commented on his long hair when we saw him last week, etc. The individual addressability of the mobile phone means that we can develop and maintain a detailed knowledge of one another that goes beyond our knowledge of their broad-stroke routines. It is clear that as people become more socially distant it is not possible to keep control of their activities at a detailed level. However, for the closest sphere of friends and family, the mobile phone gives us immediate access (Ling et al., 2012).

In the case of a couple, the immediate access provided by the mobile phone means that they could use this channel to enact their daily routines. In the words of Berger and Kellner (1964), they can use the device to construct a sense of nomos (1964).[2] This use of the device can be seen in the comments of a young couple with two small children, Trond and Tina. They describe how they commonly call one another during the day to touch bases, have a chat, plan the evening meal and to eventually cover for one another in case something has arisen.[3]

Interviewer: Do you call your wife often?

Trond (36): We talk at any rate at least once a day, if not two.

Interviewer: And you agree on getting the children and things like that? Or?

Trond: That is fixed if nothing unexpected has happened.

Tina (39): We agree . . . What we are going to have for dinner (laughter) . . .

Trond: . . . we agree on that (laughter).

Tina: . . . shopping and. . . .

Trond: . . . small things, practical messages like that.

Interviewer: What is going to happen and who is going to do what?

Tina: Yeah. (Couple with a 6- and a 9-year-old child).

Trond and Tina describe how the daily phone call gives them the chance to have a short talk, decide on the dinner menu and to deal with any issues that need discussing. The interlacing of their comments in the citation shows that they have a tightly bound common sense of the interaction. One partner starts a sentence and the other completes it. They are able to laugh at the banality of the calls, but there is also the sense that they simply enjoy talking with one another. It is clear that the calls like this were not possible before the mobile phone. However, the mobile phone allows us to interweave this type of interaction into our daily lives whenever there is the need or the inclination.

There is also the sense in the case of Trond and Tina that there is a mutual trust in one another. The work of Ampro Lasén shows that there are also other dimensions to this type of social cohesion. There is, for example, the exercise of power and the demand that the other is available to avoid suspicion (Lasén, 2011). In her analysis of mobile phone use among young couples in Spain the device becomes a way for the partners to keep track of one another. If they are unaccountably not able to answer the phone, there is cause for suspicion. She writes:

> The study of how mobile phones are used to strengthen cohesion in couples, the affective bonds, and the coordination of its members, as well as identify the couple itself, helps to reveal the dual nature of social cohesion: the achievement of trust, sharing, solidarity and identity in an interdependent relationship involves not only the establishment of a network of mutual obligations, negotiations, and latent and explicit conflicts, but also control and power relationships. Alongside the usual terms employed to define social cohesion, such as harmony, trust, shared experiences, sense of belonging, togetherness, mutual support, and loyalty, lie subjugation and power dynamics as well. (2011, p. 83)

Lasén's insight provides us with the point of departure for moving from the notion of the mobile phone as a convenience to the idea that it is an expectation. This is the slow congealing of Weber's "light cloak" into the iron cage (Weber, 1930, p. 181). The adoption of a technology of social mediation and its achievement of a critical mass in the group or in society changes the structure of expectations between individuals.

There is a coercive element associated with the use of the mobile phone (or for that matter other technologies of social mediation). On the one hand, the notion of ubicomp assumes that we will be beguiled into use. We will like the things that we get and we will enjoy them. It does not take up the opposite side that we will be coerced into using them.

We can begin to see that the web interpersonal expectations congeal into a structural need. People find that coordinating via the mobile phone is more convenient and flexible. This system exists in parallel with time/place-based systems, but given the agility of this micro-coordination, people soon learn that they can simply agree to meet at an undefined place in an open time space (Ling and Yttri, 2002). The texts that are exchanged might be: "Do you want to meet for coffee on Saturday?" As Saturday approaches the partners might further specify the time and place, that is, "I will be downtown in the morning. What about you?" To which their interlocutor might say, "Perfect! Let's text when and we can agree on a location." If both partners have their phones, it will be possible to make the final negotiations. However, if one forgets their phone, if they don't hear the ringing or if the battery is dead, the agreement is stranded. In this process, we see that our way of managing the mutual expectations with regard to the meeting has become generalized. There is a structural element associated with the way we agree to meet. It is not simply the ubiquity of the equipment, but it is the ubiquity of our mutual expectations that are pushing the use of these devices. With the development of smart phones, cloud computing and location-based services (Gordon and Silva, 2011), we can also use a variety of mapping and tracking services to find one another.

There is a collective demand (collective in the specific sense of ALL of our family members or ALL of our best buddies) that we are available via the mobile phone. This can be abstracted to being a broader expectation that people are generally available via their mobile phones. This often means that people who are more socially remote, such as our dentist or the car repair place, also begin to expect/demand that we have a mobile phone. This eventually develops into the idea that certain services are only available to those who have adopted a technology.

The Development of Ubiex

It is useful to consider how different social mediation technologies—that is, technologies and systems that mediate social interaction such as the mobile phone, the group calendar or the clock—arise, develop and become embedded in society. There is a special process whereby social mediation technologies are integrated into society. These technologies have a critical mass of users either in society as a whole or in specific groups and further our use is governed by group-based reciprocal expectations that enable but also set conditions for the maintenance of our social sphere (Ling, 2012). While there is personal agency in the mix, the use of these technologies is not necessarily our personal choice. These technologies/systems have an assumed role in social interaction. They facilitate interaction and help us to orient ourselves. Adoption and use of a fully configured social mediation technology is not, however, a free choice. As noted in this introduction, clocks and timekeeping illustrate this. Time and timekeeping helps us to navigate our social obligations. We do not, however, have the freedom to ignore timekeeping.

Social mediation technologies evolve and become embedded in society. They gain a critical mass that we in turn bless with our legitimation (Ling, 2012). Further, they reform the social ecology in their own image and most importantly, we develop reciprocal expectations regarding use. As technology becomes more ubiquitous and taken for granted it moves from being an oddity to becoming expected. It may, to some degree, disappear into the background as suggested by Weiser. In this case, it may only become obvious when it fails or when its status is threatened by the rise of a new competing approach. This was seen, for example, when a telephone exchange was destroyed in Manhattan in the 1970s (Wurtzel and Turner, 1977). The phases in this process might be characterized as (1) the hobbyist or tinkering phase, (2) the abstraction of functionality, (3) networking and the rise of a critical mass, (4) a Klondike (5) saturation and (6) regeneration.

Hobbyists and the Tinkering Phase

During the early development of a technology or a system, the users are often individuals, small specialized groups or hobbyists.[4] There is not necessarily the sense that the system will be generally adopted by the broader public. The users have a particular need or perhaps see working with the technology as a type of diversion. Early users of clocks, for example, were monks who needed a system to rouse them to pray at different points of the day. The clocks were the refinement of different horologia and other gizmos with the addition of a weight-based power train used to ring a bell after a certain time. In this early phase, each clock was unique and each had its own way of solving the different mechanical problems of regulating the device. They eventually gained a technical sophistication and certain regularity.

Early developers of clocks, radios, steam engines, automobiles and other devices were often hobbyists or tinkers (Bar et al., 2007; Wallis et al., 2012). The use areas for these devices were limited to small and specialized groups. There was a limited sense that the device would be broadly used. The clock was for the cloister. The radio was a curiosity for wealthy dilettantes. Early landline telephone users, for example, often bought two terminals, one for the office and one for the factory. The same was true of early fax machines.[5]

There is no broader need to legitimize the use of the clock, the car or the fax since it is not intended for many users. At this point, the technology has not had any impact on the broader social ecology. The existing paradigm of use is soundly rooted in place with all its attendant legitimization structures. While Henry Ford was tinkering in the workshop behind his home on 58 Bagley Street in Detroit to build his Quadricycle, the system of steam and horse-based transportation was firmly in place just outside his door. Livery stables were widely established and provided a system to fuel and maintain horse-based transportation. The cities were organized around transportation provided by either steam trains for long distance trips or horses (or walking) for more local commutes.[6] Automobile-based transportation was not to be seen on the horizon.

Abstraction of Functionality

If a technology shows itself to be useful (and/or if significant power structures push for adoption), there can be the abstraction of functionality. The clock in the abbey can be used to alert the monks of the prayer cycle, but it can also be used to remind the shopkeepers when it is time to open their stores or when the city gates should be opened.[7] The car can be used to transport goods. The fax machine can be used to send material between the one in the office and another at the factory could also be used to send material to other people who also had fax machines. The potential of the gadget becomes generalized, at least within the group of users.

It is at this point that we see what might be called the seeds of domestication. People beyond the hobbyists and tinkers start to think about how to use the technology into their everyday lives. They start to think about how the device will be placed, when they can (or should) use it and what constitutes proper use. In other words, they start to bring it into their routines. At this point, the device or system is often not well developed and crude in its functioning. The early users however employ it almost in spite of its clumsiness. Indeed, those who are partial to the previous regime use its poor functioning in counter-legitimation arguments. The aspersion "get a horse" that early drivers tolerated when their model T's broke down is an example of both the crudeness of the technology and the counter-legitimation process. However, by the dint of its usefulness, and in particular its usefulness to central persons in society, a technology begins to embed itself into the flux of daily life.

At this point in the process, the system in question is only starting to change the social ecology. Taking the example of cars, as they started to be commercialized around the early part of the 1900s, society was solidly entrenched in another transportation regime. However, as the decade progressed, livery stables started to sell gasoline, people started to understand the workings of the cars and to become mechanics, and the first tracings of today's car culture were stirring. There were no, however, reciprocal expectations that have since developed. At this point, there is the budding of the idea that the use of technology is generalizable. This is not to say that there is the reciprocal expectation of use. On the contrary, use is still seen as exclusive and perhaps cockishly avant garde.

BROADER NETWORKING AND THE PERCEPTION OF A CRITICAL MASS

Following the early development and also following the adoption by a vanguard of users, technology may begin to gain the sense that it is in more general use; that there is the development of what people perceive to be a critical mass of users. As the diffusion of a technology gains pace the device or service is refined. A number of producers may begin to move into the market and economies of scale may begin to be evident. As Henry Ford (not to mention Bill Hewlett and Dave Packard) moved from their respective garages to larger production facilities and a more commercialized stance, the products became more refined and user friendly. These production processes also meant the price of cars and electronics, respectively began to fall. Thus, the different social mediation technologies began to be within the reach of less elite users.

The democratization of a technology means that different rationales for use begin to arise. This is based on the growing perception that there is a critical mass of users. For an interactive service to take off, we need to perceive that there is a critical mass of others with whom we will interact. If we have the sense that we are the only users, or that the pool of users is not extensive, we will retain the idea that it is not useful in the broader commerce of the group.

As noted above, early users of, for example, the original telephone or the fax had the sense that they were isolated users who bought their devices in pairs. In addition, there may be the problem of conflicting standards. As a technology of social mediation becomes increasingly mainline, as the cost of production drops and as eventual standardization issues are sorted out, we develop the notion that others are also users. That is, the notion of being elite or cutting edge is replaced by the idea that the device is increasingly widespread. To use the example of the fax machine and the telephone, it eventually was not surprising that others were users. On the contrary, it became ordinary that others had access to these devices and eventually there was the expectation that they have access (Markus, 1987). Thus, instead of being surprised when others had a fax machine; people became surprised when they did not. These mutual expectations can operate at the broad social level, or within certain social groups.

It is important to note that there is no particular mathematical threshold or tipping point for these transitions. Rather the critical mass is determined by the perception of active and potential users. It may be that when one in five people use the networked office calendar system we gain the sense that there is a critical mass. It is the perception of criticality that is important. Aside from the early adopters, many people will hesitate to go through the process of learning the calendar system until there are a certain number of our colleagues who are scheduling their meetings that way (Palen, 1998). The early adopters (or possibly eccentrics) are the people who take on the heavy work of breaking the path. They work the kinks out of the system and they learn the tricks and shortcuts.[8] As others join them—perhaps enforced by an edict from the boss in the case of, for example, calendar systems—it becomes easier to justify our own use and indeed there is increasing value to the whole group as people come onboard. For the early users there is little reward and much work. There is also a missionary-like incentive to recruit others who, as they follow, make the system more central to the group. Those who do not use it are increasingly out of the loop.[9]

As social mediation technology is perceived as having a critical mass there is an increasingly common notion that it makes a valuable contribution to the group (Markus, 1987). The quirkiness of the early devices is replaced by increasingly functional systems and people learn the tricks of use. In addition, its common use contributes to the functioning of the group. This does not mean that all people gladly begin to use it. Entrenched alternatives and other notions of what is appropriate limit adoption by many group members. That said, critical mass—or more precisely the perception of critical mass—both facilitates group interaction and also constrains the alternatives open to individuals in the group.

In some cases, particularly with communication and coordination technology, we are able to enjoy group benefits of the system by simply buying (or subscribing) and learning how to use the fax, mobile phone, or the Facebook account. In other cases, other technologies become a type of requirement to operate in a particular context. This is discussed more in detail below.

As a technology moves through criticality the motivation for use changes. It moves from being an interesting diversion to being useful and then for some people to becoming increasingly necessary. Because of this there is also a transition in the legitimation narrations. The discussion regarding the technology becomes focused on its usefulness or practicality. People show pride in their ability to buy it and to master its use. There is perhaps a broader discussion of the technology in the press. In addition, to the degree that it facilitates the functioning of different institution, there is the beginning of a broader movement to enforce use. It is also important to note that there is a parallel process of anti-legitimation narrations carried on by those people who are entrenched in the devices of the previous system.

Another change associated with the transition toward a critical mass is that we begin to see adjustments in the broader social ecology. These are done to accommodate the new technology. The diffusion of the car is perhaps the most distinct case. As it became more common to have a car, shops started to make adjustments for car-bound shoppers. Road signs were developed to facilitate car-based vacations, etc.

When the mobile phone was introduced, there was no same need to rethink how this type of mediated communication should take place. There were adjustments to be sure. The fact that we call to individuals and not places has changed how we coordinate daily life and how we deal with emergencies.

The key to this period, however, is that there is the budding sense that we expect others to use the technology in question. During the late 1980s it was increasingly awkward that a client or a customer did not have a fax machine (Markus, 1987). At this phase, the alternative systems are usually available. A document can be mailed or sent with a courier. However, the existing systems began to pale out when compared to the functionality of the new. When the older systems began to fail, the upstart technology could gain a foothold.[10]

It is also at this phase that tastes begin to change and we adjust our lifestyles in order to accommodate the new device or system. It is during the transition from this phase into the next that use becomes mainstream and that broader groups adjust their attitudes and their daily regimen to the use of the system.

Klondike

As technology of social mediation moves beyond the criticality boundary, its use becomes accessible to all people in a given milieu and it becomes increasingly taken for granted. The device becomes more refined and cost is no longer seen as a barrier to use. The mobile telephone, for example, moved from being luggable to being pocket-sized and the price dropped such that many people could afford one. In addition, it gained a cultural cachet. It was no longer nerdy, but it was seen as fashionable (Fortunati et al., 2003; Fortunati, 2005) and indeed as being increasingly de rigueur. We routinize the position of the object or system.

There can be a type of Klondike-like transition when the technology becomes widely available, when the price of purchase falls and when there is a common notion that the time is into purchase. During this phase there are popular discussions in the press of, for example, which version of fax machines, mobile phones or calendar systems are best. In addition, there is informal tutoring of new users just as holdouts receive unsolicited suggestions that they might also join the pool of users. All of this expands the number of users and it further embeds the technology into our common sense of use. Unlike Weiser's sense that we gracefully accept the adoption and use of the new technology, it can be brutally forced on us, come what may. People who were not interested in jumping onto the bandwagon in earlier stages are cajoled and exhorted into use by those who have already adopted.[11]

The characteristic of the social mediation technologies is that the more people who adopt them, the more utility for those who have already adopted. Indeed, if only a handful of people use the office calendar, there is little common benefit. However, if everyone uses it, there is a bonus for the whole office. This is not to say that adoption is seen only as a positive. The time spent learning the system, the frustration with the new system when personally the old system functioned just fine in the eyes of the individual, the eventual cost of purchase, subscription costs (in the case of mobile phones) and the cost of maintenance means that users need to place consumption into a broader sense of their personal or household economy. Some people will reserve themselves from the system for just these reasons. The social pressure to adopt, however, often carries with it the majority of potential users during this period. Those who do not adopt are also increasingly seen as being frumpish, stubborn and opposed to the general interests of the group.

It is during this phase that the technology seriously begins to change the social ecology by outcompeting the alternatives. The fax machine begins to replace physically sending letters. The electronic office calendar system begins to replace paper-based calendars. The mobile phone makes the landline phone redundant. Indeed, an employer may no longer supply employees with paper-based calendars or they may limit postal delivery in favor of faxes.

As the broader social ecology changes, there is an increasing pressure or coercion on those who, for whatever reason, are not a part of the new system. This means that the holdouts need to work out their own systems for maintaining the previous regime. People who do not use the office calendar system may need to buy their own calendars for example. This is seen clearly with the adoption of the automobile. The broader system that supported car-based transportation began to direct resources toward gas stations and repair shops and away from livery stables and buggy whips. Roads were repurposed to allow for car travel and there was greater separation between home, work, shopping and other activities. As the transition to the car continued, the very structure of the city began to morph. Suburbs developed and highways were built. Stores moved out of the city core and car-based shopping areas developed (Jackson, 1985). In many respects, people own a car because the social ecology has changed and the new structure dictates that the individual has the device (Elliott and Urry, 2010; Ling, 2012). People living in city centers or in cities where there are alternative forms of transport are not coerced into having a car. However, people living in the vast suburbs of many cities buy a car because it is simply not possible to find work, buy food and carry on with daily life without one (Ehrenreich, 2008; Hjorthol et al., 2007).

In a similar way, there can be a strong economic motivation for having a mobile telephone. Following from the work of Jensen (2007) in south India, fishers were basically forced to buy a mobile phone after the system became available. Without it they were not able to gather important market information as to where there was a demand for their fish (2007). This transition happened quickly, usually within 5 to 10 weeks of the mobile telephone system becoming available, approximately 75% of the boats had one. This suggests that there was a type of economic coercion associated with use.

It is during this phase of diffusion that we begin to see the potentials as well as the coercion associated with ubiquitous diffusion. Those who, for whatever reason, cling onto the previous system, become a brake on the functioning of the group. With technologies of social mediation, there is the reciprocal expectation that others will also use the time, money and energy to adopt the technology since it facilitates the functioning of the group. It is here that we start to see the functioning of what I have called the Katz principle, named after my colleague James Katz who coined it. Simply stated, this principle describes the impact of social mediation technologies when it notes that "you are a problem to me if you do not use one." If I assume that others in my circle of friends are users of the mobile phone and they are not, it disrupts my ability to coordinate and interact with them.[12]

As this phase of diffusion progresses, we assume others will have mastered the particular technology of social mediation. We are perhaps bemused by people who do not make the transition. We think of them as quaint rustics or perhaps principle-bound luddites. For most people, however, we can safely assume that they are users of the social mediation system in question. Before scanners and email arrived, we assumed that we could fax documents to others. We assume that our family members and friends are available via the mobile phone. We know that when we agree on a time for a meeting that our fellow meeting goers will also use basically the same time reference system as we do. That is, there is the establishment of a ubiquitous set of expectations as to how the technology fits into society and how it mediates our interaction.

Saturation/Stragglers

As use of the system of fax machines, online calendars, mobile phones or other mediation technology becomes universal in the group there is the problem of the holdouts. There are the people who are unwilling, or perhaps unable, to adopt the technology. These people need to buck the legitimation processes and policing that enforce use. They need to develop a counter legitimation that is sturdy enough so as to withstand these onslaughts. They might claim that they cannot use the computer, they are poor with technology or that there is some health-based reason that they cannot comply. They might also appeal to more Arcadian values of simplicity and direct interaction. Their resistance does, however, often conflict with what has become the accepted way of doing things. The existence of these stalwarts often means that special systems need to be devised.

The example of how teens make agreements using the mobile phone and Facebook means that there are not always fixed agreements regarding the time and place for a meeting. These are negotiated through a series of exchanges that iteratively refine the time and place of the meet-up. This form of micro-coordination often assumes that all the partners are individually available via a mobile phone. Those people who, for whatever reason, do not have one are outside of the loop. If they are to be included, alternative systems of communication need to be used. This may be a simple extra process or it may involve elaborate mechanisms. It is, however, increasingly superfluous as the new social mediation technology becomes dominant. It is not simply that it is a test of wills between the people who have adopted and those who have not. The new system becomes what Weber called an iron cage (1930) and what Durkheim called social facts (1951). The system of social mediation has a facticity of its own that is not possible to ignore. This means that a certain sense of morality is associated with use or nonuse. It means that the people who refuse are necessarily encapsulated since they have declared their opposition to the flux of society.

It is at this stage of adoption that we can speak of ubiquitous expectations. A technology of social mediation is used, for all practical purposes, by all the members of a group or a society. There is a thick legitimation structure as well as a social ecology that supports use. Finally, and perhaps most importantly, there is a web of reciprocal expectations associated with use of the technology. This is not to say that the system cannot be replaced. It is the fate of many electronically mediated technologies that they are vulnerable to new innovations. The transition from the landline to the mobile phone, the fax to scanners and email and potentially the transition from texting to social networking show that while there is a facticity to many social mediation systems, their position may be susceptible to new developments.

CONCLUSIONS

Weiser has given us the notion of ubicomp. Riding on the wings of increasingly pervasive digital technology, we have seen how computing power has moved from the desktop (or the mainframe computer) into a wide variety of devices that populate our daily lives. Weiser's vision has helped us to see how these technologies are being used in the service of the individual. This is an important insight, but it remains largely at the level of the individual.

The goal of this chapter is to take this insight to the level of social interaction and indeed social expectations. There are technologies that are of use largely to the individual. An MP3 player, a digital meat thermometer, a digital pen or a lava lamp are all technologies that contain computing power, but are largely in the sphere of the individual. It is only peripherally of interest to me if a colleague or a family member has one of these items.

There are, however a set of technologies—not always only digital technologies—that have as their main function the mediation of interaction between individuals. They can facilitate communication, transportation or coordination. Their use often starts out as the whim of a tinkerer and slowly becomes seen as being useful. After a technology gains a foothold and starts to diffuse, people begin to sense that it has a critical mass. As these technologies move through criticality, they gain a legitimation structure and they begin to change and adjust the social ecology. Most importantly, we begin to expect them of one another and we begin to structure our interactions based on the assumption that others are also users. Like Weber's iron cage that starts out as a light cloak that we feel we can cast off as wished, the technologies become engrained in our daily interaction and we are, in many ways, coerced into using them. They become Weber's stahlhartes Gehäuse that has been translated as a steel hard encasing. We do not have the option to not use them.

Thus, it is not good to think of ubiquitous technology only as artifacts that facilitate our daily lives. On the contrary, rather than being calming, they can be a way in which power is exercised. In order to be a competent coworker we need to fill in our electronic calendar system. To be a good parent we need to have our mobile phone charged and have the sound turned on (you, dear reader, are hearing the voice of experience here). We need to understand timekeeping and have access to a clock in one form or another. We need to be able to transport ourselves to jobs, social events, shops that are widely separated across the suburban landscape.

Mobile communication is rapidly becoming a technology of social mediation. By placing it into the context of its social use, rather than seeing it as a personal device, we gain insight into how it is changing our quotidian routines. In its use for the mediation of text and voice, it is helping us to readjust the way that we coordinate our lives. Taking this to the next step of development, the mobile Internet is also giving us new ways to interact with one another and it is also giving us enhanced access to information through either accessing third-party servers or by accessing our own information in cloud-based applications. By examining the ubiquitous expectations that are also a part of this development, we can see that there are not only personal motivations for having a mobile phone, but that there are undeniable social motivations that are being played out in lifestyle transitions.

In short, as technology becomes intertwined into the ebb and flow of society we do not have the option of ignoring it. We gain freedom by having a car, we gain peace of mind by having a mobile phone through which we can call our family. We gain a sense of order by having a calendar with our appointments. However, we are in many ways compelled to have these. We accept them for the advantages they provide, but we are not free to ignore them. This is not because they provide a ubiquitous calmness, but because we live in a social structure where there are ubiquitous expectations.

NOTES

1. Ubicomp can be thought of as sensor networks. This so called "smart dust" can be used to monitor weather conditions, track the progress of different shipments (Kahn et al., 1999). Other versions include the embedding processors into fabric, paint and other materials so as to allow for omnipresent data collection. Castells said "Computer networks will be, literally speaking, the fabric of our lives" (Castells, 1996, p. 53).

2. Nomos is the opposite of what Durkheim called anomie (Durkheim, 1997). Where anomie is the loss of norms, nomos is their construction and maintenance.

3. This material comes from a paper I wrote in 2006 discussing the use of the mobile phone within the family (Ling, 2006).

4. This is obviously not always the case. The development of the Internet came from the needs of the military and only later was it adapted to nonmilitary use.

5. http://www.technikum29.de/en/communication/fax.shtm.

6. There was also the use of bicycles in some cases (Bijker, 1997).

7. This was indeed the case. The clock replaced a whole set of bells in medieval Italy (one for the church, one for the town hall, one for the market, etc.) with a simple system of regulation (Glennie and Thrift, 2009).

8. It is at this point that their social shaping has its freest hand.

9. There are some technologies such as the mobile phone that piggyback on earlier technologies. It was the landline telephone that broke the path for voice communication in competition with the telegraph. The landline telephone introduced the idea that we could talk with others who were not physically present (Fischer, 1992; Standage, 1998). Emergencies (Ling, 2004). The landline phone, however, facilitated the process through which we adopted the mobile phone.

10. This is the case of the fax in the United Kingdom where a postal strike in 1987 along with the decreasing cost of the fax machine gave the latter a foothold (Skelton et al., 1–95).

11. The transition can be arrested by unexpected events. The adoption of the telephone and the car were halted because of the depression of the 1930s and also due to

the Second World War. These events halted the ability to produce products. However, once past these events, the rapid adoption trajectory continued (Fischer, 1992).

12. Analysis indicates that the mobile phone is extensively used within the most intimate social sphere. Indeed this is found in a variety of social settings (Ling et al., 2010).

REFERENCES

Bar, F., Pisani, F., Weber, M., 2007. Mobile technology appropriation in a distant mirror: baroque infiltration, creolization and cannibalism. Seminario sobre Desarrollo Económico, Desarrollo Social y Comunicaciones M'oviles en América Latina.

Bell, G., Dourish, P., 2006. Yesterday's tomorrows: notes on ubiquitous computing's dominant vision. Personal and Ubiquitous Computing 11 (2), 133–143 (doi: 10.1007/s00779-006-0071-x).

Berger, P., Kellner, H., 1964. Marriage and the construction of reality. Diogenes 45, 1–25.

Bijker, W.E., 1997. Of Bicycles, Bakelites, and Bulbs: Toward a Theory of Sociotechnical Change. MIT Press, Cambridge.

Castells, M., 1996. The Rise of the Network Society. Blackwell, Oxford.

Durkheim, E., 1951. Suicide. The Free Press, New York.

Durkheim, E., 1997. The Division of Labor in Society. The Free Press, New York.

Ehrenreich, B., 2008. Nickel and Dimed: On (not) Getting by in America. Holt Paperbacks, New York.

Elliott, A., Urry, J., 2010. Mobile Lives. Routledge, London.

Fischer, C., 1992. America Calling: a Social History of the Telephone to 1940. University of California, Berkeley, CA.

Fortunati, Leopoldia, Katz, J.E., Riccini, R., 2003. Mediating the Human Body: Technology, Communication and Fashion. Lawrence Erlbaum, London.

Fortunati, Leopoldina 2005. Mobile phones and fashion in post-modernity. Telektronikk, 3(4–2005).

Glennie, P., Thrift, N., 2009. Shaping the Day: A History of Timekeeping in England and Wales 1300–1800. Oxford University Press, Oxford.

Gordon, E., Silva, A. de S.e., 2011. Net Locality: Why Location Matters in a Networked World, 1st ed. Wiley-Blackwell, Oxford.

Hjorthol, R., Jakobsen, M.H., Ling, R., Nordbakke, S., 2007. Det mobile hverdagsliv: Kommunikasjon og koordinering i moderne barnefamilier. In: Lüders, M., Prøitz, L., Rasmussen, T. (Eds.), Personlige Medier. Livet mellom skjermene. Gyldendal Akademisk, Oslo.

Jackson, K.T., 1985. Crabgrass Frontier: The Suburbanization of the United States. Oxford University Press, New York.

Jensen, R., 2007. The digital provide: information (technology), market performance and welfare in the South Indian fisheries sector. The Quarterly Journal of Economics 122 (3), 879–924.

Kahn, J.M., Katz, R.H., Pister, K.S.J., 1999. Next century challenges: mobile networking for & ldquo; Smart Dust & rdquo; Proceedings of the 5th annual ACM/

IEEE international conference on Mobile computing and networking, MobiCom '99 (pp. 271–278). New York, NY, USA: ACM. doi:10.1145/313451.313558.

Lasén, A., 2011. Mobiles are not that personal: the unattended consequences of the accountability, accessibility and transparency afforded by mobile telephony. In: Ling, R., Campbell, S. (Eds.), Mobile Communication: Bringing us Together or Tearing us Apart? Mobile Communication Research Series. Transaction, New Brunswick.

Licoppe, C., 2004. Connected presence: the emergence of a new repertoire for managing social relationships in a changing communications technoscape. Environment and Planning: Society and Space 22, 135–156.

Ling, R., 2004. The Mobile Connection: The Cell Phone's Impact on Society. Morgan Kaufmann, San Francisco.

Ling, R., 2006. Flexible coordination in the Nomos: Stress, emotional maintenance and coordination via the mobile telephone in intact families. In: Kavoori, A., Arceneaux, N. (Eds.), Cultural Dialectics and the Cell Phone: Essays in Social Transformation. Peter Lang, New York, pp. 61–84.

Ling, R., 2012. Taken for Grantedness: The Embedding of Mobile Communication into Society. MIT Press, Cambridge, MA.

Ling, R., Bertel, T., Sundsøy, P., 2012. The socio-demographics of texting: an analysis of traffic data. New Media & Society 14 (2), 280–297.

Ling, R., Bertel, T., Sundsøy, P.R., 2010. Texting among same-aged individuals: An analysis of traffic data. International Communication Association pre conference on mobile communication. Singapore.

Ling, R., Bjelland, J., Canright, G., Engø-Monsen, K., Sundsøy, P.R., forthcoming. Small and even smaller circles: The size of mobile phone based core social networks in Scandinavia and South Asia. Journal of Intercultural Research.

Ling, R., Yttri, B., 2002. Hyper-coordination via mobile phones in Norway. In: Katz, J.E., Aakhus, M. (Eds.), Perpetual Contact: Mobile Communication, Private Talk, Public Performance. Cambridge University Press, Cambridge, pp. 139–169.

Markus, L.M., 1987. Toward a critical mass theory of interactive media: universal access, interdependence and diffusion. Communication Research 14 (5), 491–511.

Palen, L., 1998. Calendars on the New Frontier: Challenges of Groupware Technology. University of California, Irvine, Information & Computer Science, Irvine.

Silverstone, R., Haddon, L., 1996. Design and domestication of information and communication technologies: technical change and everyday life. In: Silverstone, R., Mansell, R. (Eds.), Communication by Design: The Politics of Information and Communication Technologies. Oxford University Press, Oxford, pp. 44–74.

Silverstone, R., Hirsch, E., Morley, D., 1992. Information and communication technologies and moral economy of the household. In: Silverstone, R., Hirsch, E. (Eds.), Consuming Technologies: Media and Information in Domestic Spaces. Rutledge, London, pp. 15–31.

Simmel, G., 1903. The metropolis and mental life. In: Levine, D.N. (Ed.), Georg Simmel: On Individuality and Social Forms. University of Chicago, Chicago, pp. 324–399.

Skelton, C., Lynch, T., Donaldson, C., Lyons, M., 1995. Modelling Product Life Cycles from Customer Choice. British Telecommunications Laboratories, Ipswich.

Standage, T., 1998. The Victorian Internet. Weidenfeld and Nicolson, London.

Tolmie, P., Pycock, J., Diggins, T., MacLean, A., Karsenty, A., 2002. Unremarkable computing. Proceedings of the SIGCHI conference on Human factors in computing systems: Changing our world, changing ourselves, CHI '02. New York, NY, USA: ACM, pp. 399–406, doi:10.1145/503376.503448.

Wallis, C., Qiu, J., Ling, R., 2012. Mobile handsets from the bottom up: appropriation and innovation in the global south. In: Gates, K. (Ed.), Media Studies Futures. Blackwell, London.

Weber, M., 1930. The Protestant Ethic and the Spirit of Capitalism. Routledge, London.

Weiser, M., 1991. The computer of the 21st century. Scientific American, Scientific American, (September), 94–104.

Weiser, M., 1993. Some computer science issues in ubiquitous computing. Communications of the ACM 36 (7), 75–84.

Weiser, M., & Brown, J.S. (1996). Designing calm technology. Power-grid Journal, 1. Retrieved from: <http://citeseerx.ist.psu.edu/viewdoc/summary?doi=10.1.1.135.9788>.

Wurtzel, A.H., Turner, C., 1977. Latent functions of the telephone: what missing the extension means. In: De Sola Pool, I. (Ed.), The Social Impact of the Telephone, MIT Press, Cambridge, pp. 246–261.

Chapter 3

On Living in a Techno-Globalized World

Questions of History and Geography

David Morley

INTRODUCTION

My starting point is that much of media and communication studies, as presently constituted, suffers from a drastically foreshortened historical perspective, the absence of which is all the more critical now, as we enter the world of the digital media. Thus I argue that media studies needs to place contemporary developments, such as the constitution of cyberspace—with which much contemporary work is concerned—in a much longer historical perspective. As we now enter an era of digitalization, technical convergence, individualized and interactive media systems, all these issues become all the more urgent. As Lynn Spigel (2004) has put it, the more we speak of futurology, the more we need to take a longer historical perspective on these issues. In this context we must neither be "suckered" into believing all the hype about how much the new media are going to change the world, *nor* being too cynical by insisting that we have "seen it all before." The key question here is how to see contemporary changes in media cultures in historical perspective.

We are often told that, under the impact of the new technologies of our globalized age we live increasingly in a borderless world, characterized by unprecedented rates of mobility, and by the experience of time–space compression, resulting from the speed of communications and transport links. To this extent, we are offered an image of a brand new world of "all at onceness," which is remarkably similar to Marshal McLuhan's 1960s image of the "global village" of "instantaneity." New technologies and global cultural flows now transgress the boundaries around our nations, localities and homes, in so far as the media continually flood us with images of hitherto unknown people and places, and however, we should not mistake these media's "reach" for their "power." They may be almost ubiquitous, but theirs is often a rather "thin" presence, diluted by local contexts of reception. Thus global cultural forms still have to be made sense of within the context of local forms of life.

In some versions of the story of globalization, we are offered what I would characterize as an abstracted sociology of the postmodern, inhabited by an un-interrogated "we," who "nowadays" live in an undifferentiated global world. Moreover, the presumption is that our lives are increasingly determined by the effects of the "new media." From my own point of view, this is particularly problematic, as the technologically determinist nature of the claims made for the (seemingly) "automatic" effects of the new media fly in the face of the last 30 years of audience research, which has demonstrated the very complex ways in which different media technologies are, in fact, interpreted and mobilized by their users. It is curious that the discussion of new media often returns us to an outmoded discourse of "automatic" media effects whether from a dystopian or utopian point of view. My argument is that what we need is a perspective on how a variety of media technologies, both new and old, are fitted into, and come to function within, a variety of different cultural contexts.

That kind of "contextualist" approach to questions of technological change is defined by Jennifer Bryce (1987) as one in which, rather than starting with the internal "essence" of a technology and then attempting to deduce its "effects" from its technical specifications, one begins with an analysis of the interactional system in a particular context and *then* investigate how any particular technology is fitted into it.

Clearly, no technology has straightforward impacts—not least because one has to begin with the question of which people (differentially) see the relevance (or irrelevance) of any given technology to them and how they ignore, or mobilize and use it, in the specific cultural context of their own lives. This approach evidently shares much with the anthropological concept of consumption as a form of "domestication" (or indigenization) of technologies—by which the objects are effectively "customized" by being fitted into local patterns of use. This emphasis is also similar to that of the "circuit of culture" model developed by Stuart Hall and his colleagues at the Open University (du Gay et al. 1996), which likewise argues that context is no "optional extra," which we might study at the *end* of the analytic process, but rather, is best seen a "starting point"—which has determining effects on both production and consumption.

GROUNDED THEORIES OF TECHNOLOGY IN CONTEXT

In relation to the broader questions of technology and context, my general point concerns the need to focus *not* on digitalization in general, or on cyberspace in the abstract, but rather on the particular types of cyberspaces which are instituted in specific localities, under particular cultural, economic and political circumstances. This is to follow the example set by Danny Miller and Don Slater (2000), in their study of the Internet in Trinidad—as a way of studying how the worlds of the virtual and the actual are differently integrated across the globe in specific contexts. To take a further example of the importance of *how* a particular technology is instituted in a particular context, elsewhere Slater examines the contrasting cases of two villages in western Africa. In the first village, a large amount of "aid" money was acquired and a customized hall was built to house some of the latest high-speed modem computers in a purpose-built fully air-conditioned environment. Unfortunately, the hall was built in a slightly out-of-the-way location, which did not connect well with the customary pathways along which the local villagers traveled, and so this powerful technology was rather underused. In the second case, a returning Western "volunteer" left behind, in the village he'd worked in, a clapped-out laptop with a dodgy modem, giving it to the owner of a local cafe, who set it up on one of the tables in the back of his bar. Although this was a much less powerful piece of technology, the cafe where it was sited had, directly outside it, the bus stop and the taxi rank where the local people from the other villages around passed through, on their way to market. As a result of it "fitting" much more effectively with the established communications systems of the community in which it was sited, this less powerful technology had much more consequential effects than the purpose-built computers in the air-conditioned environment in the other village.

For these kind of reasons, I am very much in favor of returning to the classical definition of communications, which would reincorporate the study of physical movement within its remit. Thus, rather than focusing exclusively on the transmission of messages, which is what we tend to think of these days, when we say "communications," we might better consider how these questions can be articulated with questions of transport. This would then be to incorporate the study of the movement of people, and commodities, and technologies alongside that of information—and contemporaneously, place the study of the emerging virtual realms alongside that of the material world (cf. Morley, 2011).

THEORIES OF TECHNO-GLOBALIZATION; QUESTIONS OF REGIONALITY AND PERIODIZATION

Theories of globalization often emphasize the role of communications technologies in the process of time–space compression and de-territorialization. In this model, the virtual world is sometimes then argued to have so thoroughly "replaced" the physical world that geography is declared to be dead. However, all of this is evidently based on rather simplistic ideas about the transformative effects of digital technologies. Coming, as I do, from a cultural studies tradition which prioritizes "grounded" theory and emphasizes specificity in empirical investigations, I find myself unsympathetic to the kind of abstracted "One-Size-Fits-All" analyses of globalization-thru-technology which dominate the field today.

Rather than generalist schemes, which try to reduce the whole of history to one Big Story, we may perhaps be better served by some differentiations between the stories and perspectives of a variety of regions, areas and periods. However, if we are to attempt to produce a more concretely regionalized perspective on globalization, there is a difficulty with the definition of the units of analysis to be used. In this context, we shouldn't simply take geographical "areas" as the units of our cartography (nor presume that, within each one, we will find only one set of exclusive or "dominant" properties). Rather, we should take the relevant cultural, political and economic forms as the basic units of analysis and then look to see where they are to be found, without assuming that they are "naturally" bound to any particular geography.

The further problem concerns just how *new* all this global mobility and technical change is. David Edgerton (2006) notes that we are often told that we are entering a new historical epoch, in which change takes place at an ever increasing rate, as result of the effects of increasingly powerful technologies. However, we can perfectly well be argued to live in an age of technological stasis, relatively speaking, compared with the speed of technological change at the beginning of the twentieth century (the era of the invention of radio, the cinema, photography, the steamship, the railroad and the aeroplane). As Edgerton points out, judging by these standards, the present does *not* in fact seem particularly innovative. Most historians tend to focus exclusively on technological *novelty*, while in reality, it is older technologies which continue to dominate our lives. Moreover, our accounts of technology are fundamentally unbalanced by a tendency to focus on invention over use, acquisition over maintenance, and inevitability over choice, when what matters more is *how* technologies are used and by whom, as they are transformed and "reinvented" in hybrid forms, and their use shifts from one context to another.

Moreover, this is often a process of *uneven* development, in which technologies which may have declined for a period in one context, reemerge later as the most important technologies of the era, in other places. Consider how the bicycle declined in the West in the late twentieth century, reemerging much later in the cities of the Third World and then returning to the West, in its new ecological incarnation. Indeed, despite the claims of the innovation-centric, technological futurism on which our present visions are based, some areas of the world have recently experienced significant technological *regression*—for instance, instead of animal-drawn farm implements in countries which previously would have used tractors; or in the transformation of the previously hi-tech shipbreaking industry into one which now takes place on the beaches of Bangladesh, where armies of men break up modern ships up for scrap, using medieval implements (Buerk, 2006). These issues also raise the eternally vexed question of historical periodization. Many technologies that we think of as "old"—the donkey cart, the wooden plough—have been maintained and used throughout the last century in many places, often alongside the aeroplane and the motorcar. As we know, old technologies often achieve new forms of symbiosis with more contemporary ones.

Thinking about the history of technology-in-use—and concentrating on the adaptation operation and maintenance of things—offers us a very different perspective from that of the innovation-centered model. First, it offers us a truly *global* history, rather than focussing exclusively on the small number of (largely Western) places where invention has been historically concentrated. By this means, we can then shift our attention away from the large-scale, spectacular, masculine, prestigious technologies of the rich white world, to also bring into focus the small-scale, mundane, feminized, and often creolized technologies of the shanty towns of the world—which are the products not of professional scientists, but of millions of untutored, but inventive amateur architects, engineers and builders. When we think of these vernacular architectures, we need to ask not simply what "standard" facilities the shanty towns lack, but rather, what (often novel) systems for the supply of necessities they have developed, in order to sustain a new kind of rapidly expanding urban existence on an enormous scale, even if only at subsistence level. These systems will often be adaptations of older, imported technologies from elsewhere, now given a new lease of life and adapted for local use (such as the tin drum, flattened into roof or wall; or the scooter crossed with the rickshaw to produce the *tuk-tuk* taxis of Thailand). More importantly these technologies at the heart of the fastest growing cities, such as Lagos and Mumbai—the places where the modernity of the future is already taking shape.

These issues bring us to the eternally vexed question of how to address the question of periodization in our work. Fernand Braudel (1984) always insisted on the need to recognize the simultaneous existence of different temporalities—just as Raymond Williams (1965) noted the coexistence of residual, dominant and the emergent tendencies within any one historical period. Thus, as Bruno Latour (1993) argues, "modern time," in any pure form, has never existed—for the eras of the traditional, the modern and the postmodern have always been jumbled up and we have always worked with a mixture of old and new technologies, simultaneously utilizing Paleolithic inventions such as hammers and nails alongside contemporary cordless electric drills. In a similar spirit, Herman Bausinger (1990) rightly argues that we need to explore the extent to which folk culture is alive and well in the world of modern technology.

THE GEOGRAPHY OF CYBERSPACE : QUESTIONS OF SPEED AND INSTANTANEITY

As I noted earlier, in recent years, a variety of technologies have been said to have transformed our lives—to the extent that we are held to be living in a new era of space/time compression, characterized by the wholesale "disembedding" of cultural, economic and political activities. Some have claimed that in this cybernetic world "we no longer have roots or origins—only aerials and terminals"; and that "we" now live in a world where geography no longer counts for much (Wark, 1994).

In these arguments about the "death" of geography, one example which is often cited is the growth of the telephone "call-centers," based in India, which now handle a lot of the "customer services" calls for a variety of British businesses. However, while these call centers no longer need to be on the geographical territory of the United Kingdom, in order to deal effectively with British customers, they are not, as it were "just anywhere." They are located precisely where they are because India can offer investors the attractive combination of a high level of indigenous English language skills alongside a low-wage economy, as the direct result of the long history of Britain's imperial presence on Indian soil. It is for reasons of exactly the same kind that French and Spanish call centers tend to be located in North Africa. The supposedly "de-territorialized" geography of our postmodern era is thus much more legible if one reads it as a set of "shadow" geographies created through the complex history of imperialism.

Morever, despite the widespread dissimulations of "placelessness" practised in these call-centers, cyberspace still has a very real geography. The relative density of Internet connections per sq kilometer varies enormously as between different geographical locations, and access to these technologies still depends very much on where you are located, in both geographical and social space. The distribution of these new technologies frequently mirrors established structures of power, and flows of Internet traffic tend to follow the routes laid down by previous forms of communication. Further, as Matthew Zook (2005) demonstrates, the economy of the information age is actually rooted in very particular places. Despite the Internet's ability to transcend space, the great majority of the world's dot.com companies are clustered in particular parts of a small number of urban conglomerations.

Rather than imagining that we stand on the cusp of an era in which the virtual geography of cyberspace will somehow entirely "replace" the physical geography of the world, we would perhaps do better to accept that the virtual is now increasingly becoming a banal overlay to the world of the actual. Moreover, I would suggest that speaking of "cyberspace" in general may be less useful than recognizing that *cyberspaces* are not only various themselves, but are best characterized by reference to the particular ways in which the virtual and the actual are mixed together in different locations, according to localized cultural rules and priorities.

The same issues arise in relation to the significance of what has recently been described as the analytic "shift from stasis to mobility" within the social sciences. While sympathetic to the contribution of the "new mobilities" paradigm developed, in the United Kingdom, by John Urry and colleagues (Urry, 2008), I also take note of Peter Adey's (2006) useful caution that "If mobility is everything then it is nothing"—which clearly implies the need for a carefully *differentiated* analyses of access to mobility for people in varying social and geographical locations. Thus, in the context of claims that "we" live in a culture of speed, we evidently need to also consider the opposite issue—the question of *waiting*—which is, of course, the fate of the poor, or indeed of all those who lack the qualifications (financial or otherwise) which give access to the relevant "fast-track" or priority lane. If, for the middle classes of San Paulo, their "mobility needs" are increasingly met by helicopters which fly to and from the pads on the rooftops of their apartment buildings, their maids must often travel 4–5 hours each way, by bus, to and from the outlying *favelas* in which they live. In a similar spirit, Alan Sekula (1995) notes that it may take up to a year for illegal migrants to get from China to Canada by ship, as they must spend days waiting, at various strategic points in their journey, in order to evade customs and border controls. It is these issues to which Doreen Massey (1994, 2005) points when she insists that we attend to what she calls the power-geometry of contemporary mobilities.

ANTHROPOLOGICAL AND COMPARATIVE PERSPECTIVES

Let me turn now to the question of Euro-centrism in our field. Interpretive paradigms derived from the West have often been imported wholesale and applied elsewhere without being appropriately tailored to the local situation. Moreover, most Western media theory is both self-absorbed and parochial, with the result that universalistic theories about the media are advanced on the basis of evidence derived usually from the same few Euro-American settings. Thus our (supposedly universal) models of the world's media are in fact "unduly influenced by the experience of a few, untypical countries." Indeed, it is clearly absurd to "universalize" the particular experience of places such as the United States and Britain, as if these affluent, stable democracies, with their Protestant histories and imperial entanglements, could possibly be seen as representative of the world at large.

The anthropologist Brian Larkin's starting point (Larkin, 2008) is to ask what media theory would look like, if it began from how the media actually work in a place like the contemporary Nigerian City of Kano, rather than in Europe or the United States? His concern is to examine "*where* the insights of [Western] media theory have force" and where their analytical assumptions turn out to be socially or culturally specific, rather than based on any "essential properties" of the technologies of which they claim to speak.

Larkin uses the Nigerian experience to defamiliarize our presumptions and thus reveal the *particularity* of "what goes without saying" in the West. By highlighting processes which are played down in analyses that assume the universal normality of what are, in fact, the specific sociopolitical configurations of the West, he explicates how these differential conditions might challenge many of our taken-for-granted assumptions.

His fundamental point is well-made in a simple example, where he explicates the quite different significance of an electrical blackout in New York and Kano. In the former, it is experienced as a terrifying disaster—in the latter, as a routine/momentary annoyance, before people switch over to their private generator. While western media theory routinely assumes the universal availability of well-functioning technological infrastructures, the people of the "megaslums" of the world live in thoroughly "privatized" worlds of everyday technological improvization, where "making do"/survival strategies, based around fragile and flexible local networks, are always necessary, given the unreliability of all structural forms of provision.

To argue all this is simply to recognize that, empirically speaking, the assumed "truths" of media studies only pertain to media operating within very particular types of sociocultural, legal and economic frameworks. Thus, to conclude by returning to the issue of "contextualism," before confronting the question of contemporary changes, technological or otherwise, we must be sensitive to the very many ways in which contexts produce "media cultures" in the first place.

Thus, in relation to the dynamics of the global variations within what has been called the "post-broadcast" era, Graham Turner and Jinna Tay (2009) note how easy it is to get carried away by the overwhelming trends in Western media markets, the *similarity* of which has perhaps encouraged Western scholars to underestimate the *contingency* of the relations between television, nation and culture to which they are accustomed. Thus, they argue for the need to recontextualize the "default-settings" of Anglo-American Television studies—and to shift its focus, so as to better recognize the extent to which local, national and regional histories still determine the contemporary development of global television. This also involves recognizing the variety of platforms, formats and distinctive cultural forms currently in play in different geo-linguistic markets. Above all, we must recognize that *not* all of them are developing in the same *direction*—let alone, at the same rates—as are the Western media on which most of our theoretical models are based, and to this extent, we should resist any siren calls to extrapolate improperly "universalized" perspectives from the development of the contemporary Western media.

REFERENCES

Adey, P. (2006) "If Mobility is Everything, Then it is Nothing" in *Mobilities* Vol 1.1

Bausinger, H. (1990) *Folk Culture in a World of Technology*, Bloomington: Indiana University Press.

Braudel, F. (1984) *The Perspective of the World (Civilization and Capitalism* Vol III) New York: Harper and Row.

Bryce, J (1987) "Family Time and Television Use: A Contextualist Approach" in T Lindlof (ed) *Natural Audiences*, Norwood NJ: Ablex . . .

Buerk, R. (2006) *Breaking Ships*, New York: Chamberlain Bros Publshing.

Edgerton, D. (2006) *The Shock of the Old*, London: Profile Books.

du Gay, P., Hall, S., Janes, L. and Mackay, H. (1996) *Doing Cultural Studies: The Story of the Sony Walkman* Milton Keynes: Open University Press.

Larkin, B. (2008) *Signal and Noise* Durham, NC; Duke University Press.

Latour, B. (1993) *We Have Never Been Modern*, Hemel Hempstead: Harvester Wheatsheaf.

Massey, D. (1994) *Space, Place and Gender*, Cambridge: Polity Press.

Massey, D. (2005) *For Space*, London: Sage.

Miller, D. and Slater, D. (2000) *The Internet: An Ethnographic Approach* London: Berg Publishing.

Morley, D. (2011) "Communications and Transport: The Mobility of Information, People and Commodities" *Media, Culture and Society* Vol. 33.5.

Sekula, A. (1995) *Fish Story*, Rotterdam: Witte De With Centre For Contemporary Art.

Spigel, L. (2004) "Introduction" in L Spigel and J Olsson (eds) *Television After TV* Raleigh, Durham, NC: Duke University Press.

Turner, G. and Tay, J. (2009) *Television Studies After TV*, London: Routledge.

Urry, J. (2008) *Mobilities*, Cambridge: Polity Press.

Wark, M. (1994) *Virtual Geography*, Bloomington: Indiana University Press.

Williams, R. (1965) *The Long Revolution*, Harmondsworth: Pelican Books.

Zook, M. (2005) *The Geography of the Internet Industry*, Oxford: Blackwell.

Chapter 4

Intercultural Communication for Development in a Buddhist Perspective

Patchanee Malikhao

When I was a child in Thailand in the 1970s, our family often went from Bangkok to visit our uncle who lived upcountry. Year after year, we heard that the village where our uncle had a pig farm and a paddy field became more and more "developed." That meant that he had a better road to his house; he did not need to generate electricity anymore, as electricity poles came to every household in the village; he had tap water instead of having to pump water from the irrigation canal to his home. Later, I also learned that my country, Thailand, was called a "developing country" as the improvements needed to be called "developed" had been clustered in the capital city, Bangkok. Moreover, our GDP was so low. In 2009, 8.1 percent of the Thais, or about 5.6 million people still lived under the poverty line. That is people who earn at most $1.25 dollars a day (http://www.indexmundi.com/thailand/population_below_poverty_line.html, accessed March 7, 2013). A friend said, "You should be happy that they eulogize our status a bit. We are not underdeveloped. We are developing."

Well, the term "developed" was meant to mean "material growth, and how much money you make."

In 2003–2007, I lived in Australia, one of the seven richest countries in the world. "You need to survive financially for the first two years because Australia is a developed country; the cost of living is expensive"—that's what we read in a leaflet advising people about how to migrate to the country. When we were there, we saw pictures in newspapers of aborigines living faraway in the outback in "reservations" who still had no electricity; who had to go on foot to school, and lacked tap water. It came as a shock that we lived in a nice neighborhood with electricity, tap water, regular public transportation, Internet and cable TV access, gas lines etc. This contrast made me rethink the term "development." In my view, the natives (the "original" inhabitants of the land) have the right to get a fair share of the development we were enjoying. That is what scholars call "social justice." Income distribution and a fair share of infrastructure and services should be one of the indicators to measure "development" too.

While living in Amherst, Massachusetts, I have observed shattered, run-down, and deserted factories in many towns. The centers of mass production have shifted to other parts of the world where labor is cheap and the regulations are lax or nonexistent. I am witnessing how both African Americans and Caucasian Americans have become homeless and unemployed during the recent economic crisis. I see them begging for dollars at a few crossroads in our town. Sad news breaks daily: mass massacres from shooting sprees in many places; foreclosures of homes; huge deficits as a consequence of launching wars in Iraq and Afghanistan; and severe budget cuts that will affect the education system, airlines, social welfare services and more. Of course, we have electricity, tap water, water closets, Internet access, a central heating system, and a phone. However it is alarming that 46 million of Americans still live below the poverty line. In 2012, a single person under 65 years who earns less than $11,945 per year is considered poor (https://www.census.gov/hhes/www/poverty/data/threshld/index.html, accessed March 7, 2013). On average, a poor American earns about 33 dollars a day, which is almost 30 times more than a poor Thai. Development is quite relative, isn't it? The level of satisfaction and the quality of life (or lack thereof) may be more or less the same, but the amount of money one earns is so different.

Does development mean safety of life, a safety net for the underserved, and income distribution? Can the people in those almost deserted towns get organized and revitalize their own communities?

OUT OF CONTROL?

In the 1960s, development focused only on the economic dimension: growth in terms of GDP and per capita income. At that time, the framework of development was under the *modernization paradigm*. Empirical indicators such as the level of literacy, the coverage of railroads, roads, electricity, etc., were good indicators of how developed a country was. Later there was a concern about fair income distribution among those who benefitted from modernization and growth: the middle class and those who had fewer opportunities: the grassroots. Then it was argued that development should include social justice as one of the development indicators. Social justice refers to more income distribution and a fairer share of infrastructure and services.

Indeed, development is a discourse. It has many different meanings and definitions. As Wolfgang Sachs (2013, 5), a German scholar in development economics puts it:

> On the one hand there are those who implicitly identify development with economic growth, calling for more relative equity in GDP. Their use of the word "development" reinforces the hegemony of the economic world view . . . On the other hand, there are those who identify development with more rights and resources for the poor and powerless. Their use of the word calls for de-emphasizing growth in favor of greater autonomy of communities.

Regarding development theories, the modernization paradigm, which endorses economic growth, has been popular since the 1950s. It emphasizes development as a path that traditional countries should follow and step up the ladder to get to the holy grail of living like Western people. That assumption has proven to be inadequate. The slumping down of the European and US economy; the busting of the inflated housing market in Japan and the United States; the droughts and strong UV levels that cause skin cancer in Australia; the suicidal cases in Japan and Korea, etc. should not be dream goals for any developing country.

Now, it has become clear that Westernization should not be the ultimate goal of development. Booming economies that lead to environmental disasters or global climate change should not be considered a desirable development goal. The People's Republic of China is a case in point. Recently, the Chinese government admitted that there are a number of cancer villages where water is no longer drinkable due to chemical waste dumps in waterways. Smog from factory smoke stacks cause respiratory problems (http://www.the-scientist.com/?articles.view/articleNo/34500/title/China-Admits-to--Cancer-Villages-/, accessed March 7, 2013). At present, citizens in Beijing and other major cities in China and in other developing countries are suffering from more than threshold-levels of pollution in the name of modernization and growth. Instead of battling poverty, diseases and ignorance, the governments of many developing countries are implementing economic and environmental policies that aggravate the poverty-stricken grassroots and induce new sorts of cancer and respiratory diseases out of ignorance, negligence or corruption.

MEDIATIZATION AS MODERNIZATION AND GLOBALIZATION

Hjarvard (2013, 7) argues that:

> mediatization should be viewed as a modernization process on a par with globalization, urbanization, and individualization. The modernization process causes

> urbanization as economic growth takes place more in big cities where human resources are needed. Stark differences between the level of development in the centers and peripheries in the world system (North-South), and in a nation-state of developing countries (capital/big cities-rural areas) are the consequences of geo-economic and political factors.
>
> The study of mediatization is an interdisciplinary discipline using critical sociology to a long term study of the relationships of culture and society transformed by the role of the media in a large scale. (Hjarvard, 2013, 3)

Mediatization is a new area of research; it is not the same as mediation because mediation concerns a smaller scale of communication process, not the structural change of the society at large (Hjarvard, 2013, 2–3). In the postmodernity or late modernity, the media, due to advanced communication and information technologies, have become the propeller of the culture industry. The media institution impacts are observable in the social interactions and cultural practices. The impacts of which are so bountiful that we cannot exclude the role of the media from structural culture change.

In the 1970s, the so-called Latin American dependistas proposed another paradigm, the so-called "Dependency Paradigm," criticizing multinational corporations and neo-colonialism that moved natural and human resources around the globe in the name of "globalization." The dependistas were criticized as well because they blamed outside actors and factors too much, and were less concerned about their own problems of internal colonization in which the wealthier exploited the poor in one's own country.

Globalization was not a new concept in the 1970s though. According to historians, the globalization process started long ago in archaic times. Transnationalization, colonialization, and postcolonialization fall in the wide category of globalization. Contemporary globalization is conceived as a free flow of cultural products, information, social practices and economic/financial exchange that go across the boundaries of the nation-states. Globalization is not similar to modernization as it is not a homogeneous and unidirectional movement: it is spreading unevenly. That means some areas are more globalized than others.

Mediatization, as a part of globalization, characterizes the postmodern culture industry. The productions and products of the communication and cultural industry record and reproduce cornucopia of social interactions, representations and organization systems in diverse media formats. Mediatization brings about new forms of contexts in a profound way.

Individualization can be traced back to the study of Carl Jung's work on individuation (http://www.mindstructures.com/carl-jung-individuation-process/, accessed March 27, 2014). He stated that:

> it is a process by which individual beings are formed and differentiated; in particular it is the development of the psychological individual as a being distinct from the general, collective psychology.

According to Jung, the ego is a subject of one's consciousness and a subset of the self which includes both the consciousness and unconsciousness of the total psyche. Globalization increases this individualization process. Consequently, the German Sociologist Ulrich Beck proposed his individualization theory, arguing how individuals in the late modernity take their lives in their own hand (due to migration and economic opportunities) and thereby create a new identity of social life (http://damienlanfrey.net/web/index.php?option=com_content&view=article&id=12:notes-on-ulrich-becks-qindividualizationq&catid=13:modernity, accessed March 27, 2014).

Individualization goes hand in hand with mediatization. With the breakthrough of the new media as a consequence of the digitization revolution, new formats of self-expressions have become popular as the audience gains recognition of his/her private, social and public achievements (Hjarvard, 2013, 150). Hjarvard (2013, 11) explains that the mediatization process affects an individual autonomy and social belonging in such a way that the individual gains more autonomy by relying deeply on the external world in the act of connecting to the large social networks available. He calls this phenomenon, *soft individualism*. Elliott and Lemert already observed a new kind of individualism in 2006. They propose that globalization has a profound impact on the individual level. They defined this *new individualism* as a highly risk-taking, experimenting and self-expressing individual underpinned by new forms of apprehension, anguish and anxiety. High levels of individualism can lead to *narcissism*. Twenge and Campbell (2009, 19) state in their book, *The Narcissism Epidemic*, that the central feature of narcissism is a very positive and inflated view of self and this value is growing rapidly in the American culture fueled by the mass media, including the new media, and changes in parental approaches to upbringing that emphasizes self-expression. Symbolic representations of the new American culture of self-expression or participating audience/amateur journalist are the emphasis on celebrities in the media, the success of Facebook as social networking sites, the uploading of personal videos on YouTube, twitter (microblogging and text-based social networking or SMS on the Internet via its own website) and blogging (Twenge and Campbell, 2009). In many cases the audience can be a target or a commodity when profit-making and commercial values are built into the media system of advertising (Jackson, Nielsen and Hsu, 2011, 63). The self or ego of the audience will in these cases be coupled with commercial products to increase self-confidence, self-respect, self-esteem, etc.

In the mediatization process, the audience is participating more because the media technology allows them to be active, not only as consumers but also as producers. In journalism that means that the demarcation between professional journalists, mass communicators and citizen journalists has become thus somewhat blurred. For some this is considered a positive development, others are less convinced.

PARTICIPATION FOR HAPPINESS

The multiplicity paradigm as proposed by Jan Servaes in the1990s, emphasizes participatory communication and planning strategies to achieve one's desired goal of development. Development goes hand in hand with social change, preferably social change which is sustainable. If it is not sustainable development, our natural resources will become exhausted and the environmental impacts, such as the greenhouse effects, climate change, the dislocation of plates due to the melting of the ice in the North Pole, the big holes in the ozone layer in the Southern hemisphere, etc. will drive us into extinction.

Development cannot be achieved without sustainable social change! That is why the Bhutanese government announced in 2010 that Bhutan is not going to consider the growth of GDP as the single indicator for development. The Bhutanese use the term GNH or "gross national happiness" instead. Though, it is difficult to fathom the level of happiness, it is obvious from the examples given above that having more money does not necessarily lead to more happiness. The Bhutanese GNH implies good governance, sustainable socioeconomic development, cultural preservation, and environmental conservation. It seems they are applying the multiplicity paradigm in their framework of sustainable development. This is more congruent to the kind of development the public want. No one wants to have cancer. No one wants epidemics. People want a clean and green environment, clean water and food, good education for their children, good and affordable health care, good elderly care and childcare. In sum, they all want peace and happiness.

THE BUDDHIST WAY

P.A. Payutto, a renowned Thai Buddhist monk, proposed that sustainable development in a Thai Buddhist perspective is comprised of three components: economy, ecology, and evolvability. In his view, the middle way or the balance of demands and resources has to be integrated (Payutto, 1996,

168–173). His new term, "evolvability," means the ability of a human to develop oneself to live in harmony with nature, not to conquer the nature, nor destroy it. In his view, *human development* has to come first.

Humans need to be trained to have the right attitude toward the natural environment and conserve one's own natural resources. That is what we called "being educated." Educated citizens do not need to hold a degree from any institution. They must be lifelong learners with minds that possess what Buddhists call four sublime states of mind: metta (compassion), karuna (loving-kindness), mutita (sympathic joy) and ubekka (equanimity). They must possess these characteristics: first, the right occupation meaning not to exploit other humans nor the environment; second, a moral conduct to maintain self-contentment; and, third, the right wisdom to accept differences among mankind and understand that there are different paths to development with the aim to conserve the environment (Payutto, 1996, 183–184). This view of Payutto strongly advocates the cultural perspective of development, that is, putting people at the center of development should be on the top of our development agenda.

There is a saying in Chinese that the reverberating energy of picking a flower can be felt among the stars. This means that we should see and foresee the relationships of beings and nature in a holistic way. Being uneducated, in this author's view, means seeing things in fragments, aiming at exploiting other humans and nature at an individual level or at a national level, considering only monetary profits, having a fixed mindset/worldview/beliefs, and not being able to appreciate others who are different in culture, social status, education, race, and ethnicity. Therefore, we urgently need to reeducate ourselves and plan to educate the new generations.

Development from within is needed as part of a moral ethics curriculum to uplift national pride. This should be passed on from one generation to the next. Being educated from within coins with the term selflessness or sunyata in Buddhism and its importance will be explained next.

When Prince Siddharta Gautama, who passed away 543 years before Christ, discovered how to attain Nirvana (supreme bliss), he proclaimed himself as Buddha, which means the "awakened" or the "enlightened" one. According to the Buddhist doctrine, the perception of "self" brings sufferings to us because we all are ignorant (avijja) in the true nature of things (May, 1984, 95). That is the fact that there is nothing we can hold on to as we all age; go through phases of change, get sick; and eventually die either of old age, sickness or accidents. The world keeps changing and nothing is here to stay. Impermanence or transience (anicca) is evident. For most Buddhists, it is known that the Lord Buddha taught that there is no substantially self (anatta) and the doctrine of anatta had become a dogma and a component of Buddhist identity (May, 1984, 93). By adhering to the thirst or craving for existence of self and sense experience (in Pali, tanha), suffering (dukkha) arises (May, 1984, 9).

Suffering arises because of one's clinging to the five aggregates or Panca-khandha that give us the individual illusory of self (or atta): matter (rupa), feeling (vedana), perception (sanna), mental formation (sankhara), and consciousness (vinnana). Then, come the clinging and attachment to "self." The "self" or ego is being fed by tanha (craving): craving for sensuality (kama tanha), craving to be . . . (bhava tanha), and craving not to be . . . (vibhava tanha).

According to the doctrine of causal genesis or dependent origination (Paticcasamuppada), the ultimate cause of "self" can be traced through 12 conditioned factors that we fabricate our "self." The way to cease the "self" is that we ought to see that the world is full of complexity. There is no linear thinking or direct causality. Together with this law, the ways to the cessation of "self" was explained by Buddha as the *four Noble Truths*. Masao Abe stated the essential of what Buddha taught as follows:

> The four Noble Truth, the fundamental teaching of Buddha run as follows: that existence is suffering; that the cause of suffering is craving or thirst; that by the extinction of craving nirvana may be attained; that the means for the attainment of nirvana is the practice of the Eightfold Noble Path: right view, right intention, right speech, right conduct, right livelihood, right effort, right mindfulness, and right concentration. (Abe, 1989, 205)

The concept of sunyata or voidness or naught appeared in the prajnaparamita or the perfection of wisdom as the antithesis of "self" (Thomas, 1951, 96). Buddhadasa famously stated that "Nirvana here and now" meaning that one can attain nirvana in this lifetime as one can be trained to be a selfless person. Working with the voidness of self, is appreciated as one does not put own bias or predisposition in seeing the truth or in the intervention of social interactions.

With moral conducts and selfless minds, educated individuals will not dump waste in waterways nor support the dumping of chemical waste by any corporation either within the nation-state boundary or from the First World country in other developing countries; will not sell medicines that do not meet the safety standards of the First World to people in the Third World; will not repack hormone or chemically tainted meat to sell in developing countries; will not use toxic chemicals to prolong the shelf life of fresh produce and meat; will not mix substances inadequate for human consumption in ready-for-consumption food; will not feed cattle or poultry with contaminated food or inconsumable substances or changing diets of farm-raised animals from natural to synthetic diets to lower costs; will not encourage the use of genetic engineering for food production and commercialization without profound research on the consequences to human body; will not use genetically modified product in food and not inform the consumers, etc.

It is now proven that the world is related and interrelated. What we don't want may come to us inevitably. For example, toxins emerged from the incineration of plastics in vegetables and milk, mercury and lead or even radioactive substances in seafood and fresh water fish, antibiotic and chemicals in farm-raised fish and prawns, radioactive clouds that can move across the continents, radioactive substances in fresh produce, GMO substances hidden in consumable products, contaminated fresh produce that cause food-borne diseases, and new forms of viruses, and more.

COMMUNICATION COMPETENCIES

Now it comes to communication to enable social change for the better. Good governance, civic society, participatory communication and all cannot be achieved without good communication, both mass and interpersonal. Servaes (2013, 371) proposed three streams of action: first, the media must be activated to build up advocacy for policy decisions; second, networking among interest groups and alliances, individuals, political forces, academic and non-academic organizations, business, industry, religious groups etc. is needed; public demand and movements of citizens to push development issues and agendas are needed. Participation and power in its nature and kinds are analyzed in detail in Servaes (2013).

As participatory communication concerns communication with people from different language (perhaps), attitudes, values, belief systems, disbelief systems and behaviors, intercultural communication is needed as a prerequisite to advocate participatory communication. Marshall Singer clarified that we experience everything in the world not "as it is" but only as "the world comes to us through our sensory receptors" (Singer, 1987, 9). It is not wrong to say that everyone's world is unique and different because of the influence of one's own culture[1] and the meanings of symbols we interpret. People who practice selflessness will succeed more in understanding others. Dialectical communication would be good for them to communicate as this requires intersubjectivity, meaning communicators can take turn subjecting to one another's authority and respect.

Martin and Nakayama (2004, 62) recommend a dialectical approach to intercultural communication: first, it is important to remember that cultures change and so do individuals; second, one should view various aspects of culture in a holistic perspective; and third, holding contradictory ideas simultaneously is not uncommon. They came up with six dialectics that characterize intercultural communication:

1. Cultural–individual dialectic. It is important to know that communication is sharing and it concerns both group members and the culture of the group.
2. Personal–contextual dialectic. While communicating with a person, never neglect the context to which that person belongs.
3. Differences–similarities dialectic. We must keep in mind that dealing with people from different cultures requires the realization of their differences to reduce prejudice and predisposition. There must be some similarities in the way humans communicate and build up experiences.
4. Static–dynamic dialectic. One should notice that some cultural patterns of ethnic relations remain constant while others are dynamic.
5. History/past–present/future dialectic. One needs to focus on the past and the present while doing intercultural communication.
6. Privilege–disadvantage dialectic. People may be simultaneously privileged in some respects and disadvantaged in others. This includes the ability to speak foreign languages.

We need to realize that participation is not just a fad. Actors who are competent in intercultural communication and have evolvability will certainly participate effectively under the framework of multiplicity that recognizes felt needs, environmental concerns, self-reliance, respect of culture, and conservation of natural resources. Self-evaluation of a community for social usefulness should be considered to complete the circle of acting, observing/measuring, reflecting, improving, and learning (Servaes, 2013, 376).

CONCLUSION

This author has come to the conclusion that although development has many dimensions—sociocultural, political, economic, and environmental—, human development should come first. Mediatization and communication technology increase the perception of self and realizing the virtue of being selfless from a Buddhist phenomenology will help cultivate educated-from-within communicators. Participatory communication and advocacy for sustainable development and the evaluation thereof needs selfless participants who possess evolvability and who are interculturally competent.

NOTE

1. For culture, Singer means: "Technically speaking, group-related, learned perceptions (including verbal) and nonverbal codes, attitudes, values, and belief and disbelief systems plus accepted and expected codes of behavior, taught by groups with which we identify, are what constitute culture." (Singer, 1987, 9).

REFERENCES

Martin, J. N. & Nakayama, T. K. (2004). *Intercultural Communication in Contexts.* McGraw Hill: Boston.

Payutto, P. A. (1998). *Sustainable Development* (third ed.). Bangkok: Buddhatham Foundation.

Sachs, W. (2013). Liberating the World from Development. *New Internationalist.* 460, March, p. 25.

Servaes, J. (2013). "Communication for Sustainable Social Change is Possible but not Inevitable." In Servaes, J. (ed.), *Sustainability, Participation, and Culture in Communication*. Bristol: Intellect, pp. 369–388.

Part II

ARGUMENTS

Chapter 5

Technological Innovation and Social Change

Colin Sparks

INTRODUCTION

Enthusiasm for the developmental potential of new media and information technologies is very widespread. All sorts of claims are made for the ways in which it will transform the lives of the world's poorest people and solve the apparently intractable problems of social exclusion, hunger and endemic disease. It is tempting for the historically minded scholar to dismiss much of this talk as uninformed tosh and try to remind the enthusiasts of the hopes placed in earlier technologies, like hybrid rice and satellite communication (Contractor, Signhal and Rogers, 1988). Similar claims were made for their transformative potential back in the last century. These technologies certainly produced changes but many of these were different to those their propagandist anticipated. Hybrid rice entrenched the social power of rich farmers rather than poor peasants and landless laborers, while satellite television turned out to be most potent in spreading entertainment rather than enlightenment (Pendakur and Kapur, 1997; Kumar, 1995; Page and Crawley, 2001).

Of course, there has been a substantial drop in the level of absolute poverty, adult life expectancy has risen and infant mortality has fallen in much of the world, but still the intractable problems of rural poverty remain endemic in many countries. According to the World Bank, in 2010 around 1.2 billion people were living in *extreme* poverty, which they define as an income of less than $US1.25 per day (World Bank, 2013a). With the more "generous" threshold of $US2 per day, there are still 22 countries in which more than half of the population remains in poverty and 16 in which more than 75 percent of the population are poor (World Bank, 2013b). Inequalities between different parts of the world remain gigantic, even on the most basic registers. Life expectancy in the high-income countries is still 10 years greater than in low-income countries, and while 2.7 children out of every thousand born alive will die before their fifth birthday in the high-income countries, for low-income countries the figure is 58.7 (World Health Organisation, 2013).[1] Technological changes have certainly contributed to these improvements, but neither old nor new technologies have acted as a panacea for the problems of development.

In this chapter I want to look briefly at some of the claims made for the transformative potential of new information and communication technologies and then examine in rather more detail three of the underlying theoretical propositions that sustain such views. The chapter begins by looking at the strengths of the case for technology as a motor for social change and then goes on to discuss how technologies spread in society and their interactions with the social structures in which they are embedded. It then looks at the record of successful development in a historical perspective and discusses the relationship between democracy and development. Finally, the chapter re-examines the potential of information and communication technologies in the processes of social change, and in particular their potential for empowerment rather than development.

THE CASE FOR TECHNOLOGICAL REVOLUTION

There has been considerable enthusiasm for new information and communication technologies for development (ICT4D) for more than two decades and it remains pervasive. There is a well-established scholarly journal dedicated to the subject and, here and elsewhere, almost every article on the subject begins by stating some version or another of the claim that these new technologies have transformative potential (Diaz Andrade and Urquhart, 2009; Duncombe, 2011; Ashraf and Malik, 2011; Islam and Grönlund, 2011). Scholarly studies are, of course, naturally cautious in their predictions but outside of the academy one finds more robust opinions. For example, Eric Schimdt of Google wrote after his 2013 trip to Africa that:

> The demographic dividend in Africa of young people is their greatest hope, in my opinion. Today high rates of unemployment show an economy underperforming to its true potential. This new generation expects more, and will use mobile computing to get it. (Schmidt, 2013)

Other, more scholarly, voices have long expressed the same view that the urgent problems facing the world's poor can finally be resolved now that a new technology is becoming available. More than a decade ago, UN's *Human Development Report* cataloged the four ways in which the potential of "rapid advances in two technologies—digital storage and processing of information

(information) and satellite and optical fibre transmission of information (communications)" had the potential to transform the prospects for development (United Nations Development Programme, 2001, 32). In their view, political participation, institutional transparency, income levels and health would all receive an enormous boost if only these new technologies could become widely, if not universally, available.

It is important to begin any discussion of these claims by recognizing that they have real substance. As Figure 5.1 demonstrates, mobile telephony is growing extremely rapidly in Africa, as it is throughout the developing world. It dwarfs fixed line connections (estimated at 1.4 per one hundred of the population in 2013). Internet connectivity in Africa was estimated at 16.3 percent of individuals in 2013 but only 6.7 percent of household had access at home, while 10.9 percent of individuals had active mobile broadband accounts (International Telecommunication Union, 2014). Internet penetration in Africa is still low, but it is growing, and its main vehicle is mobile telephony. All of the evidence, from Cairo to the Cape, suggests that young people in Africa are discontented and wish for a better life, and that they are likely to employ mobile telephony as an integral part of their lives, so we may conclude that, at least on this level, Schmidt is right in his prognostication.

There is another important sense in which the enthusiasts for these technologies are also right. As is very well known, mobile telephones and the Internet, however accessed, are by their nature dialogic devices. This differentiates

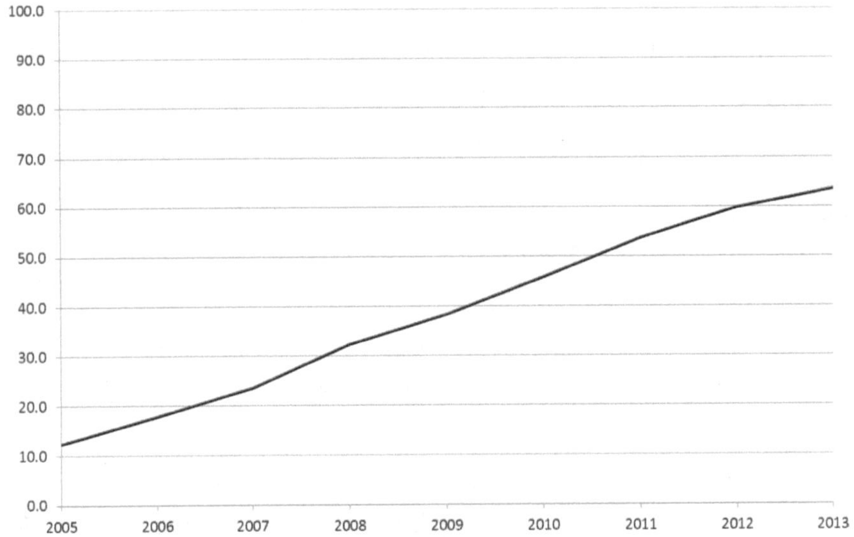

Figure 5.1 Mobile-Cellular Subscriptions per 100 Inhabitants in Africa. *Source*: International Telecommunication Union, 2014. Figures for 2012–2013 are estimates.

them sharply from the previous communication technologies, which had an inherent monologic bias. The premise of the classical paradigm of development communication, which is still alive and well in a whole raft of institutions, particularly those concerned with health communication, was that there was a one-way transfer of useful information from the expert to the nonexpert. At the very most, the nonexpert recipient was allowed to give feedback (the dotted line on all those visual representations of communication models) as to whether or not the program was working (Sparks 2007).

Various critical attempts to provide alternatives to this paradigm, most notably those based on participatory communication, have always had to wrestle with the inescapable fact that the technologies in question are not inherently either dialogic or democratic. Newspapers, radio, and television are and always have been monologic technologies designed for one to speak to many and they are only capable of very partial adaptation to the needs of democratic communication. What they fit best is the elite competition models of "democracy" as exemplified in the writings of people like Lipmann and Schumpeter, not to mention the practice of most contemporary advanced societies. Even the community radio station and the local news sheet, which have less pronounced distinctions between producers and audiences and which have thus better potential to realize the kinds of democratic involvement favored by participatory theories of communication, necessarily involve the selection of particular voices and the exclusion of others. As a consequence, lurking behind the rhetoric of democratic participation we often find the de facto domination of projects by particular elite groups.

New ICTs promise to change all this. In the classic paradigm, the archetypical peasant audience listened to the radio station, or gathered around the state-installed television set, and constituted an audience for programs designed by remote experts and transmitted over a satellite to a thousand villages across a subcontinent. After watching such programs, it was thought, the archetypical peasants would be filled with new (scientific) knowledge in the place of their traditional and erroneous beliefs. They would go forth and change their ways for the better. All of the evidence, unfortunately, is that any substantial learning and any lasting change was at best very marginal. With the new technology, the sons and daughters of the archetypical peasants can go to a telecenter (provided, it is true, either by business, the state or a big foundation) and decide for themselves how to use it. They may seek education, or the latest prices for their main cash crop, or information about the medical condition of their aged parents, or the electoral manifesto of their favorite party. They might just decide to play World of Warcraft. The point, however, is that *they* make the decisions, based on *their* needs, *their* desires and *their* choices. Few would claim that every last user of every last communication technology will be transformed by the experience but the potential to realize the participatory dream of the population as conscious subjects of the development process rather than the objects of social engineering decided by far-away elites is certainly present.

The gap between potential and reality, however, is notoriously wide in developmental activities, and although this potential is very real there are a number of obstacles in the way of the kinds of outcomes that Schmidt and other enthusiasts envisage. As one observer noted, the development literature is haunted by the memory of failed projects: "the literature [on information systems for developing countries] manifests an acute anxiety about failure" (Avergou, 2008, 137). Here, we only have time and space to look at three of the common obstacles encountered by projects designed to use ICTs in the context of democratic and sustainable development: the issue of diffusion; the social life of technology; and the very concept of development itself.

DIFFUSION

The study of the diffusions of innovations is one of the (few) areas where scholars engaged in social enquiry can look physicists in the eye and strut around claiming to be real scientists. The scholarly literature, with its most elegant and convincing articulation in the work of Everett Rogers, abounds with normal distributions, neatly divided into groups measured with the accuracy of one decimal point (Rogers, 1995). There are smooth S-curves that plot the gradual completion of the diffusion process. Compared with the middle-range theories that are the best that can often be observed in the chaos of social reality, this is immensely impressive. The problem with all of this is that while some technologies are adopted in ways that at least approximate to theory, others aren't. It takes quite some stretch of the imagination, for example, to describe the diffusion of the mobile telephone in Africa presented in Figure 5.1 as an S-curve. Figure 5.2 provides another example, from the developed world, in the take up of multichannel television in the United Kingdom. Again, it is very hard to claim that social reality approximates to the demands of theory: it requires both real determination and vivid imagination to categorize this as an S-curve.

In order to make any sense of the ways in which technologies spread throughout any given society, it is necessary to step back from the physics-envy and recognize that these processes, like any other social processes, are rather more complex and messy than is suggested by the nice scientific concept of "diffusion." The first thing to note is that the physicists will be very careful to specify under what conditions the theory they are advancing can be expected to hold, and social scientists should be even more concerned to make any such limitations on the validity of their idea. As is obvious, not all technological innovations actually do diffuse in any significant way:

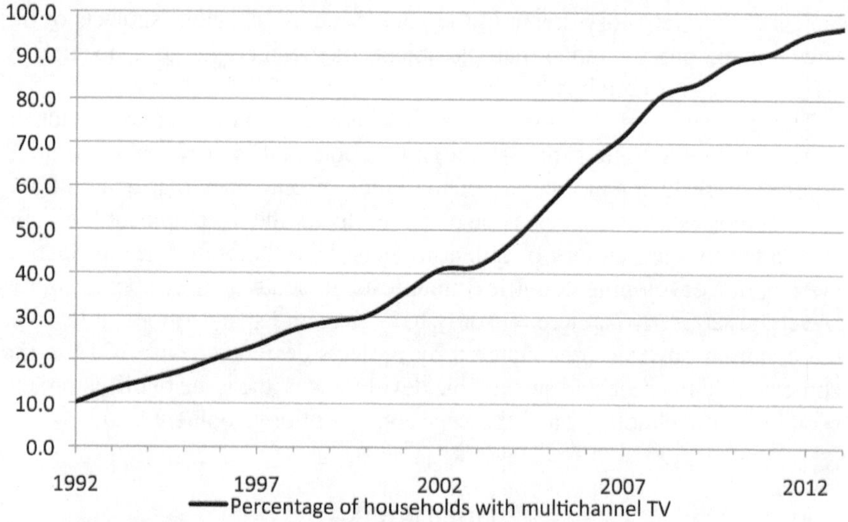

Figure 5.2 Multichannel TV in the United Kingdom. *Source*: Broadcast Audience Research Bureau, 2014.

stereophonic sound systems clearly did spread widely in the developed world but quadrophonic systems did not, at least in their original form (Postrel, 1990). Even those technologies that do spread do not always spread to the whole of society. Some spread to more or less everybody: in Hong Kong today, for example, there are more than two mobile phone chips in operation for every last man, woman and child in the Special Administrative Region (Hong Kong Trade and Development Council, 2013). Others bump up against all sorts of obstacles—economic, social, cultural, and legal—that means they only ever reach a small proportion of a given population. One well-known example much studied in the development and health communication literature is the struggle to ensure the regular use of condoms amongst sexually active individuals: it has long been known that in many parts of both the developed and developing world cultures of masculinity provide disincentives to the spread of what is hardly a new technology (Gómez and Marín, 1996; Marín, 2003; Sheeran, Abraham and Orbell, 1999).

The fact is that the origins of these intellectual problems lie in considering the questions as ones of the diffusion of technologies, rather one of the adoption of social practices. Different technologies do indeed have different affordances and understanding what a particular technology permits and what it prohibits is one extremely important dimension in the complex of social, economic and cultural factors that lead to a particular social practice. As we discussed above, new information and communication technologies, and in particular the Internet, have affordances which are potentially radically different to those of the earlier communication technologies that have been utilized in development programs. It is social processes, not technology, that determine whether these affordances are realized, by whom and in what ways.

A new social practice can be independent of technological change, closely related to a technological change, or the consequence of a group of quite different technologies. The uptake of multichannel television in the United Kingdom, cited above, is a case in point. The key affordance provided by what is termed "multichannel television" is the ability to receive more channels than those provided by the older analog terrestrial system; five in the case of the United Kingdom. This social reality is now almost universal in the United Kingdom and it this social process that has been adopted, not any particular technology. The technologies which have achieved this are various, involving both technical distribution mechanisms and models of revenue generation. Technically, these factors included: analog direct-to-home satellite, digital direct-to-home satellite, analog cable, digital cable, and digital terrestrial television. Economically, they included: subscription satellite services, free-to-view satellite services, subscription cable, free-to-view digital terrestrial and, briefly and unsuccessfully, subscription digital terrestrial television. While the social practice has certainly "diffused," it has done so through a range of different technologies based upon different economic models.

From this perspective, the classic model of the diffusion technologies can be seen as a special case of a more general problem of the adoption of social practices. We can identify four conditions that have to hold before we can expect the actual pattern of adoption to fit closely to the predictions of theory:

1. The technology in question must have one, unambiguous, affordance which allows adopters to address an issue that is present in their daily lives. The technology may also have other affordances, but from the point of view of its diffusion it is the central one that is significant. This affordance may, of course, be quite different from the one around which the technology was designed.
2. There must be a clear and unambiguous benefit from the changes in social practices that the technology facilitates. This benefit must be sufficient as to outweigh any associated disadvantages, at least in the minds of the population. Again, it is not necessary to hypothesize that the benefit with which the technology endows its users is the one for which it was originally designed, or that the users' computations of the balance of utility is either rational or fully informed.
3. The population in question must possess a relatively high level of uniformity, both with regard to the benefit that the technology would bring and with regard to their social and economic ability to acquire the new technology and to use it in the appropriate manner. Although there are strong

economic forces which push toward the cheapening of products as a consequence of mass adoption, in many cases there are countervailing factors, for example patents in the case of many medicines, which impede this process. Similarly, there can be powerful social differentiations which mean that only some groups within a population are in a position to acquire a technology while others, for whatever reason, find themselves unable to utilize it.

4. There must be no external impediments that prevent the social practices that the technology permits. An external impediment can either be negative, for example, a culturally sanctioned prohibition on a particular practice, or positive, for example, the need to abandon some other cherished practice in order to adopt the new one. To flirt with the language of economics for a moment, the opportunity cost of adopting a technology must be positive, and preferably strongly positive.

The diffusion of hybrid seed corn in Iowa certainly appears to have fitted these criteria fairly well, but the number of cases in which these conditions hold strictly is likely to be relatively limited. It is much more likely that the real-world outcomes of the diffusion of technologies will be better described by a theory which recognizes the limits imposed by these conditions. Rather than beginning theorizing this process of social change from a situation whose parameters are likely to be abnormal, it is surely better to try to develop a theory capable of accounting for both normality and abnormality.

THE SOCIAL LIFE OF TECHNOLOGIES

The literature on the use of ICT for development is more or less universally convinced of its potential and equally convinced of the social problems that are encountered in attempting to implement programs based upon its use. The much-studied programs to build telecenters in southern Indian states provide a substantial body of evidence that the existing social structures contain a range of factors that limit the take-up of the services. Participation in and usage of the facilities, even when consciously designed to avoid the most obvious problems, are strongly marked by gender, land-ownership, caste and educational level (Dossani, Misra and Jhaveri, 2005; Kendall and Singh, 2006; Best and Maier, 2007).

This should hardly come as a surprise to anyone familiar with the history of attempts to promote social development, but the implications of these findings go far beyond the sites upon which they are based. There is an underlying theme in much of this literature that echoes the classical paradigm of modernization: these problems exist because of the premodern social structures of the countries in question and one of the objectives of development is precisely to transform them into more modern societies in which such problems can be effectively challenged if not resolved.

This argument can only be effective if it can be shown that the problems of differential access do not exist in those more "modern" societies that are being held up as examples. In this case, part of the problem is that much of the evidence from South India is drawn from Kerala, in which a succession of leftist movements and governments has achieved some significant successes in "modernizing" the state with respect to combatting gender inequality and caste discrimination. What is more, the evidence appears to be that social differentiation in access to and ownership of technologies remains a feature even of very "modern" societies. For example, in 2010 in the United States, households with an annual income of $100,000 or more were at least twice as likely to have home access to the Internet as those on incomes of $25,000 or below—the 40 million or Americans who live around or below the official poverty line. Overall, as Figure 5.3 shows, home Internet access correlates very closely ($R^2 = 0.9759$, as it happens) with household income band (National Telecommunications and Information Agency, 2011).

This sharp differentiation of access to the Internet according to income is observable in one of the most advanced countries in the world, and one in which the spread of the Internet has been a reality for around 20 years. The fact that it might be an enduring feature of the patterns of access rather than a temporary result of unequal diffusion (early adopters are notoriously from the better off sections of society) is thus one that must be taken seriously. This conclusion is reinforced by a further National Telecommunications and Infrastructure Agency (NTIA) study, published in the same year, which

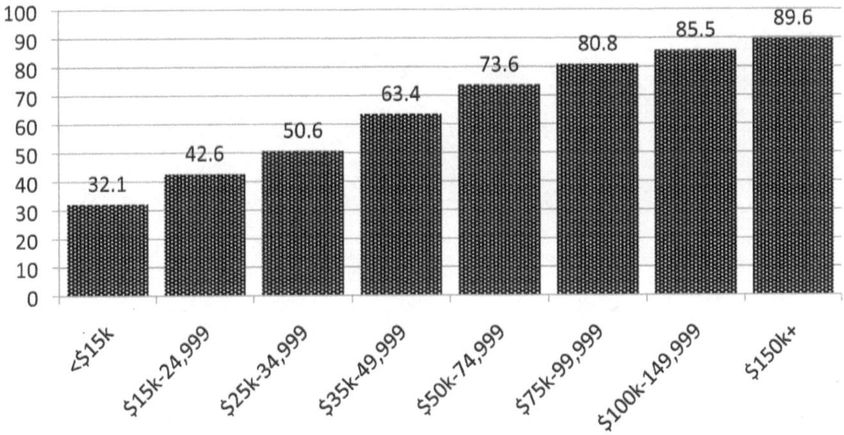

Figure 5.3 Home Internet use in the United States in 2010. *Source*: National Telecommunications and Infrastructure Agency, 2011a.

investigated the reasons that the stubborn 29 percent of Americans who do not have Internet access at home give for their situation. As Figure 5.4 shows, nearly half (47 percent) claim that they do not need the Internet. Expense is still a real issue, but the largest single group is made up of people who claim, rightly or wrongly, that it does not afford social practices that they would find valuable. Similar views have been expressed elsewhere with respect to a range of ICTs. One elderly Finnish respondent told a researcher that he did not use a computer because: "A computer? Why don't I use one? Well, it's not much use when it comes to shovelling snow and it's just in the way when carrying firewood" (Hakkarainen, 2012, 1206).

The US figures, however, are for the relatively crude measure of physical access which is in some countries much closer to complete availability, so this may just be a feature of adoption in a market society which can be overcome by judicious government intervention of the kind practiced in northern Europe. Research in the Netherlands, where there is very much wider availability, shows that while the issue of crude physical access is only of marginal importance, there are powerful social factors which determine what kinds of uses the technology will be put to. Notably, it is argued that the patterns of usage do not simply reflect social inequality but actually function to help reproduce it. Those with high income and education will use the Internet further to enhance their knowledge and skills, thus reinforcing their privileged position. Those with low income and education will use the Internet for entertainment and diversion. These patterns demonstrate that there is emerging

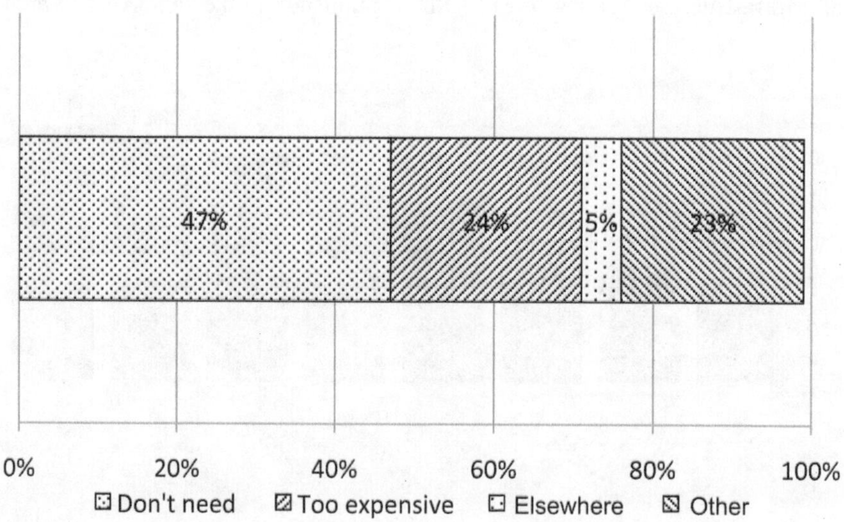

Figure 5.4 Reasons for US Citizens not having Home Internet Connections. *Source*: NTIA, 2011b.

a "structural usage gap." This gap is between different social groups, some of whom habitually "take advantage of the serious Internet activities they engage in, while others only use the Internet for everyday life and entertaining activities" (van Deursen and van Dijk, 2010, 908).

The evidence from "modern" countries thus suggests very strongly that the patterns of advantage and disadvantage that are pervasive in the Indian research are not ones that are a simple function of developmental level. While the precise details of social differentiation in usage and access might change, there is every indication that they are endemic to a differentiated society. The influence of social inequality upon the take up of particular technically dependent practices is so great that one European study concluded that overcoming the various dimensions of digital division was only possible if European societies were to adopt policies that aimed at producing "a social system that promotes the economic development and social welfare of its citizens by reducing inequality in all its various aspects." (Guerrieri, Bentivegna and Meliciana, 2010, 139).

The problem, however, is not simply that technologies encounter populations with different economic, social and cultural endowments, but they are designed and implemented in particular ways which in their turn reflect the influence of the social structures. Technologies are not simply the neutral products of societies: they embed the social relations within which they are conceived and developed. Any technological innovation encourages particular kinds of behavior and discourages others and this has implications far beyond the mere facts of its utilization. As one scholar studying the implementation of information systems, not specifically in a developmental context, put it:

> If we understand rationality in a substantive sense to refer to socially constructed, institutionally sedimented beliefs of what matters for the agents involved in organizing enactments and why, information systems innovation, bound to the economic rationality of modern management, is an attempt to enforce a particular rationality over others in the organizational reform. (Avergou, 2002, 236–7)

All of the evidence is that, in the context of recent developmental efforts, the substantive rationality that is being implemented in these projects is that of market economies and indeed one may almost conclude that in practice the concept of "development" has become more or less synonymous with the "economic rationality of modern management."

The South Indian examples involve systems explicitly designed to ease the working of markets and some, indeed, are financed by large agricultural businesses, which see the advantages in ensuring more consistent production quality and in reducing the price of their inputs through removing the layers of intermediaries that stand between them and their suppliers (Gollakota, Pick and Sathyapriya, 2012, 186). A similar market-oriented logic is present in schemes initiated by noncommercial organizations like foundations and by governmental organizations. Kendall and Singh, in a study of a network initiated by a nonprofit foundation, took as their sole measure of "success" the revenues accruing to the franchisees (Kendall and Singh, 2006, 3). The World Bank, a major and influential figure in development, set the tone in its *World Development Report 1998–99* which argued that only through privatization and much lighter regulation would it be possible to sustain development and realize the potential of the new ICTs (World Bank, 1998). They returned to this theme on many occasions in the subsequent years, for example, in their 2002 report *Building Institutions for Markets* and, as its own summary of its history states "the Bank embraced privatization as a necessary step for countries seeking the advantages of a market-based system . . ." (World Bank, 2001; Yusuf, 2009, 72).

As Yusuf notes in his historical account of the World Bank's strategy for development, and as we discuss below, there are some problems with the practical implications of this approach. It is likely, however, that ICTs designed with this approach in mind will be intended to promote social structures that are patterned after developed market economies. In the case of digital technologies, as we saw, the evidence is that, very far from eroding social inequalities, these technologies fit comfortably into societies one of whose major characteristics over the last decades has been increasing inequality. The introduction of technologies that reproduce inequalities into societies that are marked by inequalities is, to say the least, likely to lead to the continuation of inequalities. Indeed, what evidence there is suggests that this is the case: Gollakota and her colleagues, for example, found, although did not discuss in detail the fact, that users of the telecenters they researched had landholdings roughly twice the size of those who did not use the technologies (Gollakota, Pick and Sathyapriya, 2012, 197).

All technologies, and certainly the new information and communication technologies, are doubly determined by the societies in which they are both produced and consumed. The production of these technologies necessarily responds to the kinds of social relations within which they are conceived and attempts to address the problems and potentials of those societies. In the case of the new information and communication technologies, they were overwhelmingly produced in highly developed market economies characterized by high degrees of social inequality and were designed to be used by institutions and individuals who had been, more or less adequately, socialized into the expectations of such an order. As ICT4D, these same technologies, in both their physical and their social design, are introduced into societies with different institutions and cultures but also invariably marked by high degrees of social inequality. The likelihood is, then, that they, like earlier innovations, will be taken up in ways which reproduce and exacerbate those inequalities.

These technologies are complex and capable of multiple uses, some of which are quite contradictory. Even a technology designed with one simple function in mind can be used in different contexts for radically different ends. The AK47 is a technology with one simple function, but it has been used for many different social purposes: by the Soviet military to crush a popular revolt in Hungary in 1956; by Vietnamese peasants to defeat US invaders twenty years later; by the military wing of the African National Congress (ANC) in its attempt to overthrow the apartheid regime another twenty years later; not to mention by a host of other movements from pirates to partisans ever since. There is a world of difference between the technical complexity and social utility of that example and the sophisticated technologies we are discussing, so it is not surprising to find that the latter have been used in even more contradictory ways. Despite this complexity, however, the evidence suggests that the pattern of social adoption that is underway at this moment is very similar to that of previous technologies deployed for developmental goals: it will act to reinforce the existing social and political structure and may well exacerbate inequalities (Mody and Borrego, 1991, 164).

THE ELEPHANT IN THE ROOM

The idea of ICT4D implies that the technologies in question will be used to achieve some form of "development." This category is notoriously problematic and definitions have encompassed everything from crude measures of industrial output through life expectancy and gender equality to more problematic measures like human happiness. The United Nations Human Development Report annually produces a thick volume which attempts to measure and rank countries for their level of human development. Most contemporary accounts of development, and attempts to promote development, pay at least lip-service to some kind of participatory dimension in which the groups who will be part of the process are consulted as to their wishes and desires. Sustainable development, it is persuasively argued, can only be achieved if it is undertaken with the full consent of the people involved (Servaes, 2013).

This is an attractive proposition, which fits well with the democratic predilections of many observers, the present author included. Too many projects aimed at development and implemented from above, from grand catastrophes like the Soviet collectivization of agriculture or Mao's Great Leap Forward, to much less dramatic and mundane attempts to impose social change on unconsenting people, have had ruinous results for the individuals and communities involved. History, logic and humanity seem to point convincingly to the need to begin with the desires and ambitions of the people concerned and to find a path to development that helps them realize those goals. Of course, in practice, many projects fall short of this ideal, but even major promoters of development like the World Bank have long proclaimed participation and involvement as essential aspects of their activities.

The problem with this view of the world is that it ignores both the history of successful development and one particularly clear contemporary example. Historically, the development of the developed countries was anything but a participatory activity. It involved the eviction and dispossession of the peasantry, the brutal exploitation of the new proletariat, a massive and protracted international trade in African slaves, the destruction of indigenous industries in conquered territories and systematic genocide against whole populations, both directly through murder and less directly through the imposition of corrosive social and economic arrangements that produced disease, famine and despair. This long and bloody process, conducted without any semblance of democracy and famously labeled the primitive accumulation of capital, was the necessary precursor to the prosperity of those models of modernity that are held out as a goal for development. After all, the very prosperous modern city in which these lines are being written owes its origin to the military conquest of foreign territory in order to build a base from which an officially sanctioned drug trade could force its way into a country whose rulers were struggling to keep their population free of dangerous narcotics.

The material reality of successful "development" is not, however, simply a historical passage upon which we can look back with regret from the comfort of a more civilized present. The unfortunate truth is that the most successful example of development in the contemporary world has used contemporary versions of many of the same processes to great effect. That example is China, which, along a number of axes, has achieved developmental success on a scale unprecedented in human history. One crude and obvious measure of that achievement can be seen if we compare the numbers of persons in China living below the World Bank's poverty line of $US1.25 per day in purchasing power parities. In 1981, the first year for which comparative figures are readily available, there were 835 million people in China living on or below that line; in 2009 the figure was 157 million. In other words, more than 650 million people have been lifted out of the most abject poverty (World Bank, 2014). These millions have not, of course, all become wealthy, although some of them have become extremely rich, but the material circumstances of all of them have improved very substantially.

For this writer, and for many other authors who have advocated participation as a key to development, these are unpalatable facts. Of course, the improvement in people's living standards is a cause for celebration, but the methods by which it has been achieved cast doubt upon some of our most cherished ideas. Very far from the active involvement of the people of China in determining their futures, this colossal social transformation has been achieved without a single shred of democracy. All of the important decisions have been taken by the top leadership of the Communist Party, which consists of a self-perpetuating oligarchy, whose members owe their position to secretive intra-party struggles rather than any democratic mandate and whose families are busy looting the country for their personal gain. There has been plenty of dissent—peasants opposing the illegal seizure of their land, workers demanding higher pay and better conditions, citizens protesting at the corruption of officials, national minorities fighting for genuine self-determination, intellectuals arguing for freedom of speech or democracy, whole populations determined not to suffer any further environmental degradation—but all organized dissent has been crushed with a greater or lesser degree of brutality. This is not a model of development that matches any of the aspirations of generations of dedicated individuals who have tried to promote the cause of the world's poorest people. But it has, in its own terrible way, worked. It may not continue to work in the same way for ever, but there will be no going back to the bad old days of underdevelopment and universal poverty.

We can illustrate the scale of the problem by contrasting the developmental changes that have taken place in China with those in India. While of comparable size, and beginning from much the same baseline, India has had a rather different trajectory. Despite sharing many problems with China—endemic corruption, widespread state repression of minorities, cavalier disregard for the environmental cost of development and so on—India has been, for almost all of its independent history, a political democracy and many of the developmental initiatives undertaken there have had at least a veneer of democratic involvement. The outcomes, as displayed in Table 5.1, have been markedly different. The data available is patchy and open to question, and it does not contain series for many of the key variables that would illuminate the argument. The series that do exist are often fragmentary and it has only been possible to construct this table using very approximate comparisons. As a consequence, its detail should be regarded with some scepticism, although the overall picture is certainly clear.

On all of the measures, China outperforms India, both absolutely and in terms of the gap between men and women. Children in China are twice as likely to be born with trained medical assistance than in India, less likely to be malnourished by a factor of ten, and have roughly a decade more of life to look forward to. Both of their parents are likely to be much better educated, they are twice as likely to grow up literate, and the women will marry nearly five years later. These are some of the standard measures of development, and against them China performs very much better than India.

Table 5.1 Some Comparisons of India and China

Indicator	*India (date)*	*China (date)*
Percentage of births attended by skilled health staff	52.3% (2008)	99.1% (2008)
Percentage of children under five less than a healthy weight	43.5% (2006)	4.5% (2005)
Female life expectancy at birth	67.7 years (2011)	76.4 years (2011)
Male life expectancy at birth	64.3 years (2011)	73.8 years (2011)
Female/male gap	3.4 years	2.6 years
Female years of education (25+)	3.2 years (2010)	6.9 years (2010)
Male years of education (25+)	4.3 years (2010)	7.5 years (2010)
Female/male gap	−1.3 years	−0.6 years
Adult female literacy (15+)	50.8% (2006)	92.7% (2010)
Adult male literacy (15+)	75.2% (2006)	97.5% (2010)
Female/male gap	−24.4%	−4.8%
Ratio of young literate females to males (15–24)	84.1% (2006)	99.9% (2010)
Female age at first marriage	20.2 years (2006)	24.7 years (2010)

Source: World Bank 2014. Figures have been rounded.

What these figures do not mean is that China is in some way a better place or that it has finally solved the problems of development. Even the achievements noted above have their downsides. Greater medical attendance at birth can be accompanied by pervasive and frequently unnecessary interventions, for example, and increased longevity leads to growing problems in care of the elderly. The gaps on a more inclusive measure are nowhere near as wide as in some of the individual instances. The 2013 *Human Development Report* puts both firmly within the category "medium human development," but even on this account China stands at 101, while India is 136 out of 184 countries reviewed (United Nations Development Programme, 2013, 145–46).

The evidence does, however, suggest that development and democracy are less closely associated than many writers have wished. Development, in this context, is really a question of finding a niche in the world market and subordinating the mass of society to developing industries capable of exploiting it. That was the basis for the original development of Europe and the United States, and it has been the basis for China's development. India, rather less successfully and rather more democratically, is exploring a similar trajectory.

The aims of most development theorists, however, were usually rather broader than simply increasing per capita gross domestic product. Even those working within the classical paradigm tended to see the objective as one of enriching people's lives as well as feeding and housing them. At its best, in the more rigorous versions of participatory communication, democracy was more than an optional extra, or a necessary condition for efficient material advancement. On the contrary, it was one of the major objectives.

The evidence very strongly suggests that this association is one that is in urgent need of reconsideration.

THAT OTHER POTENTIAL

We are faced, then, with a paradox. The ICTs have an enormous capacity to increase the possibilities for genuine dialogue in a society but they are being developed and deployed on assumptions that are blind to this potential. To the extent that they are linked to development, it seems that they act to reinforce existing social divisions and produce new ones. To the extent that there is evidence of development, that evidence points to the absence of any substantial democratic dimension in most, if not all, of the cases (Western Europe, the United States, Japan, the USSR and China). Development, then, is not a social process that requires democratic participation to be successful.

That, however, is not the end of the story. All the societies we have considered, developed and developing, are marked by social differentiation and some degree of social conflict. Different groups have different needs and different aims and, to the extent that they are able, they organize to achieve those ends. This suggests that any technology with the affordances of ICTs will, if introduced in to such a society, be used in different ways by different people. In particular, to the extent that there are social conflicts within a society, it will become one of the resources in such conflicts.

It is here that we can find the real core of the democratic potential of ICTs: they represent an additional technology that can be used in the struggle for democratic ends. That is not the entire story, of course, since these very same technologies can be powerful instruments for the surveillance of the population and the deployment of sanctions against dissidents, but it is an important reality for which there is already very substantial evidence.

At least as far back as the EDSA II movement that overthrew President Estrada in the Philippines in 2001 there has been strong evidence of the use of ICTs, in that case mobile telephones, in the organization of political movements (Coronel, 2001). More recent examples of the political potential of these networks, in both the developed and the developing world, are the big political protest movements of the last few years—the Arab Spring, the Occupy movement, the Los Indignados movement and many others, which owe at least part of their impact to the use of online resources. The shape and structure of these new social movements, and the role played in them by technological developments, is a major new research theme in political communication (Harlow, 2012; Gustaffson, 2012; Mercea, 2011; Wolfsfeld, Segev and Sheafer, 2013; Hussain and Howard, 2013; Bennett and Segerberg, 2012).

Again, there is a danger of abstracting these technologies from their social context, and indeed of overestimating their importance to the events in question. There is no doubt, particularly in Egypt, that there were other forms of communication that played an important role. Al Jazeera broadcasts continuous feeds of the protests in Tahir Square and the mosques were clearly also major centers of communication and mobilization. What is more, there were alternatives present even during the revolution itself. The supporters of the old regime used the same technologies to communicate their ideas and to mobilize their supporters, and at critical points the state attempted to deny electronic resources the opposition by closing down the mobile networks. Although the technologies in question have the affordance of permitting democratic opposition to organize, the same affordance provides the same opportunities to antidemocratic forces, and they have not proved a guarantee against the reestablishment of the old undemocratic regime under new and equally repressive leadership.

Nevertheless, it is this potential that provides the key link between democracy and development. Development has taken place and is taking place, but so far all of the great successes have been achieved at an enormous human cost, which some would argue was sufficiently high as to throw in to question the value of the entire exercise. This kind of development, as is well known, produces at least a period of great social tension and accompanying instability. From the Chartists in Britain in the first half of the nineteenth century to China today the processes of urbanization and industrialization have produced political challenges to the prevailing undemocratic orders. The traditional organizational tools of these movements were primarily the printing press and, to a lesser extent, broadcasting, when that became available. The truly revolutionary potential of ICTs is that they simplify and cheapen the organizational tasks involved. Forty years ago, producing a factory newspaper was a long and complicated task even in the most advanced of economies. Today, with simple computing and desktop printing power, it is very much easier. These new technologies do not produce the ideas and values that such a project requires. The origins of the forces that drive such initiatives lie where they have always lain—in the relations of production inside the factory or office. What the technologies do is to give the activists new ways of carrying out the old tasks of organization and agitation.

This potential to aid the struggle for human liberation is not peculiar to the developing world: the developed world faces its own crises and has its own social divisions. Some of the issues at stake in the richer countries are different from those of the developing world, but many are surprisingly similar, in nature if not in scale, and of course the danger of environmental catastrophe affects all countries to a greater or lesser degree. Just as the problems do not differ greatly, neither do the solutions nor the methods needed to achieve them. In all countries, rich and poor alike, the only way to begin addressing them is radically to transform the primary goals of society from those centerd on the blind accumulation of wealth to those designed to embody society's control over its own destiny: in other words, a renewal and enormous extension of the scope of democracy to mean something more than the periodic selection of (remote, wealthy, often white and usually male) individuals to take the decisions whose consequence will determine all our futures.

ICTs have the potential to play a central role in that process. It is in this field that their dialogic capacities are not merely desirable but essential. Technology is only part of the story. It does not guarantee that any dialogue will be rational, or consensual, or that it will reach conclusions that this writer, or any other, thinks are desirable. What technology can do is to make that dialogue of the many a concrete possibility.

NOTE

1. There are, of course, grotesque inequalities in the advanced countries, as well. In 2011, nearly a quarter (24.2 percent) of the population of the 27 countries of the European Union were living in conditions that put them "at risk of poverty or social exclusion" and this figure has been growing, albeit relatively slowly, since 2008 (Eurostat, 2013b). We do not have space to explore the implications of these facts in this chapter.

REFERENCES

Ashraf, Md. Mahfuz, and Bushra Tahseen Malik. "View from practice. Gonokendra model: a response to "information poverty" in rural areas of Bangladesh." *Information Technology for Development* 17, no. 2 (2011): 153–61.

Avergou, Chrisanthi. *Information Systems and Global Diversity*. Oxford: Oxford University Press, 2002.

Avergou, Chrisanthi. "Information systems in developing countries: a critical research review." *Journal of Information Technology* 23 (2008): 133–46.

Bennett, Lance, and Alexandra Segerberg. "The logic of connective action." *Information, Communication and Society* 15, no. 5 (2012): 739–68.

Best, Micahel, and Sylvia Maier. "Gender, culture and ICT use in rural south India." *Gender, Technology and Development* 11, no. 2 (2007): 137–55.

Contractor, Noshir, Arvind Signhal, and Everett Rogers. "Metatheoretical perspectves on satellite television and development in India." *Journal of Broadcasting and Electronic Media* 32, no. 2 (1988): 129–48.

Coronel, Sheila. "The media, the market and democracy: the case of the Philippines." *Javnost/The Public* 8, no. 2 (2001): 109–26.

Diaz Andrade, Antonio, and Cathy Urquhart. "The value of extended networks: social capital in an ICT intervention in rural Peru." *Information Technology for Development* 15, no. 2 (2009): 108–32.

Dossani, Rafiq, D.C. Misra, and Roma Jhaveri. *Enabling ICT for Rural India.* Stanford, CA: Asia Pacific Research Centre, Stanford University, 2005.

Duncombe, Richard. "Resaerching impact of mobile phones for development: concepts, methods and lessons for practice." *Information Technology for Development* 17, no. 4 (2011): 268–88.

Eurostat. "People at risk of poverty or social exclusion." February 26, 2013b. http://epp.eurostat.ec.europa.eu/tgm/refreshTableAction.do;jsessionid=9ea7d07d30e7a36e58e56ea746a7bfb1e7b3799b2176.e34OaN8PchaTby0Lc3aNchuMc30Re0?tab=table&plugin=1&pcode=t2020_50&language=en (accessed April 30, 2013).

Gollakota, Kamala, James Pick, and P. Sathyapriya. "Using technology to alleviate poverty: use and acceptance of telecentres in rural India." *Information Technology for Development* 18, no. 3 (2012): 185–208.

Gómez, Cynthia, and Vanoss Barbara Marín. "Gender, culture and power: Barriers to HIV-prevention strategies for women." *The Journal of Sex Research* 33, no. 4 (1996): 355–62.

Guerrieri, Paolo, Sara Bentivegna, and Valentina Meliciana. *Analysis of e-Inclusion impact resulting from advanced R&D based on economic modelling inrelation to innovation capacity, capital formation, productivity, and empowerment.* Brussels: European Commission, DG Information Society and Media Directorate, 2010.

Gustaffson, Nils. "The subtle nature of Facebook politics: Swedish social network site users and political participation." *New Media and Society* 14, no. 7 (2012): 1111–27.

Hakkarainen, Pävi. "'No good for shovelling snow and carrying firewood': Social representations of computers and the Internet by elderly Finnish non-users." *New Media and Society* 14, no. 7 (2012): 1198–215.

Harlow, Summer. "Social media and social movements: Facebook and an online Guatemalan justive movement that moved offline." *New Media and Society* 14, no. 2 (2012): 225–43.

Hong Kong Trade and Development Council. "Telecommunications Industry in Hong Kong." October 11, 2013. http://hong-kong-economy-research.hktdc.com/business-news/article/Hong-Kong-Industry-Profiles/Telecommunications-Industry-in-Hong-Kong/hkip/en/1/1X000000/1X003VIX.htm (accessed January 20, 2014).

Hussain, Muzammil, and Philip Howard. "What best explains successful protest cascades? ICTs and the fuzzy causes of the Arab Spring." *International Studies Review* 15, no. 1 (2013): 48–66.

International Telecommunication Union. "Development Statistics." 2014. http://www.itu.int/en/ITU-D/Statistics/Pages/stat/default.aspx (accessed January 17, 2014).

Islam, M. Sirajul, and Ake Grönlund. "Bangladesh calling: farmers' technology use practices as a driver for development." *Information Technology for Development* 17, no. 2 (2011): 95–111.

Kendall, Jake, and Nirvikar Singh. *Internet Kiosks in Rural India: What Influences Success.* New York: NET Institute, 2006.

Kumar, Keval. "Communication approaches to participation and development: challenging the assumptions and perspectives." In *Participatory Communication: Working for Change and Development*, by S. White, S. Nair and J. Ashcroft, 76–92. New Delhi: Sage, 1995.

Marín, Barbara Vanoss. "HIV prevention in the Hispanic community: sex, culture, and empowerment." *Journal of Transcultural Nursing* 14, no. 3 (2003): 186–92.

Mercea, San. "Digital prefigurative participation: the entwinement of online communication and offline participation in protest events." *New Media and Society* 14, no. 1 (2011): 153–69.

Mody, Bella, and Jorge Borrego. "Mexico's Morelos satellite: reaching for autonomy." In *Transnational Communications: Wiring the Third World*, by Gerald Sussman and John Lent, 150–64. Newbury Park, CA: Sage Publications, 1991.

National Telecommunications and Information Agency. *Exploring the Digital Nation: Computer and Internet Use at Home.* Washington, DC: National Telecommunications and Infrastructure Authority, 2011.

Page, David, and William Crawley. *Satellites Over South Asia: Broadcasting Culture and the Public Interest.* New Delhi: Sage, 2001.

Pendakur, Manjunath, and Jyotsna Kapur. "Think Globally, Program Locally: Privatisation of Indian National Television." In *Democratising Communication?: Comparative Persepctives on Information and Power*, by Mashoed Bailie and Dwayne Winseck, 195–217. Cresskill, NJ: Hampton Press, 1997.

Postrel, Steven. "Competing networks and proprietary standards: the case of quadraphonic sound." *The Journal of Industrial Economics* 39, no. 2 (1990): 169–85.

Rogers, Everett. *Diffusion of Innovations.* Fourth Edition. New York, NY: The Free Press, 1995.

Schmidt, Eric. "10 Observations on Technology in Africa from Eric Schmidt of Google." January 23, 2013. http://www.ictworks.org/2013/01/23/10-observations-on-technology-in-africa-from-eric-schmidt-of-google/ (accessed January 17, 2014).

Servaes, Jan. *Sustainability, participation and culture in communication. Theory and praxis.* Bristol: Intellect, 2013.

Sheeran, Paschal, Charles Abraham, and Sheina Orbell. "Psychosocoal correslates of heterosexual condom use: a meta-analysis." *Psychological Bulletin* 125, no. 1 (1999): 90–132.

Sparks, Colin. *Development, Globalization and the Mass Media.* London: Sage Publications, 2007.

United Nations Development Programme. *Human Development Report 2001: Making New Technologies Work for Human Development.* New York: Oxford University Press, 2001.

———. *Human Development Report 2013. The Rise of the South: Human Progress is a Diverse World.* New York, NY: United Nations Development Programme, 2013.

van Deursen, Alexander, and Jan van Dijk. "Internet skills and the digital divide." *New Media and Society* 13, no. 6 (2010): 893–911.

van Dijk, Jan, and Alexander van Deursen. "A social revolution online? The digital divide shifts to gaps of usage." Phoenix, Arizona: Paper presented at the 62nd

Annual Conference of the International Communication Association, May 24–28, 2012.

Wolfsfeld, Gadi, Elad Segev, and Tamir Sheafer. "Social Media and the Arab Spring: politics comes first." *The International Journal of Press? Politics* 18, no. 2 (2013): 115–37.

World Bank. *Data.* 2014. http://data.worldbank.org/ (accessed January 22, 2014).

———. "The State of the Poor: Where are the Poor and Where Are they Poorest?" *www.worldbank.org.* April 17, 2013a. http://www.worldbank.org/content/dam/Worldbank/document/State_of_the_poor_paper_April17.pdf (accessed March 19, 2014).

———. "World Development Indicators: poverty rates at international poverty line." *wdi.worldbank.org.* 2013b. http://wdi.worldbank.org/table/2.8 (accessed March 19, 2014).

World Bank. *World Development Report 1998–99: Knowledge for Development.* New York, NY: World Bank and Oxford University Press, 1998.

World Bank. *World Development Report 2002: Building Institutions for Markets.* New York NY: World Bank and Oxford University Press, 2001.

World Health Organisation. "Global Health Observatory Data Repository." *apps.who.int.* 2013. http://apps.who.int/gho/data/node.main.1098?lang=en (accessed March 20, 2014).

Yusuf, Shahid. *Development Economics Throught the Decades.* Washington DC: The International Bank for Reconstruction and Development/The World Bank, 2009.

Chapter 6

Cultural Values and Digital Networks as Predictors of Sustainable Democratic Development

Marko M. Skoric, Yong Jin Park, and Yalong Jiang

> *Freedom is fostered when the means of communication are dispersed, decentralized, and easily available, as printing presses or microcomputers. Central control is more likely when the means of communication are concentrated, monopolized and scarce, as are great networks. But the relationship between technology and institutions is not simple or unidirectional, nor are the effects immediate.* (Pool, 1983, p. 5)

The above quote comes from one of the seminal works on communication technology and political development, Ithiel de Sola Pool's *Technologies of Freedom* (1983). He branded his research zeitgeist "soft technological determinism" indicating that technologies neither determine nor are completely neutral with regard to sociopolitical outcomes. In this chapter, we advocate a softer version of this paradigm, suggesting that long-lasting cultural traits, largely defined by religious traditions and political history, play a role in shaping the structure of communication systems, which carries specific implications for political development. According to this perspective, certain cultural values act as catalysts of technological and political changes, with technological systems mediating the relationship between culture and political development.

While the characteristics of communication technologies seemingly have predictable implications for the development of political institutions and their functioning, the adoption and use of new technologies does not happen in a cultural vacuum (Steers, Meyer and Sanchez-Runde, 2008). Therefore, understanding the nature of interactions between technology and culture is necessary when developing any models of sociopolitical development. Technologically deterministic models of social change suggest an independent (from society and culture) and decisive impact of new technologies and associated practices on society, while contextualism sees technologies as being inseparable from the social world and reflective of its concerns, value and beliefs (Nye, 2006). We believe that the latter perspective offers better explanatory power, while allowing for a reciprocal influence between culture and technology, as well as their mutual and transformative impact on society.

CULTURE AND TECHNOLOGIES

Throughout history cultural variables have influenced the development, regulation and uses of communication technologies. Cultural norms determine how one communicates in relation to personal (gender, age, social status) and situational characteristics (time, place and medium of communication). Widely held values and beliefs shape the flows of information as well as preferences toward certain modes of communication. From the ancient Egyptian civilization, to the present day China, culture has played a strong role in determining the character of technologies used, and proscribing who can communicate, as well as what can be said, written or painted. Furthermore, these culture-shaped communication networks and practices have exerted a strong influence on the character of political institutions.

There are several historical examples that illustrate how cultural factors have shaped the diffusion and uses of communication technologies since the ancient times. For instance, the communication technology of freedom for the ancient Greeks was the alphabet. Although this technology had already existed in Phoenicia, it only brought freedom when it was widely diffused among the ordinary citizens of the Greek city-states. Literacy stabilized the language and culture across different classes and helped establish a sense of cultural unity, creating a decentralized network of democratic poleis that fought together when under a Persian attack (Mann, 1986). The Greeks nurtured a culture of individual rights and freedoms, and emphasized egalitarianism, at least among all free men. These values promoted the widespread adoption of alphabet and spread of literacy, which in turn yielded additional democratic (and practical) benefits for their society. Similarly, although movable type was invented in China in the eleventh century and adopted and improved in Korea a century later, it did not become a technology of freedom until it was diffused in the West during the Protestant Reformation (Pool, 1983). The printing press enabled production of vernacular translations of the Bible which further eroded the power of the highly centralized and hierarchical organization—the Catholic Church—by taking income away from the church scribes and by removing the right to interpret God's words from the exclusive realm of the Church (Eisenstein, 1968).

TECHNOLOGIES OF FREEDOM

Ithiel de Sola Pool was among the pioneers in the academic community emphasizing the importance of the technological transformations for human freedom and development of democratic political institutions. Traditionally, the literature on development and communications largely focused on traditional one-to-many media such as television and radio, with many of Ithiel de Sola Pool's contemporaries actively engaging in efforts to promote modernization, economic development and democracy in less-developed countries via the mass media (Deutsch, 1961; Lerner, 1958). Some scholars suggested that issues of political development and modernization could be thought of as problems of communication, noting the quintessential role of the free and independent press in the rise of American democracy (Pye, 1963). Their hope was that the introduction of then new technologies, television and radio, would promote nation-building efforts in developing societies by disseminating persuasive information in a top-down fashion (Rogers, 1976).

Although early research suggested that the mass media made the strongest contribution to development when paired with interpersonal and group communications (Mathur and Neurath, 1959; Pye, 1963; Schramm, 1981), scholars surprisingly did not focus their attention to horizontal communication networks such as telephone at the time. This relative lack of scholarly interest in horizontal, person-to-person technologies persisted at least until the late 1970s when Ithiel de Sola Pool's work *The Social Impact of Telephone* was first published (Pool, 1977). By then, it became clear that the old paradigm of top-down, communication-aided development had largely been exhausted (Rogers, 1976).

We take Pool's lead and focus on two characteristics of new ICTs that differentiate them from traditional vertical media, namely their widespread accessibility and their decentralized nature (Pool, 1983). In addition to providing direct person-to-person communications, digital communication technologies are characterized by the decentralization of information collection, creation and processing, paired with a significant increase in individual control over the communication process. Still, Pool's optimism about the liberating impact of new communication technologies did not solely rest on the character of technologies. He also emphasized the importance of the U.S. formal as well as informal institutions, the Bill of Rights and the political culture favoring individual rights and pluralism, respectively (Pool, 1983, p. 251). Thus, a study of the relationship between ICTs and political development should also examine the dominant cultural tradition of a society in question, including its religious traditions and its political history. Indeed, any study of political communication necessitates an examination of global cultural differences as the patterns of political communication widely vary between different cultures (Graber, 2005).

MODELING CULTURAL VALUES AND INFORMATION AND COMMUNICATION TECHNOLOGIES

In addition to having a direct influence on a democratic institutions and practices, the impact of culture can also be indirect, that is, mediated by communication networks. In our theoretical model, culture shapes the structure, regulation and uses of communication systems, which has direction implications for the nature of communication patterns within a society. A cleverly designed communication system should reflect the structure of the society where it is deployed, whether it is authoritarian, oligarchic, commercial or militaristic (Pool, 1990).

The arrival of digital communication networks characterized by the widespread accessibility and decentralized nature has signaled a shift toward a more user- and culture-centric communication world. Since the late 1980s, new technologies have increasingly allowed users to localize, customize and personalize media content, technologies and services (Abramson, Arterton and Orren, 1988; Negroponte, 1995). Most importantly, the low cost and ease of use associated with the new bread of ICTs holds a promise of widespread, if not universal, access to global flows of information. Therefore, it is reasonable to expect that ICTs will be more reflective of dominant cultural values than it has been the case with traditional mass media in the past.

We argue that specific cultural conditions promote the development of horizontal digital technologies, characterized by their widespread accessibility and decentralized nature, and foster their democratizing uses. These conditions have been shaped by religious heritage and political history of a society and constitute an important part of its political culture. There are several characteristics of political culture that could potentially play key roles in this process. In our previous work (Skoric and Park, 2014), we examined whether the patterns of authority relations prevalent in a society were also reflected in the structure of communication networks, arguing that hierarchical cultures should be less supportive of nonhierarchical communication technologies than the more egalitarian ones. In addition, we tested whether political cultures placing high value on freedom of choice, self-expression and participation created demands for communication technologies that offered greater control over the communication process. The above dimensions of cultural variability have been shown to be of great importance when it comes to the issues of political and economic development (Inglehart and Welzel, 2005), and we believe that this is partly reflected in cultural shaping of digital communication networks.

In our analysis, culture is viewed as a widely shared societal system of attitudes, values and knowledge that is transmitted from generation to generation (Inglehart, 1990). Research in history, psychology, sociology and political science has provided a solid basis for estimating the importance and prevalence of certain values, beliefs and norms in societies around the world, also allowing researchers to analyze their sociopolitical consequences (Fukuyama, 1995; Harrison and Huntington, 2000; Putnam et al., 1993). In recent decades, the emergence of comparative, multi-country surveys (i.e., World Values Survey, Eurobarometer, etc.) has greatly increased our ability to empirically test hypotheses about the role of culture in economic, technological, and political development (Inglehart and Baker, 2000; Inglehart and Welzel, 2005; Servaes and Heinderyckx, 2002; Skoric and Park, 2014).

MODELING DIGITAL NETWORKS AND DEMOCRATIC INSTITUTIONS AND PRACTICES

It is important to note that centralized control over communication networks does not necessarily imply a democratic deficit, but it hardly represents a fertile ground for blooming of civic activism, public deliberation and citizen education that characterizes "strong democracy" (Barber, 1984). The one-to-many nature of the mass media does not allow for reciprocal communication, and is thus antidotal to participatory forms of communication (Enzensberger, 2000). In support of this view, a range of media malaise theories has emerged in the literature since the 1970s (see Norris, 2000), blaming the mass media for disconnecting citizens from politicians (Patterson, 1993), increased cynicism toward politics (Cappella and Jamieson, 1997), civic disengagement (Putnam, 2000), and fragmentation of society (Turow, 1997). The above studies have focused exclusively on the U.S. case, but similar claims have also been made about other postindustrial democracies, usually focusing on trends toward commercialization and tabloidization, resulting in impoverishment of the public sphere (Dahlgren, 1995), corrosion of trust (Lull and Hinerman, 1997), and widening of knowledge gaps and increased cynicism (Kaase, 2000). Historically, Nazi Germany and the Soviet Union are vivid examples of how hierarchical and highly centralized communication systems can under certain conditions be successfully put to totalitarian and oppressive purposes.

Institutional Democracy

Since the 1990s, research has indicated that in authoritarian and semi-authoritarian countries, the mass media systems are usually well developed while horizontal networks tend to be purposefully neglected (Hills, 1993). Selian (2002) reported that the year-to-year growth of telephone main lines per 100 inhabitants was significantly correlated with measures of political rights and civil liberties. More recent research shows that the Internet could function as an agent of democratic change, but that its democratizing role was most visible in the societies that were already developed and at least partially democratic (Groshek, 2009). It is hard to dispute that the Internet has lead to a more egalitarian and democratic access to information and communication flows around the world, with more people having access to the global flows of information than ever before, even when taking into account Internet filtering and censorship. For instance, today social media networks directly link almost a quarter of the mankind in real-time—any two (out of more than a billion) Facebook users in the world are linked by less than five degrees of separation (Backstrom et al., 2012).

Furthermore, mobile technologies are likely to bring about even more significant democratizing benefits. There are in fact suggestions (e.g., Campbell and Park, 2008) that the personal nature of the mobile telephony heralds the transition away from "mass" to direct "point-to-point" communication that can potentially bypass censorship. Studies (e.g., Bhavnani et al., 2008; Geser, 2004; Sinha, 2005) have examined the mobilizing power of mobile phone diffusion in developing nations, that is, how different nation-states appropriate technological affordance to leapfrog via the mobile to support social, economic, or technological advancement. The logic follows that the horizontal nature of mobile access may disrupt or spur new flow of personalized information to the extent that it encourages the development of democracy (see Park, 2014).

Corruption

Corruption is one of the most pervasive social phenomena, plaguing societies since the ancient times, regardless of their political system, religion or culture. While definitions of corruption are numerous, most of them focus on politicians' or government officials' tendency to obtain wealth or power illegally, that is, private gain at public expense (Lipset and Lenz, 2000). Corruption can be viewed as a measure of integrity of political elites, that is, to what extent elites respect the rule of law or abuse power for their own benefit, denying citizens the services they are entitled to receive. Elite integrity has been identified as a vital factor that differentiates between formal and effective democracy (Inglehart and Welzel, 2005).

Corruption takes both economic and democratic tolls. First, as a rule, corruption raises the transaction costs and thus breeds inefficiency and waste. Among other consequences of corruptions, researchers have cited reductions in foreign direct investment (FDI) and total investment, as well as increases in income inequality (e.g., Gini coefficient) and levels of poverty (Lambsdorff, 1999). Furthermore, corruption fosters culture of mistrust, cynicism and powerlessness. Corrupt government officials remove the decision-making process from the public into the private realm, diminishing transparency and accountability.

Scholars have cited numerous factors as important correlates of corruption. These factors include the quality of judiciary, economic competitiveness and freedom, educational level, press freedom and length of democratic tradition (see Lambsdorff, 1999). Research shows that corruption is also related to the Freedom House Index of Democracy, more strongly to its civil liberties indicator (Lipset and Lenz, 2000). There is rather strong evidence linking certain cultural values to corruption (e.g., La Porta et al., 1997). For example, following Banfield's (1958) work on amoral familism in Italy, Lipset and Lenz (2000) argue that traditionalism with an emphasis on familism and coupled with a relatively narrow radius of trust and cooperation are the main cultural factors predicting levels of corruption. Such a cultural syndrome is likely to exert its influence on corruption via shaping of ICT structure and use, effectively reducing the potential of ICTs to promote transparency and government accountability.

How can modern ICTs help fight corruption and promote transparency? Research shows that ICTs can be effective tools of promoting openness and transparency and consequently reducing corruption and waste (Bertot, Jaeger, and Grimes, 2010). There are essentially two ways of promoting government transparency via ICTs:

a. top-down: through implementation of e-government initiatives which enable greater citizen access to government information, promote the opening of government processes and procedures, and allow citizens to observe and participate in the policy-making processes;
b. bottom-up: through social and mobile media-enabled citizen journalism and whistleblowing which can frequently circumvent established systems of censorship and control.

Still, the success of the ICT-enabled anticorruption initiatives is dependent on a range of factors, including dominant cultural beliefs about the value of transparency as well as the levels of technological accessibility and literacy (Bertot, Jaeger, and Grimes, 2010). More generally, research shows that many e-government initiatives in developing countries have not been successful, often due to the lack of understanding of the local context and vested interests of numerous stakeholders involved in the implementation of such programs (Dada, 2006).

RESEARCH HYPOTHESES

Based on the conceptual contours discussed above, we conjecture the two routes that are linked to the sustainable democratic development:

First, we propose that societies where hierarchy and authority are emphasized, where people distrust each other, have few liberty aspirations and little sense of personal agency and freedom, are less likely to develop and see widespread use of modern ICTs.

Second, this pattern of technological underdevelopment and underutilization of the existing digital networks will diminish opportunities for political development of these societies, as reflected in lower scores on the measures of institutional democracy and corruption perceptions.

Our fundamental premise is that the contextual nuances of cultural values may either encourage (or curtail) a particular structure and use of digital networks, which in turn helps incubate the nation-state environments in which democratic institutions and practices can become sustainable. In this study we aim to replicate and refine the findings from our earlier work (Skoric and Park, 2014) and test the model using an updated measures of cultural values, ICT development, institutional democracy and corruption.

MEASURES

Cultural Indicators

Empirical indicators of cultural values were taken from the World Values Survey (fifth wave, from 2005–2008), the largest comparative survey in the world, covering 56 societies (WVS, 2008). The traditional vs. secular-rational dimension consists of the following items, emphasizing traditional values and their opposites: (a) God is very important in respondent's life; (b) It is more important for a child to learn obedience and religious faith than independence and determination; (c) Abortion is never justifiable; (d) Respondent has a strong sense of national pride; and (e) Respondent favors more respect for authority (see Inglehart and Baker, 2000).

The liberty aspiration factor includes three items: (a) Seeing that people have more say about how things are done at their jobs and in their communities; (b) Giving people more say in important government decisions; (c) Protecting freedom of speech. Respondents assigned a priority to each of the three liberty items, ranging from first, to second, to no priority, which made possible to create a six-point index, on which zero indicates the lowest and five the highest priority placed on liberty (Welzel and Inglehart, 2005).

Freedom of choice and personal agency is measured using a question "please indicate how much freedom of choice and control you have over the way your life turns out," with responses ranging from 1 to 10 on a Likert-type scale. In the present study, the values for all the above cultural variables are means calculated at the national level.

The ICT Development Index (IDI) was developed by the International Telecommunication Union (ITU) in 2008 as a composite index combining 11 indicators into one benchmark measure with a purpose of monitoring and comparing developments in ICT globally. It is a modified version of the ITU's Digital Access Index that was used in our previous study (Skoric and Park, 2014). The IDI contains following three subindices:

- Access subindex (40%): Capturing ICT readiness, and including five infrastructure and access indicators (fixed-telephone subscriptions, mobile-cellular telephone subscriptions, international Internet bandwidth per Internet user, percentage of households with a computer, and percentage of households with Internet access).
- Use subindex (40%): Capturing ICT intensity, and including three ICT intensity and usage indicators (percentage of Internet users, fixed (wired)-broadband subscriptions, and active mobile-broadband subscriptions)
- Skills subindex (20%): Capturing ICT capability or skills as indispensable input indicators; it uses three proxy indicators (adult literacy, gross secondary enrolment and gross tertiary enrolment), and is given less weight when compared to the other two subindices.

To measure institutional democracy, we adopted the Democracy Index developed by the Economist Intelligence Unit, that measures a state of democracy in 167 countries. The index is based on 60 indicators grouped in five different categories: electoral process and pluralism, civil liberties, functioning of government, political participation, and political culture. Our measure summed all the scores from the Democracy Index from 2008–2010 for each country; the higher the score, the more democratic a country is.

To measure corruption, we used the Corruption Perception Index provided by Transparency International, which scores 178 countries and territories on the basis of how corrupt their public sectors are perceived to be by their citizens. Our measure summed all the scores from the Corruption Perceptions Index from 2008–2010 for each country; the higher the number, the less corrupt a country is.

The measure of Gross National Income (GNI) per capita was taken from the World Bank's database (World Bank, 2008). To create our control variable we summed up the GNI values from 2006 and 2007, year during which most the World Values Survey data was collected.

RESULTS

Table 6.1 presents the means and standard deviations, while Table 6.2 presents bivariate correlations between all the variables in the model.

Table 6.3 presents the results from a series of hierarchical linear regressions, with the Democracy Index (DI08-10), Corruption Perception Index (CPI08-10) and the ICT Development Index (IDI07) entered as criterion variables. The three cultural dimensions were entered as predictors variables, with an aim of testing both the direct and mediating path leading to democracy and corruption measures.

In Table 6.3, starting from the left, the first column shows that the traditional vs. secular-rational dimension is negatively associated with the DI (Beta=−0.372, $p < 0.01$), while freedom of choice and personal agency (Beta=0.258, $p < 0.05$) and GNI06-07 (Beta < 0.297, $p < 0.05$) are positively associated with the DI. After adding the IDI07 into the model, the second column shows that the negative associations with the traditional vs. secular-rational dimension from first column disappeared, indicating a mediating relationship. The results from the second column suggest that greater

Table 6.1 Means and Standard Deviations

	IDI07	*DI08-10*	*CPI08-10*	*Traditional*	*Liberty*	*Freedom*	*GNI06-07*
N	152	211	211	56	52	56	211
Mean	3.40	8.32	9.82	0	0.82	6.95	20121.30
SD	1.91	6.13	6.96	3.00	0.17	0.95	28338.79

Note: ICT Development Index (IDI07), Democracy Index (DI08-10), Corruption Perception Index (CPI08-10); Gross National Income GNI06-07.

Table 6.2 Bivariate Correlation between Variables

	DI08-10	*CPI08-10*	*Traditional*	*Liberty*	*Freedom*	*GNI06-07*
IDI07	0.579**(152)	0.784**(152)	0.223**(152)	0.390**(50)	0.240(54)	0.753**(211)
DI08-10		0.676**(211)	0.427*(211)	0.458**(52)	0.553**(56)	0.342**(211)
CPI08-10			0.311**(211)	0.390**(52)	0.378**(56)	0.664**(211)
Traditional				−0.433**(52)	0.097(56)	−0.129**(56)
Liberty					−0.190(52)	0.355*(45)
Freedom						0.220(56)

*p<0.05, **p<0.01, ***p<0.001.

emphasis on liberty (Beta=0.188, $p < 0.05$), freedom of choice (Beta=0.195, $p < 0.05$), and higher ICT development (Beta=0.660, $p < 0.001$) are all predictive of more democratic institutions and less corrupt government practices.

The third column shows that the traditional vs. secular-rational dimension is negatively associated with the CPI (Beta=−0.346, $p < 0.05$), while GNI06-07 (Beta=0.458, $p < 0.001$) is positively associated with the CPI. After adding the IDI07 to the model, the fourth column shows that all the significant associations in third column disappeared, indicating a mediating relationship. From the results in fourth column we could claim that higher ICT development index (Beta=0.545, $p < 0.01$) is associated with reduced corruption perceptions.

Finally, the last column suggests that of all three cultural dimensions entered as predictors, only the traditional vs. secular-rational dimension was associated with the IDI07, showing a moderately strong negative association (Beta=−0.493, $p < 0.001$). In addition, GNI06-07 was positively associated with the IDI07 (Beta=−0.493, $p < 0.001$).

All significant relationships from the regressions are presented in Figure 6.1 (for the Democracy Index '08-10) and Figure 6.2 (for the Corruption Perception Index '08-10).

Table 6.3 Multiple Regressions for Democracy Index (DI08-10), Corruption Perception Index (CPI08-10) and ICT Development Index (IDI07)

	DI08-10 (Beta)		*CPI08-10 (Beta)*		
	Model 1	*Model 2*	*Model 1*	*Model 2*	*IDI07 (Beta)*
Traditional	−0.372**	0.047	−0.346*	−0.077	−0.493***
Liberty	0.194	0.188*	0.067	0.062	−0.009
Freedom	0.258*	0.195*	0.104	0.052	0.095
GNI06-07	0.297*	−0.005	0.458***	0.209	0.457***
IDI 07		0.660***		0.545**	
Adjusted R^2	0.571***	0.695***	0.536***	0.616***	0.699***
N	50	50	50	50	50

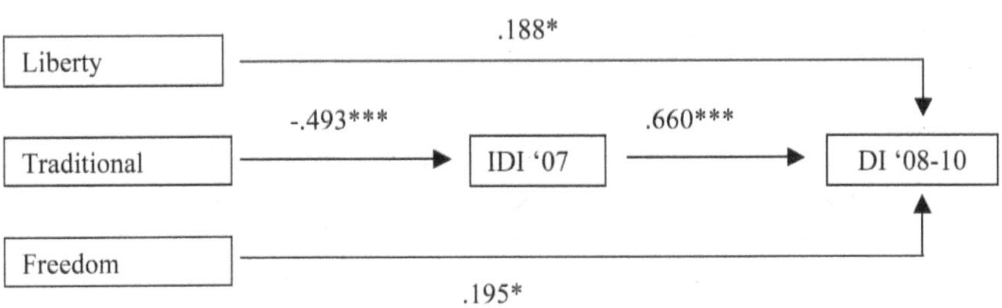

Figure 6.1 Standardized Regression Coefficients for the Relationships between Three Cultural Variables and Democracy Index (DI)2008-10, with the Relationship between Traditional vs. Secular-Rational Dimension and DI2008-10.

Table 6.4 Two Mediating Paths Tested Using Sobel's Method

	z-score
Path in Figure 6.1	
Traditional → IDI2007 → DI'08-10	−3.340***
Path in Figure 6.2	
Traditional → IDI2007 → CPI'08-10	−2.738**

*p<0.05, **p<0.01, ***p<0.001.

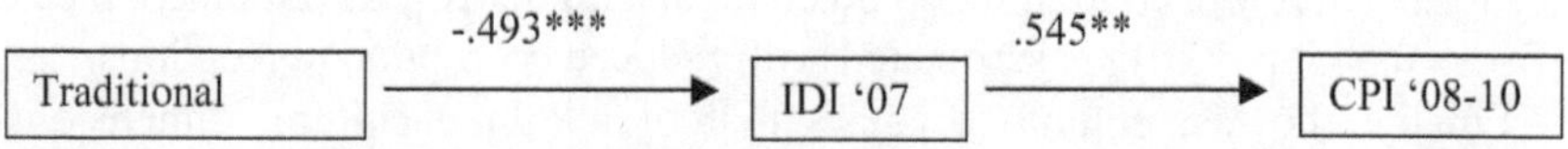

Figure 6.2 Standardized Regression Coefficients for the Relationships between the Traditional vs. Secular-Rational Dimension and CPI2008-10.

Lastly, we tested whether the indirect paths from cultural values to institutional democracy via the IDI are significantly different from zero, which would indicate that the predictions regarding the mediating role of the IDI were supported (Table 6.4). The path leading from the traditional vs. secular-rational dimension to institutional democracy via the IDI is significantly different from zero (Sobel test=−3.340, $p < 0.001$) as is the path leading to the Corruption index (Sobel test=−2.738, $p < 0.01$).

DISCUSSION

In our current analysis, we partially replicated the findings from Skoric and Park (2014), demonstrating a moderately strong relationship between the traditional vs. secular-rational dimension of cultural variation and the IDI. Furthermore, the findings suggest strong relationships between the IDI and both the Democracy Index (DI) and the Corruption Perception Index (CPI), even after economic development is being controlled for. Still, unlike Skoric and Park (2014), we find no evidence linking the freedom of choice and personal agency dimension with the IDI; instead, this cultural dimension is directly linked to the Democracy Index (DI). This pattern of findings is almost identical for the liberty aspirations factor, which was found to have no relationship with the IDI, but still weakly associated with the DI.

The most robust finding in our study is the relationship between the ICT development and the measures of institutional democracy and corruption, which fully replicates our previous research. This is especially important since the measures of ICT development, institutional democracy and corruption used in this study are either slightly modified (IDI) or significantly different (DI, CPI) from the measures used in the Skoric and Park (2014) study. Furthermore, we replicated the finding in a different time period, which increases the generalizability of our findings. Given that, we suggest that the relationship between the structure and use of ICTs and the development of democratic institutions and practices is neither a measurement artifact, nor a phenomenon limited to a specific stage of the ICT development.

Still, it is also important to note the difference between the current findings and those from the earlier study (Skoric and Park, 2014). In our study only the traditional vs. secular-rational dimension was linked with the ICT development as neither the liberty aspirations nor freedom of choice and personal agency dimensions were found to be linked to the IDI; instead, we found that they were directly related to the Democracy Index (but not to the Corruption Index). This is interesting as it indicates that over time some cultural values may become less important for the ICT development and use, while retaining their direct importance for institutional democracy. This study suggests that as the technologies mature and become commodified, it is possible that the emphasis on personal freedom and choice is no longer a significant catalyst for the ICTs adoption and use.

Although our findings suggest a strong relationship between ICT development and the quality of democratic institutions and practices, we do not support a technologically deterministic perspective of political development. First, the technological systems in our model are shown to be shaped by dominant cultural values (in this case traditionalism vs. secularism). Second, the IDI captures more than just ICT infrastructure and access to ICTs, as the measures of ICT intensity, usage and skills comprise 60 percent of its value. Therefore, we argue that the contextualist perspective which sees technologies as being inseparable from the social world and reflective of its cultural values (Nye, 2006) provides a more suitable research paradigm for future research. This is not to say technologies are neutral or inconsequential. In fact, Ithiel de Sola Pool (1983) was right to emphasize the importance of their democratizing characteristics, namely their widespread accessibility and their decentralized, person-to-person nature, but it is also important to note that their structure and use is still significantly shaped by cultural forces.

Do our results allow us to make causal claims about the nature of relationships uncovered? Is the proposed mediation model the most logical one, given the data sources used?

In terms of temporal order, World Values Survey (WVS) data precedes the technology variables, as the WVS fifth wave was collected between 2005 and 2007, while our ICT and democracy/corruption variables date from 2007 and 2008–2010, respectively. Therefore, the proposed model follows the temporal sequence of events, potentially allowing for casual, mediating paths between variables. Moreover, we view cultural values as relatively enduring and unlikely to change within a short period of only few years—it would be difficult to imagine how cultural values could quickly change as a result of the deployment of ICTs. If such cultural shifts are to occur, they are likely to happen after significantly longer periods of technology use. Moreover, as one of the pioneers of the communication and development field, Wilbur Schramm, argued that any attempts to single out a specific causal link in the relationship between communication and development are futile, as this interaction should be viewed as constant and cumulative (Schramm, 1981, p. 50). Therefore, our model views culture as the catalyst for both technological and democratic development, allowing for a reciprocal influence between the variables, but specifying technology as the meditator in the present model.

CONCLUSION

Our findings highlight the importance of including cultural variables in any comparative examination of the political implications of ICTs, as they show that culture explains a substantive portion of variance in the ICT development and use even after economic factors are being controlled for. Our evidence shows that cultural values matter and have an enduring impact on both technological and political development. Furthermore, the study emphasizes the value of empirically testing these relationships, that is, by generating specific hypotheses, subjecting them to rigorous testing, and finally replicating and refining the results. We believe that survey-based, nationally representative indicators of cultural values represent a strong asset when conducting such analyses. In addition, given the increasing availability of Internet and mobile access around the world (already reaching saturation in many developed economies), the emphasis should be shifted to examining the actual ICT usage patterns, rather than on the simple enumeration of the levels of penetration and quality of access (e.g., percentage of subscribers, bandwidth). This approach should also produce a more systematic and nuanced understandings of how social and economic structures lead to the differences in the quality of access and use (see Mertens and Servaes, 2010).

For policy-makers and practitioners, our study provides specific information about the cultural preconditions favoring the development and use of modern information and communication technologies as well as the nature of their linkages with sustainable democratic development. In societies where religion, family ties, deference to authority, and avoidance of political conflict are emphasized, the ICT development and use are likely to be stifled, which in turn creates the conditions unfavorable to the development of democratic institutions and practices. In such context, the diffusion of modern ICTs will be slower and their role in promoting political development likely minimized.

Still, ICTs can make a difference when it comes to promoting democratic outcomes. In many authoritarian countries, citizens increasingly use new technological means to challenge the status quo and push for democratic reforms, while citizens of liberal democracies use ICTs to fight government corruption and abuses of power. Indeed, activists, opposition leaders and engaged citizens around the world have effectively employed social and mobile media platforms for political expression, information dissemination, organization, and mobilization. They have used the power of ICTs to provide information that is not available through traditional channels, such as news of strikes, protests and political arrests. Internet applications and technologies have also been utilized to facilitate whistleblowing and promote greater transparency in governance. Many governments have responded to this challenge by either deliberately stifling the development of new communication networks, or when this development has been allowed or even encouraged, by attempting to control, limit and censor these new flows of information. Some of the methods employed range from widespread surveillance of digital communications to very sophisticated techniques of monitoring and censoring user-generated political content in authoritarian countries such as China (King, Pan, and Roberts, 2013). Interestingly, as demonstrated by the recent Edward Snowden revelations, extensive surveillance of citizens' digital communications do not seem to be confined to the nondemocratic countries, although the U.S. surveillance program has not thus far been linked with any attempts to limit and censor information online. The future of such surveillance programs is likely to be dependent on the reactions from both formal (Congress, courts) and informal (political culture) institutions, in line with the contextualist perspective which sees technologies as being inseparable from cultural and societal forces.

In conclusion, we started from Ithiel de Sola Pool's (1983) version of "soft technological determinism" in which technologies neither determine nor are neutral when it comes to social and political development, and then expanded our theoretical framework to account for a decisive and long-lasting impact of culture in the process of political development. Our study shows that certain cultural values create conditions favorable to democratic institutions and practices by promoting the development and use of widely accessible and decentralized communication technologies. The existence of such communication systems is strongly linked with institutional democracy and the integrity of political elites, providing an environment conducive of sustainable political development.

REFERENCES

Abramson, J. B., Arterton, F. C., & Orren, G. R. (1988). *The electronic commonwealth: The impact of new media technologies on democratic politics*. New York: Basic Books.

Backstrom, L., Boldi, P., Rosa, M., Ugander, J., & Vigna, S. (2012). *Four Degrees of Separation*. Proceedings of 4th ACM International Conference on Web Science (WebSci), 2012.

Banfield, E. C. (1958). *The moral basis of a backward society*. Glencoe: Free Press.

Barber, B. (1984). *Strong democracy: Participatory politics for a new age*. Berkley: University of California Press.

Bertot, J. C., Jaeger, P. T., & Grimes, J. M. (2010). Using ICTs to create a culture of transparency?: E-government and social media as openness and anti-corruption tools for societies. *Government Information Quarterly, 27*, 264–271.

Bhavnani, A., Won-Wai Chiu, R., Janakiram, S., & Silarszky, P. (2008). *The role of mobile phones in sustainable rural poverty reduction*. World Bank Report—ICT Policy Division Washington, DC: World Bank.

Campbell, S. W., & Park, Y. J. (2008). Social implications of mobile telephony: The rise of personal communication society. *Sociology Compass, 2*(2), 371–387.

Cappella, J. N. & Jamieson, K. H. (1997). *Spiral of cynicism: The press and the public good*. New York: Oxford University Press.

Dada, D. (2006). The failure of e-government in developing countries: A literature review. *The Electronic Journal of Information Systems in Developing Countries, 26*(7), 1–10.

Dahlgren, P. (1995). *Television and the public sphere: Citizenship, democracy and the media*. London: Sage.

Deutsch, K. W. (1961). Social mobilization and political development. *American Political Science Review, 55*(3), 493–514.

Eckstein, H. (1969). Authority relations and governmental performance. *Comparative Political Studies, 2*, 269–326.

Eisenstein, E. L. (1968). Some conjectures about the impact of printing on Western society and thought: A preliminary report. *Journal of Modern History, 40*(1), 1–56.

Enzensberger, H. M. (2000). Constituents of a theory of the media. In S. Marris & S. Thornham, *Media Studies: A reader*. New York: New York University Press.

Fukuyama, F. (1995). *Trust: The social virtues and creation of prosperity*. New York: Free Press.

Geser, H. (2004). Towards a sociological theory of the mobile phone. Retrieved from http://itu.dk/courses/KMP/E2007/Pdf/Geser_lektion6.pdf

Graber, A. D. (2005). Political communication faces the 21st century. *Journal of Communication, 55*(3), 479–507.

Groshek, J. (2009). The democratic effects of the Internet, 1994–2003: A cross-national inquiry of 152 countries. *The International Communication Gazette, 71*(3), 115–136.

Harrison, L. E. & Huntington, S. P. (2000). *Culture matters: How values shape human progress*. New York: Basic Books.

Hills, J. (1993). Telecommunications and democracy: The international experience. *Telecommunications Journal*, *60*(1), 21–29.

Huntington, S. P. (2000). Foreword: Cultures count. In L. E. Harrison & S. P. Huntington (Eds.), *Culture matters: How values shape human progress*. New York: Basic Books.

Inglehart, R. (1990). *Culture shift in advanced industrial society*. Princeton, NJ: Princeton University Press.

Inglehart, R., & Baker, W. E. (2000). Modernization, cultural change, and the persistence of traditional values. *American Sociological Review, 65*, 19–51.

Inglehart, R., & Welzel, C. (2005). *Modernization, cultural change and democracy: The human development sequence*. New York: Cambridge University Press.

King, G., Pan, J., & Roberts, M. (2013). How censorship in China allows government criticism but silences collective expression. *American Political Science Review, 107*(2), 326–343.

La Porta, R., López-de-Silanes, F., Shleifer, A., & Vishny, R. (1999). The quality of government. *Journal of Law, Economics and Organization, 15*, 222–279.

La Porta, R, López-de-Silanes, F., Shleifer, A., & Vishny, R. (1997). Trust in large organizations. *American Economic Review, 87*(2), 333–339.

Lambsdorff, J. G. (1999). Corruption in empirical research—a review. Transparency International Working Paper. Retrieved August 18, 2004 from http://www.transparency.org/working_papers/lambsdorff/lambsdorff_eresearch.html

Lerner, D. (1958). *The passing of traditional society: Modernizing the Middle East*. Glencoe: Free Press.

Lipset, S. M., & Lenz, G. S. (2000). Corruption, culture and markets. In L. E. Harrison, & S. P. Huntington, *Culture matters: How values shape human progress*. New York: Basic Books.

Lull, J., & Hinerman, S. (1997). *Media scandals*. Cambridge: Polity Press.

Maitland, C. (1998). Global diffusion of interactive networks: The impact of culture. *Electronic Journal of Communication, 8*(3–4).

Mann, M. (1986). *The sources of social power: A history of power from the beginning to A.D. 1760*. Cambridge: Cambridge University Press.

Mathur, J.C., & Neurath, P. (1959). *An Indian experiment in farm radio forum*. Paris: UNESCO.

Mertens, S., & Servaes, J. (2010). ICT policies on structural and socio-cultural participation in Brussels. In E. E. Adomi (Ed.), *Handbook of research on information communication technology: Trends, issues and advancements*. Hershey PA: IGI Global.

Negroponte, N. (1995). *Being digital*. New York: Knopf.

Neuman, W. R. (1991). *The future of mass audience*. New York: Cambridge University Press.

Nye, D. E. (2006). *Technology matters: questions to live with*. Cambridge, MA: MIT Press.

Park, Y. J. (2014). My whole world's in my palm! The second-level divide of teenagers' mobile use and skill. *New Media and Society* (Online First). January 27, 2014, doi: 10.1177/1461444813520302

Patterson, T. (1993). *Out of order*. New York: Knopf.

Pool, I. d. S. (1977). *The social impact of telephone*. Cambridge: MIT Press.

Pool, I. d. S. (1983). *Technologies of freedom*. Cambridge: Belknap Press.

Pool, I. d. S. (1990). *Technologies without boundaries: On telecommunications in a global age*. Cambridge: Harvard University Press.

Putnam, R. D., Leonardi, R. & Nanetti, R. Y. (1993). *Making democracy work: civic traditions in modern Italy*. Princeton: Princeton University Press.

Pye, L. W. (1963). Communications and political development. Princeton: Princeton University Press.

Rogers, E.M. (1995). *Diffusion of innovations*. New York: Free Press.

Schramm, W. (1981). *Mass media and national development: The role of information in developing countries*. Urbana: University of Illinois Press.

Selian, A. (2002). ICTs in support of human rights, democracy and good governance. International Telecommunication Union Background Paper for the 2003 World Summit on the Information Society. Retrieved March 17, 2004 from http://www.itu.int/osg/spu/wsis-themes/humanrights/ICTs%20and%20HR.pdf

Servaes J., & Heinderyckx F. (2002). The 'new' ICTs environment in Europe: Closing or widening the gaps? *Telematics and Informatics, 19*(2), 91–116.

Sinha, C. (2005). Effect of mobile telephony on empowering rural communities in developing countries. Paper presented at the World Forum for Information Society, Tunis, Tunisia, November 14–16, 2005.

Skoric, M. M., & Park, Y. J. (2014). Culture, technologies and democracy: A cross-national analysis of political development. *Telematics & Informatics, 31*(3), 364–375.

Steers, R. M., Meyer, A. D., & Sanchez-Runde, C. J. (2008). National culture and the adoption of new technologies. *Journal of World Business, 43*(3), 255–260.

The Economist Intelligence Unit (2010). *Democracy Index (2008 & 2010)* [Data file and code book]. Retrieved from http://www.eiu.com/public/who-we-are.aspx

Transparency International (2010). *Corruption Perception Index (2008–2010)* [Data file and code book]. Retrieved from http://www.transparency.org/research/cpi/overview.

Turow, J. (1997). *Breaking up America: Advertisers and the new media world*. Chicago: University of Chicago Press.

World Bank (2008). *GNI per capita, PPP (current international $) (2007–2008)* [Data file and code book]. Retrieved from http://data.worldbank.org/indicator/NY.GNP.PCAP.PP.CD

World Values Survey Association (2009). World value survey 1981–2008 official aggregate [Data file and code book]. Retrieved from http://www.wvsevsdb.com/wvs/WVSData.jsp

Chapter 7

Being Meaningfully Mobile

Mobile Phones and Development

Jo Tacchi

The broad question I address in this chapter is "what makes mobility, particularly related to mobile phones, meaningful?" Or put differently "in relation to mobile phones, what does it mean to be meaningfully mobile?" The concept of meaningful mobility was introduced in relation to gender and mobile phones in India (Tacchi, Kitner and Crawford, 2012) as a way of arguing that while mobile phones can be seen as active agents and facilitators of change or development (in that case within the evolving field of gendered relationships), they must be understood within specific uses and settings in order to understand the part they play in social and economic meaning making. The ways in which as researchers we understand ideas of mobility—incorporating domains such as the technical, social, economic, political and spatial—determines what we research and the kinds of meanings we uncover or develop in relation to mobile phones and development. Here I want to explore the concept of meaningful mobilities further in order to challenge two types of determinism in relation to technology and development. First is technological determinism and the need to balance what I will call contextualized affordances. The second is a form of cognitive determinism that shapes the field of communication for development and social change and the types of research questions we ask.

I begin this chapter by thinking through some of the implications, in a broad sense, of technological determinism for media and communication for development and social change. I then explore this further through the two determinisms, technological and cognitive, which I illustrate through examples. In relation to the first determinism, I argue that there are at least two important things that need to be taken into account when considering technological determinism within a communication for social change agenda. We must consider the affordances, or constraining and enabling material possibilities of media technologies, and we must consider contexts. A balance is needed between what technologies provide in terms of affordances as the possible properties of technologies and broad and specific contextual constraints and opportunities. As examples, I discuss the implications of mobile phones and disadvantaged women in India drawing upon research that examines the use of phones by women in a slum cluster, and initiatives to improve the position of women in terms of lives and livelihoods (see also Tacchi et al., 2012) in India. Here I apply the concept of meaningful mobilities to ask what role mobile phones play in women's changing situations and agencies in spaces that are highly complex, structured, constrained and subject to uncertainty.

Building on these examples, I then move on to the second determinism, to explore how disciplinary and other fields can be seen to frame and in some cases over determine what questions are asked when we think about technology and development. Here I suggest we need to take a step back and think about the types of framing that we need to pay attention to; how development and other powerful structures and systems can approach mobile phones to frame perceptions of groups of people in society and therefore highlight and prioritize particular characteristics and differences over others. As researchers, how mobile phones and mobilities are framed points us to some extent toward certain interpretations or subject matter. Technologies and populations are framed within certain frameworks and value systems that may distract us from other possibly more interesting explorations. I conclude the chapter by discussing what these examples tell us about technological determinism and media and communication for development and social change, and suggest ways in which the concepts of contextualized affordances and meaningful mobilities (when appropriately framed) can help us to move forward.

TECHNOLOGICAL DETERMINISM AND MEDIA AND COMMUNICATION FOR DEVELOPMENT AND SOCIAL CHANGE

Technology is generally understood to be an important and determining factor of development. Notions of development that emerged following the Second World War incorporated modernization theory frameworks, with technology and progress conceived of as interlinked and linear. Developing countries in the global south were framed as not having reached the stage of development of countries situated in the global north, and framed within neocolonial constructions (Crewe and Harrison, 1998; Escobar, 1995). Development as a concept frames our thinking about much of the world (Ferguson, 1990; Servaes, 1999). It is "an enormously powerful set of ideas which have guided thought and action" since the mid-twentieth century, with development institutions spending millions of dollars each year, developing and operationalizing countless plans, workers and policies (Gardner and Lewis, 1996, 2). The introduction of "modern" technologies are part of an agenda to move societies forward on the pathway to development (Waisbord, 2001). Production technologies and industries for farmers, medical technologies and family planning, mass media programing and media infrastructure such as a free press, radio, and (more recently) Internet and mobile phone are part of central notions of technology "transfer" that are core to agendas promoted to shape and change socioeconomic situations (Gardner and Lewis, 1996).

Communication, media and development have been connected as fields of knowledge and of practice, with a range of names such as communication for development, development communication, media development, communication for social change (Manyozo, 2012; Waisbord, 2008; Wilkins, 2000; Wilkins and Mody, 2001). Manyozo (2012) suggests there are three distinct but not entirely separate main approaches within what he terms media and communication for development; media for development, media development, and participatory or community communication. Underlying theories of media and communication for development have shifted over time from technology diffusion and transfer models toward participatory communication paradigms, although Manyozo argues there is a persistent dominant binary between the two. He argues that we need to move beyond this binary and that our focus should be on participation, policy and power. Reasonably constant themes have defined the broad field since the 1970s (Servaes, 2008) with communication techniques and media seen as participatory processes for social change, and dialogue seen as key to socially inclusive processes (for example, Fraser and Villet, 1994; Rogers, 1976; UN, 1997; WCCD, 2006). Development communication as exchange, meaning creation and processes rather than the transmission of messages now dominates theoretical work, while in practice earlier modernizing paradigms and economic growth approaches still persist (Fraser and Restrepo-Estrada, 1998; Inagaki, 2007; Mansell, 2011). How do we explain the persistence of modernizing paradigms that focus on technologies and linear progress, in a participatory development agenda that prioritizes processes and dialogue? It might be argued that modernization theory hinges on technological determinist ideas, while participatory approaches lean toward social determinism.[1]

In the field of ICT4D ideas of technological innovation for economic growth tend to dominate. Contrary to the participatory approaches in media and communication for development, efficiency in the dissemination of information is more central. This places most of the agency on the technology and its developers rather than on poor and marginalized populations (Unwin, 2009). ICT is seen by some in this field as making a progressive contribution to development, but largely missing its inherent promise for transformation. According to Heeks (2010), disruptive technologies can deliver resources from North to South, and can transform development itself. Such a focus on the transformational qualities of ICT as a universal good and universal concept tend not only to obscure particularities of poor and marginalized populations but also reinforce normative modernizing models of development (Unwin, 2009). They also risk maintaining a notion of technology as a solution to the problems of development, in particular through increased efficiency in production (Heeks, 2005). While the promotion of new ICTs for development has led to innovative experiments, some also observe a rapid evolution and expansion of technological determinist responses from development agencies (Article 19, 2005, 3).

Surely we cannot ignore the inherent possibilities and implications of "transformative" technologies, but just as surely we need to understand the implications of technological determinist approaches. So, for example, ICTs as transformative technologies have been said to create new "knowledge economies" and "networked societies" (Castells, 1996; Selwyn, 2004). They are also said to have created what is termed a "digital divide," whereby those who are not part of new technological developments are disadvantaged or left behind. As Robin Mansell (2002) pointed out, there have been inadequate interpretations of the causes and consequences of the "divide," focused on infrastructure rather than people, so that actions to "bridge" the divide or to understand its consequences were seen as stemming from the ways in which social and technical relationships were understood at a position removed from actual practices. In a technological determinist frame of thinking, following Postman (1993) and Mcluhan and Fiore (1967), when people have access it is the form or the medium that shapes uses and changes cultures. Yet if this were the case, we might see far more "successes" in the media, communication, and ICT for development interventions that view technology as a tool for change. And yet mobile phones have achieved high levels of penetration not through the efforts of development agencies but through commercial business expansion mixed with deregulation and increased competition, to be "generally freedom-enhancing, and that is an appropriate point of departure for the hagiography of the mobile phone" (Sen, 2010, 2).

Coleman's (2010) review of ethnographies of digital media tells us how important context is to understanding media, but it also demonstrates that there are particular affordances and constraints implicated in these technologies, however differently experienced they might be in a range of cultural, social and political contexts. Following Williams (1974, 133), while we have to reject technological determinism, we should not substitute it for the notion of determined technology. Tenhunen's (2008) study of mobile phone use in rural India shows that technologies not only amplify ongoing processes of cultural change, which we might consider to be a form of local appropriation of technology or determined technology, they can also be seen to *influence* social, cultural and political processes. As Williams states:

> We have to think of determination not as a single force, or a single abstraction of forces, but as a process in which real determining factors—the distribution of power or of capital, social and physical inheritance, relations of scale and size between groups—set limits and exert pressures, but neither wholly control nor wholly predict the outcome of complex activity within or at these limits, and under or against these pressures. (1974, 133)

Within development there is a preference for linear, cause-effect models, with implications for structural and power relations, and hierarchies of knowledge. Mansell (2011) notes development's preference for a "one knowledge system," which fails to appreciate the political nature of knowledge, and the importance of multiple knowledges. Escobar's work stridently critiques developments' ignorance of the complex cultural constructions of local knowledges and their modes of operation and relations to social and cultural fields (Escobar, 1995, 2007). So it is not just about whether the focus is on technology, or on people or contexts—it is more than this. It is about how knowledge and experiences are understood and valued, and the implications of this. Yet externally engineered technologies play a role in our development agendas—in shaping and framing what we do and how we think about and prioritize development agendas—telecenter programs, for example, reveal our underlying yet often implicit notions and models of both technology and development (Gurstein, 2011).

DETERMINISM 1: TECHNOLOGICAL DETERMINISM AND CONTEXTUALIZED AFFORDANCES

Technologies for communication, from the printing press to the mobile phone, can be seen to effect the evolution of societies (Ong, 1982), but the question of how far they drive social change, set the conditions for social change and determine that change is contested. We might rather look for the "operative relationship" (Williams, 1974) between modern technologies and complex societies. In this way, we can avoid polarizing accounts that either lean too far toward technological determinism, or completely ignore the inherent implications of modern technologies. This is to argue that we cannot understand the relationship between technologies and social change without capturing the particular details of how they are experienced in the everyday. How media and communication practices relate to the places, spaces, relationships, routines and everyday lives of people are complex and embedded. Thus, in development communication, we might usefully think about contextualized affordances. We can explore the notion of contextualized affordances further by considering some of the ways in which the smartphone has been thought about in development, and some examples of use.

Mobile phones in development are often viewed through economic and livelihood perspectives (Osorio and Postill, 2010). This relates to a dominance of an economic growth approach to development. Yet the goals of development are not purely economic, including broader human development as well as capabilities and freedoms (Deneulin with Shahani, 2009; Sen, 1999). There is no empirically demonstrable "automatic" connection between high GDP and "the ability of people to flourish" (Alkire and Deneulin, 2009, 15); well-being and freedom are not straightforwardly linked to incomes (Nussbaum, 2011).

Mobiles for Development (M4D) as a field has a "dual heritage" (Donner, 2010). It is framed by ideas of the mobile as an enabler of choice, and framed within a broader ICT4D perspective where technologies can be designed and woven into social systems to bring about social change. Donner calls this "embedded directionality," which requires a development goal as well as an appreciation of the social structures and contexts that services and applications can be introduced into. The focus is often skewed toward livelihoods, rather than lives. Donner (2009) recognizes the blurring of lives and livelihoods in mobile phone use, including their use and value for self-expression, agency, and social connection in addition to economic activities. Yet, as he points out, "most benefits of the mobile in development processes remain unobserved and under-studied, in unorganized 'peer-to-peer' voice calls and text messages" (Donner, 2010, 9). Mobile phones are held up as a key instrument of development, important because of the scale of uptake in the global South, and the fact that they provide the first access to electronically mediated communication for millions of people. However, ethnographic research on their everyday uses and implications in developing countries points to a range of implications that challenge "monolithic visions" of their effects (Ling and Horst, 2011). Here we might then ask, in specific contexts, what are the affordances that matter, and what makes mobility meaningful?

To illustrate these issues, we can draw upon ethnographic research in an Indian slum,[2] giving examples of three women and their use of phones, including landline and smartphones. We can then compare this with the mobile phone use of women in rural Gujarat, encountered through research with the Self-Employed Womens Association (SEWA).[3] This helps to illustrate the range of ways women manage their lives within highly structured and restricted environments, and the various uses and consequences of phones and ideas and practices of mobilities within this. The first woman

I discuss used a landline rather than mobile phone, but nevertheless the story works well to illustrate these tensions and complex relations between restrictions and mobilities.

Savita and her family live in a large slum cluster in Delhi, which in 2004 included her husband and four sons (then aged between 14 and 21 years). A runaway boy of 13 also lived with them and helped them run a tea shop on the busy road they lived on, which marked one of the boundaries of the slum, and set it apart from the neighboring legal settlement. Savita's house had two rooms, a toilet and a wired telephone connection (a landline). In the context of this slum, they were doing fairly well. Apart from running the tea-shop and a catering business, three of their sons were working and contributing to the household income. Rajbeer was Savita's husband, and initially our main point of contact who took the lead in answering our questions, even when they were directed at Savita. Rajbeer had suffered a serious accident some years before and was chronically ill and unable to work outside the home.

Plans were underway for the two oldest sons to get married to two sisters from a lower-middle class family. Savita and Rajbeer were planning to invest in another *jhuggi*[4] in the same camp, to accommodate the extending family. But just two months prior to the weddings Rajbeer died. Savita's position in the family transformed dramatically, as she became head of the household. While Rajbeer was alive we were given a sense of a warm, loving and supportive relationship with his parents who lived nearby. However, soon after the two-week mourning period, when discussions about rescheduling the double wedding were underway, Savita took a stand not to postpone it contrary to the wishes of her in-laws. This was her first act of defiance and open acknowledgment of an acrimonious relationship with her in-laws. Savita's mobility—spatial, social and economic—were strictly contained in her role as a wife and daughter-in-law. But once she was head of the family, the possibilities for mobility increased. She visited her maternal relatives living on the outskirts of the city, a trip for which she would previously have had to seek permission and funds from her husband. She now controlled the businesses, made household decisions, and arranged the weddings. Her access to the phone also significantly changed.

During our regular visits before and following Rajbeer's death, our conversations were often interrupted by calls on their landline. Until Rajbeer's death, he or one of his sons or the live-in help would answer it. Savita did not answer the phone or have a "direct" conversation on it. Even if the caller inquired after her, her husband or sons would relay the conversation back and forth. In one of our conversation about the potential of the mobile phone, Rajbeer made it clear that he objected to the increasing freedoms or mobilities available to women, including working outside the home, travelling beyond their neighborhoods unescorted, and the use of mobile phones. While he saw phones as corrupting influences for women, his oldest son was one of the few local men at that time who owned one and Rajbeer saw this as important for his son's work and career. The control of telephones in this household, like most others here, lay with the men.

On one occasion, during the early period of mourning, the telephone rang. A number of Savita's female neighbors, relatives and friends were present. They observed and listened in silence as Savita received the phone call. Afterwards they asked her to repeat the entire conversation twice. Savita had been nervous accepting this call, but quickly Savita's ease and confidence to use the phone increased. She began conducting much of her business and social relationships through the phone, especially since as a widow her physical and social mobility was still restricted in some ways, as she was required to be in extended mourning for at least a year.

When her two daughters-in-law were brought within the household after marriage, Savita instituted a strict regime of control over their movements, and use of phones, similar to the one she had to abide by while Rajbeer was alive. She saw it now as her responsibility to uphold a level of family morality and virtue, which is visibly demonstrated through a range of constraints and restrictions. The notion that access to technologies can increase economic or social mobility in any straightforward way is clearly too simplistic as a way of understanding Savita's increased and indeed differently constrained mobility following the death of her husband. Two young women in an adjoining slum, Rani and Monica, further demonstrate how the social, cultural and moral landscapes within which these women live, can determine and constrain their use of mobile phones, as well as how the phones can help them in some ways to resist or defy them.

Rani, aged 26, and Monica, aged 21, are neighbors who both live in women-dominated muslim households in a neighboring camp in the same slum cluster as Savita. Rani and Monica live completely different lives, despite some similarities in their situations. Rani lives with her aging mother, younger brother, and her daughter aged 9. She also has two younger sisters who are married and live with their in-laws. Her father was a drug-addict and died a few years ago. Rani left her abusive husband when she was 18, a year after her marriage, and returned home. Their house has four rooms, with two rented out to other families. Rani's mother works as a domestic help in a nearby middle-class household. Rani's contribution to the well-being of the family has been significant. She paid to extend the home from a one-room dwelling to four rooms. She also arranged her two sister's marriages, including the financial aspects. Rani generated these funds by working as a sex worker.

She has a few regular clients in and around Delhi, none from her slum cluster which is a deliberate decision. It is important for Rani to both support her family financially, and to protect their reputation, especially for her young daughter. She had tried working as a maid in a nearby middle-class settlement, but she found the treatment she received humiliating, and in some cases threatening. She found her treatment as a sex worker (with a few rare exceptions), especially given her establishment of regular clientele, far more respectful and lucrative. She has earned enough money to improve the condition of her family, but also fears the implications for them of the kind of work she does if it becomes known. While she is aware that some do know her line of work, she has made useful connections to local crime figures and wealthy men, and so neither her family nor neighbors condemn her openly since her family depend on her earnings, and her neighbors occasionally need to seek her help when they have local issues, "My family is so hypocritical; of course they know where the money is coming from and had no issue about taking it when they needed but if there is the slightest issue, they call me a *randi*,[5] and say I bring shame to the family. Earlier I used to get affected by what they said, but now I retort back and say, who is worse a *randi* or those who eat of her living."

Rani's particular position has allowed her a rarely available economic, social and spatial mobility, and yet it is also highly precarious. Rani also has a level of self-taught expertise with electronics and can take apart mobile handsets. Both local men, women and young girls often came to her to learn the functionalities of their new phones. She entered the second-hand and extra-legal "mobile markets" selling handsets (some stolen) to her extended networks in and beyond the slum. She herself carries two mobiles, and has three mobile connections. Only one connection is in her name, which she never uses when dealing with her clients. Even though she is illiterate she sends and receives several text messages in a day. Some of them are conversational, but most of them are forwarded poetry and words of affection, which her brother or one of her educated neighbors such as Monica read to her.

We met Monica's mother one afternoon in late 2011. She runs a general store in the slum. As the only store catering to a few hundred households, it does thriving business. She runs the business with an iron-fist, and is revered and respected as *badi-khaala* (elder aunt). She is known for her astute business sense and is not shy to admit it herself: "I am an illiterate woman, but I was always keen on improving the situation for the family. If I had left it to my husband we would be still hand-to-mouth." Besides the corner store, they own four *jhuggis* and an apartment in a lower-middle class area in southwest Delhi. Three of the *jhuggis* are rented out augmenting the income from the shop. The afternoon we met her, she was approaching Rani to escort her to the Tihar Prison where her son was imprisoned having been involved in a burglary. Monica's mother would not consider Rani as an escort, and until her son was imprisoned Monica and her sister had lived in the apartment in the lower-middle class area. When her son was arrested she became worried about their safety because of regular police visits. She is especially concerned about Monica who is a college graduate and beautiful, and who attracts a lot of male attention.

After her graduation, Monica studied a course in fashion merchandizing, and found a job in an export house in the outskirts of Delhi. She commuted an hour each way, and loved her job, and the experience and exposure that it brought. However after only five months she left because, "Amma made it impossible. She would call me every half hour to inquire what I was doing, and if I did not answer the phone she would start calling the landline. It was most embarrassing. If I said I would reach home by seven, and was delayed on account of traffic or some work, she would get hysterical. The limit was when one-day she landed up at my work, because the previous night my boss—a man—had called about some work. I just could not take it." *Badi Khaala* does not regard her strict control over her daughter's movements as oppressive or controlling, it is her duty, "young girls need to be protected. There are so many distractions, and we cannot allow her to go ashtray." Her biggest fear is Monica marrying outside their religion, and her second biggest concern is a love marriage rather than an arranged marriage.

Monica on the other hand accepts that she will marry a groom of her mother's choice, but finds her mother's day-to-day oversight extremely difficult to deal with. Her smart-phone connection with the outside world stops her from "going mad." She is constantly online through her latest Samsung handset. She uses a range of social media sites including Facebook, Orkut and Twitter. She also uses Skype to keep in touch with her extended family working in the Middle East and with her online friends. She has mild flirtations on Facebook and has 400+ friends, many of them unknown to her offline "of course Amma does not know what I am up to on the phone. She thinks I am texting, and gets annoyed at times but that is it. She does not know internet, or Facebook." Monica knows that her mother would prohibit the use of the phone if she knew what she did with it, but for her it was a way of managing or circumventing the severe restrictions placed upon her in what she considered a harmless way. Offline her social life is highly monitored, online she is free.

The lack of socially sanctioned mobility is clear from the examples of the three women from the slum cluster. It was also evident in Gujarat where the SEWA was formed, as a women's self-employed textile workers union in 1972 to help women gain greater security in unpredictable working environments. Since it formed it has grown to over 1.5 million members in a range of occupations, including farming, factory work, and home-based small businesses (Bhatt, 2006). SEWA has over 600,000 members in the State of Gujarat. SEWA provides a social network for poor women, designed to help them achieve greater security in terms of their livelihoods as well as agency. The members of SEWA who took part in our research talked about how SEWA gave them a sense of belonging, solidarity, identity, recognition, security, and greater social, cultural and political agency. SEWA has helped open up social, economic and political spaces for these women, along with access to a range of savings and loan schemes. SEWA has formed a lending organization, a marketing association, a bank, a learning institution, and an insurance group. It has also encouraged the use of new technologies like computers and mobiles, and provides loans for members to purchase mobile phones.

In the Great Rann of Kutch, one of the largest salt deserts in the world, seasonal salt workers come here from across the state, and SEWA supports some of these daily wage earners to help them build sustainable livelihoods. The women working here spend months at a time far from home. A salt worker and her husband talked to us about how they leave their children with their extended family in the village. This is common practice for children of school age since the environment is extremely harsh and the work very hard. The woman talked about how the mobile phone allowed them to contact neighbors and send or receive messages about their children, giving them peace of mind that they can be contacted if there is a problem. This peace of mind may not make the backbreaking work any easier, but eases their concern for their children who they will not see for months.

Other members of SEWA talked about how mobiles allowed them to be away from home but not just in terms of easing their worry about their children. It was also a case of women being sanctioned to be outside of the home for work or for training, which previously had been very difficult because their role as wives and mothers meant staying at home. During a large meeting of more than 40 group leaders at the main SEWA center in Anand, women talked about the role of mobiles in facilitating networking with other women and SEWA groups. They also talked about the ability to work and travel outside the home, but a constant theme across the discussions and interviews was on the importance of their men "allowing" their involvement with SEWA. Permission to become involved in SEWA had to be provided by the men, and the members had a range of strategies to help reassure them and persuade them to approve this. Permission was generally given once men appreciated and saw the advantages, most often in terms of helping the family to become more financially secure.

A widow called Kapilaben, a member of the SEWA executive committee, told us how she had never held a mobile or used one before she took a loan of 3,000 rupees from her local SEWA association to buy one. A small lily farmer with just over half an acre of land, she picks the flowers each morning, and sells them in local markets. Before she had her mobile she would spend the whole day going from one market to another, searching for traders to buy her flowers. Now she uses the mobile to line up the traders as she picks the flowers, delivering them by 9 am. She has used the time and money saved to start another enterprise with a group of SEWA members, producing and selling goods through SEWA's Rural Distribution Network (RUDI).

Kapilaben had been a subsistence farmer, growing millet and other crops until her husband died and she had to find ways to earn extra income to support her family. She was trained in organic farming through SEWA, who helped her research lily farming, and provided loans to set up her business. She took out her third SEWA loan in 2003 to purchase her mobile phone. As well as income security, the kind of meaningful mobility she has achieved over the past few years also encompasses self-esteem, empowerment, and leadership, and as a widow who is head of her household, constraints on her mobility are lessened. Recognizing the benefits of the kind of agency she has developed, she is now taking out another SEWA loan to support her daughter's education.

A similar theme across the women discussed above is that of their shifting lives, especially related to and emphasized by the constraints, insecurities and vulnerabilities that these women face. While much of the discussion on phones is linked in some way to economic aspects of life—business, income generation—the complex constraints and restrictions faced by the women cannot be ignored. In one example we could say that the mobile afforded a connection that restricted economic independence—Monica's short working life, interrupted by her mother's ability to contact her through mobile and landline—at the same time as it provided an unprecedented freedom in the online realm. Studying mobile phones can reveal how their uptake can make existing social relationships and processes of change more visible (Ling and Horst, 2011).

Mobile phones can help to challenge power structures and alter communicative practices and norms to an extent, yet as Sey (2011) points out, they are not so much a mechanism for development in themselves, but rather contribute to processes of development as they highlight, extend and magnify communicative and other capabilities. They can be thought of as agents within wider social and cultural change, and indeed can highlight and amplify existing restrictions, insecurities and tensions. Underlying gender dynamics complicate our understanding of the role of mobiles (Chib and Chen, 2011). Tensions can emerge through challenges to social sanctions, as Wallis (2011) found through her study of mobile phones and labor relations for young migrant women in Beijing. Gender, age, class and place-based power relations severely constrain these women and their efforts to find better employment. Mobile phones can help their job movement, but only for some women who incorporate phones into preexisting microenterprises, or women who use phones to generate higher income for their employers, and some employers use mobiles for surveillance and to harass young female employees (Wallis, 2011).

DETERMINISM 2: CONCEPTUAL DETERMINISM, DISCIPLINARY FRAMES AND THE QUESTIONS WE ASK

As noted above, Donner has pointed out that many uses of the mobile phone such as social messaging and voice calls have been understudied by development research. Therefore we only partially understand how mobiles take part in processes of development. We can see from the examples drawn upon above that mobiles help women to remain connected with family, source labour opportunities and improve their livelihoods but we can also start to see how it feels to be a seasonal salt worker or a sex worker, or a young women monitored by her mother, the competing demands and relationships, as well as the range of mobilities that matter to women. We can start to see that meaningful mobility from the women's perspectives varies across these examples. Extending from the examples above that complicate the relationship between technologies, contexts and change, there is a need to consider the frame.

While we might agree that mobiles and mobility mean different things and have different implications and raise different questions in different contexts, how do we then go on to research and understand the implications of this? The approach and disciplinary frame determines what questions are asked, what knowledge is useful and therefore highlights and prioritizes particular differences over others. This points us toward certain interpretations and subject matter. The examples of research in the slum cluster can be contrasted with the research in Gujarat. Different questions were asked because the frame of the research in Gujarat was development; it was looking at women through a development initiative and asking what role mobiles played. The work in the slums was undertaken outside of a development agenda, to explore and understand the everyday lives and experiences of slum dwellers and how they relate to the world outside the slums, with communication and phones being just one aspect of what emerged as interesting (see Chandola, 2012, 2013).

Research undertaken within a development frame is understandably focused on instrumentalist agendas—for example, how are and how could mobiles be used to increase livelihoods. While it is clearly possible to compare and use the findings from both sets of studies above to develop arguments and concepts such as contextualized affordances and meaningful mobilities, here I want to point to the importance of nevertheless paying attention to the frame. In the context of literature about technologies and communication for development, where it is not uncommon for technological advances to become conflated with modernization paradigms of development itself, we tend to know little about how people think and feel about their lives, their relationships, families, social conditions, labor, and relationships with technologies. The underlying models of development frame the perspective and the research questions (Tacchi, Kitner and Kiran, 2014).

Buskens (2010) points out how (rightly) concerned we are in ICT4D with increasing agency and choice for the women we work with, such as the women discussed in this chapter. However, she also gives an example from Africa of a woman's struggle to succeed, as a way of pointing out that the ways in which ICT4D (and I would add development more broadly) fails to understand the complex and layered environments that women live in, different situations and locations, and the gendered socialization, expectations and aspirations. Buskens tells us how Bahati left home as a young woman and moved to the city to stay with her aunt and try to earn a living. The first job she tried did not work out for her and she turned to hairdressing. She found a position and worked hard for two years, saving what she could from her small earnings. The only way for her to make more money was to set up her own business.

Once she had saved enough, she bought a mobile phone, and was then able to build a customer base for her own business. This example is both inspiring since Bahati showed enormous strength and determination, and instructive because she succeeded against the odds. Yet the lesson Buskens wants us to understand is not that we can therefore replicate such success, by sending mobile phones to all women in Africa so that they can set up their own business like Bahati. Buskens rather invites us to think about how successful Bahati might have been without the two year's hairdressing experience, or how a married woman might compare, given the very different expectations, and, whether a woman with children would be able to dedicate all her time and energy to work outside the home. She challenges us to think beyond "fix it" notions of technologies and development. As researchers we do need to think about the agency of women in development contexts, but also think about our own agency and choices in how we think about and research development. This is because we, as development researchers and scholars, are among the people who define development discourses and practices and research processes, along with donors, and practitioners. We greatly "influence the theoretical, methodological, and normative concepts employed in the ICT4D knowledge construction processes" (2010, 20).

This is important because we can look across all of the examples discussed in this chapter and see that marginalized and poor women live in a range of different ways, within a range of different structures and systems that restrict and set gendered expectations for them. While these women's agency is experienced and adapted to within the various and particular structural and systemic environments they live in, and the tensions and competing demands upon them, development sets a further set of expectations about what women might do and how they might achieve agency and the ability to make choices. Rather than hand out mobile phones so that other women might strive to succeed and demonstrate agency as Bahati did, what if we focused instead on the structures that restrict women and their mobility, or the infrastructures that mean it takes a poor working woman two years to save for a mobile.

As researchers and scholars, Buskens considers that we too experience agency and choice *within* the structural powers in our environment. For us it is funding organizations, corporations and mainstream neoliberal development models that provide the frame of reference for our research, and our choices. The concept and practice of development "is defined by the powerful and signifies the powerful" (Buskens, 2010, 20) it is therefore essential to acknowledge the frames that we employ. ICT4D is a "reflexive endeavor" because the ways we make sense of what we research "impact the reality that we study because of the ways that power and knowledge construction interact and intersect" (Buskens, 2010, 20). Our knowledge construction is not neutral. Buskens challenges us to recognize our theoretical, methodological and normative positions and acknowledge and take account of its impact on those we research. She considers this as a form of agency; "It is in becoming aware of our perspectives, realizing them for the choices they are, and acknowledging the impact they might eventually have on the lives of the women to whom we aim to reach out that we, as researchers and scholars, can exercise our agency" (Ibid).

In international development, mobiles are understood to have effectively "leapfrogged" older communication and information technologies that required costly infrastructures that are largely considered to be the responsibility of governments. Mobiles, on the other hand, are proliferating and reaching into previously technologically unconnected areas through mainly commercial initiative. Rather than a social good, with universal service agendas pressuring governments to act, commercial operators are tapping into new markets with affordable devices and connections.

This is not to suggest abandoning instrumentalist approaches in development, but equally to recognize what might be missed in such an approach. Asking different questions and considering different frames may enrich rather than distract from the goal of understanding the human condition and attempts to improve it. Other frames and models of development do exist, and Buskens mentions the Institute for New Economic Thinking (http://ineteconomics.org), and the Venus Project (http://www.thevenusproject.com) as examples. Human development approaches and Sen's capabilities approach remind us that within development we should be thinking not narrowly about economic growth, but broadly about people in environments. They tell us to focus not just on markets but on a range of capabilities and freedoms to help people choose the kind of lives they have reason to value. This can be hard to measure and hard to research, and does not fit well within the mainstream development framework for research and evaluation (Lennie and Tacchi, 2013).

CONCLUSION

What the discussion above intends to highlight is that there are key domains of mobility related to mobile phones including social, political, economic, and geographic. It also aims to illustrate some of the key constraints on mobility including gender, class, caste, economics, social status and so on. The implications I argue are that we need to push ways of thinking about mobiles, mobilities and development beyond the economic, in ways that balance affordances and context, and that acknowledge, challenge and make explicit our frames, which requires a degree of reflexivity. This can help us to ask different questions and pay attention to underlying processes of change that might at first seem unimportant. Looking at mobiles in the context of a highly effective development organization (SEWA) highlights underlying processes of social change that have been a work in progress for more than 30 years. A focus on technologies like mobiles can draw our attention to underlying structures, and social change, can magnify them, but not wholly create them. In India, rather than focus on consumption and economic comparisons, how would it be to consider the constraints and restrictions faced by slum dwellers and other marginalized populations and the social, sensual and feelingful dimensions of how they think and feel, and the role of mobiles in this? In essence I am arguing for seeking out and recognizing local perspectives.

For Sen (2010, 2) it is important not to overemphasize the contrast and opposition between local and global knowledge, it is "very fashionable these days to praise 'local knowledge' and its great importance." I am certainly not arguing that local knowledge is *more* important that external or "global" knowledge, and agree with Sen that the way communication technologies are being absorbed across the world show how easily assimilated they are within local environments and knowledge systems. But there is, within development, a need to consider unequal power relationships between donors and development practitioners and their recipient communities, and the different conceptual frames through which we perceive what change is achievable and desirable, and what futures are worthy of aspiration.

The discussions above indicate the complexity and multiple interlinked structures, constraints, and opportunities that marginalized women in particular face. Understanding the multiple meanings of technologies within such contexts is a vast undertaking. Focusing on ideas of contextualized affordances and meaningful mobilities, the changes that technologies help to facilitate in particular places and times, that are meaningful and valued locally, can perhaps provide a suitable focus when exploring mobile phones and development and understanding how to counter tendencies toward technological and conceptual determinism.

NOTES

1. Participatory approaches are also widely critiqued and challenged (see, for example, Cooke and Kothari, 2001).
2. The author has worked in this space with Dr Tripta Chandola since 2004. The research reported here includes Dr Chandola's ethnographic work with the author (funded by the Department for International Development—DFID—from 2003 to 2005), and further ethnographic work undertaken by Dr Chandola first for her doctorate, supervised by the author, and subsequently through postdoctoral work.
3. This data is from three research projects. The first was "Moving Content: Creative Engagement in Marginal Spaces" (2007–2010), funded by Intel and led by the author with Jerry Watkins, Kathi R. Kitner, Jay Melican and Sue Faulkner, and research assistants Kiran M.S., and Tripta Chandola. The other two short projects were "Technologies of Attachments" and "Smartphones and Social Participation," again funded by Intel and led by the author with Kathi R. Kitner and Kate Crawford, Kiran M.S., and Tripta Chandola.
4. Slum dwelling/house.
5. Scrubber or whore.

REFERENCES

Alkire, S. and Deneulin, S. (2009). 'The Human Development and Capability Approach'. In Deneulin, S. with Shahani, L. (Eds.) *An Introduction to the Human Development and Capability Approach: Freedom and Agency.* London: Earthscan. pp. 22–48.

Article19 (2005). Experiencing Technical Difficulties: The Urgent Need to Rewire and Reeboot the ICT-Development Machine. *Thematic Reports*. London: 45.

Bhatt, E. R. (2006). *We are Poor but so Many: The story of self-employed women in India*. New Delhi: Oxford University Press.

Buskens, I. (2010). 'Agency and Reflexivity in ICT4D Research: Questioning Women's Options, Poverty, and Human Development' *Information Technologies & International Development* Volume 6, Special Edition, pp. 19–24.

Castells, M. (1996). *The rise of the network society*. Cambridge: Mass, Blackwell.

Chandola, T. (2013). 'Listening in to Water Routes: Soundscapes as Cultural Systems' *International Journal of Cultural Studies* 16: 1.

Chandola, T. (2012). 'Listening in to Others: Moralising Soundscapes in Delhi' *International Development and Planning Review*, 34(4).

Chib, A. and Chen, V. H. (2011). 'Midwives with mobiles: A dialectical perspective on gender arising from technology introduction in rural Indonesia' *New Media & Society* 13(3): 486–501.

Coleman, G. E. (2010). 'Ethnographic Approaches to Digital Media' *Annual Review of Anthropology*, 39: 487–505.

Cooke, B. and Kothari, U. (Eds) (2001). *Participation: The New Tyranny*. London: Zed Books.

Crewe, E. and Harrison, E. (1998). *Whose development? an ethnography of aid*. New York: Zed Books.

Deneulin, S. with Shahani, L. (Eds) (2009). *An introduction to the human development and capability approach: freedom and agency*. London: Earthscan.

Donner, J. (2010). Framing M4D: The utility and continuity and the dual heritage of 'mobiles and development' *The Electronic Journal on Information Systems in Developing Countries* 44(3): 1–16.

Donner, J. (2009). 'Blurring Livelihoods and Lives: The Social Uses of Mobile Phones and Socioeconomic Development' *Innovations* 4(1): 91–101.

Escobar, A. (1995). *Encountering Development, the making and unmaking of the Third World*, Princeton: Princeton University Press.

Escobar A. (2007). '*World and Knowledges Otherwise*': The Latin American Modernity/Coloniality Research Program. *Cultural Studies* 21(2–3): 179–210.

Ferguson, J. (1990). *The anti-politics machine: "Development," depoliticization, and bureaucratic power in Lesotho*. Cambridge: University of Cambridge Press.

Fraser, C. and Villett, J. (1994). *Communication: A key to human development*. Rome: FAO. http://www.fao.org/sd/cddirect/cdpub/sdrepub.htm

Fraser, C. and Restrepo-Estrada, S. (1998). *Communicating for development. Human change for survival*. London: I.B. Tauris Publishers.

Gardner, K. and Lewis, D. (1996). *Anthropology, Development and the Post-Modern Challenge*. London: Pluto Press.

Gurstein, M. (2011). 'Telecentres are not "sustainable": Get over it!' *Gurstein's Community Informatics* blog, accessed 2 February 2014. http://gurstein.wordpress.com/2011/05/18/telecentres-or-community-access-centres-or-public-interest-access-centres-or-community-technology-centres-etc-etc-are-not-%E2%80%9Csustainable%E2%80%9D-get-over-it/

Heeks R. (2005). ICTs and the MDGs: On the Wrong Track? I4D online.net, February. URL http://i4donline.net/feb05/perspective.pdf.

Heeks R. (2010). Do information and communication technologies (ICTs) contribute to development? *Journal of International Development* 22(5): 625–640.

Inagaki, N. (2007). Communicating the impact of communication for development. Recent trends in empirical research. The World Bank, Working Paper Series, Development Communication Division.

http://wwwwds.worldbank.org/external/default/WDSContentServer/WDSP/IB/2007/08/10/000310607_20070810123306/Rendered/PDF/405430Communic18082137167101PUBLIC1.pdf

Lennie, J. and Tacchi, J. (2013). *Evaluating Communication for Development: A Framework for Social Change*. Oxford: Earthscan, Routledge.

Ling, R. and Horst, H. (2011). 'Mobile communication in the global south' *New Media & Society* 13(3): 363–374.

Mansell, R. (2002). 'From Digital Divides to Digital Entitlements in Knowledge Societies' *Current Sociology* 50(3), 407–426.

Mansell, R. (2011). 'Power and Interests in Information and Communication and Development: Exogenous and Endogenous Discourses in Contention' *Journal of International Development.* Article first published online: 10 July 2011. DOI: 10.1002/jid.1805.

Manyozo, L. (2012). *Media, communication and development: three approaches.* New Delhi: Sage.

McLuhan, M. and Fiore, Q. (1967). *The Medium is the Massage.* New York: Random House.

Nussbaum, M.C. (2011). *Creating Capabilities: The Human Development Approach.* Cambridge Massachusetts: Belknap Press.

Ong, W. J. (1982). *Orality and Literacy: The Technologizing of the Word.* New York: Methuen.

Osorio, F. and Postill, J. (2010). 'Mobile rewards: a critical review of the Mobiles for Development (M4D) literature' EASA2010: Crisis and imagination, Maynooth, Ireland (24–27 August 2010).

Postman, N. (1993). *Technopoly: The Surrender of Culture to Technology.* New York: Vintage Books.

Rogers, E. (Ed.). (1976). *Communication and development: Critical perspectives.* Thousand Oaks, CA: Sage.

Selwin, N. (2004). 'Reconsidering political and popular understandings of the digital divide' *New Media & Society* 6(3): 241–262.

Sen, A. (1999). *Development as Freedom.* New York: Anchor Books.

Sen, A. (2010). 'The Mobile and the World' *Information Technologies & International Development* Volume 6, Special Edition, pp. 1–3.

Servaes, J. (1999). *Communication for development. One World, Multiple Cultures.* Cresskill: Hampton Press.

Servaes, J. (Ed.). (2008). *Communication for development and social change.* London: Sage.

Sey, A. (2011). "We use it different, different": Making sense of trends in mobile phone use in Ghana. *New Media & Society* 13(3): 375–390.

Tacchi, J., Kitner, K. and Crawford, K. (2012) 'Meaningful Mobility: Gender, development and mobile phones' *Feminist Media Studies* 12(4): 528–537.

Tacchi, J., Kitner, K. and Kiran, M. S. (2014), 'Drive-by Wi-Fi and digital storytelling: development and co-creation' *The Journal of Community Informatics* 10(1). http://ci-journal.net/index.php/ciej/article/view/1015.

Tenhunen, S. (2008). Mobile technology in the village: ICTs, culture, and socail logistics in India. Journal of the Royal Anthropological Institute (N.S.) 14, 515–534.

UN Resolution 51/172 (1997). http://www.undemocracy.com/A-RES-51-172.pdf

Unwin, T. (2009). 'Information and Communication in Development Practices' In Unwin, T. (Ed.) ICT4D: Information and Communication Technology for Development. Cambridge: Cambridge University Press.

Waisbord, S. (2001). Family tree of theories, methodologies and strategies in development communication. South Orange, NJ: Communication for Social Change.

Waisbord, S. (2008). The institutional challenges of participatory communication in international aid. *Social Identities, 14*(4), 505.

Wallis, C. (2011). Mobile phones without guarantees: The promises of technology and the contingencies of culture. *New Media & Society* 13: 471–485.

WCCD (World Congress on Communication for Development) (2006). Rome Consensus www.uneca.org/africanmedia/documents/Recommendations_Rome_Consensus.pdf

Wilkins, K. (Ed.). (2000). *Redeveloping communication for social change: Theory, practice and power.* Boulder, Colorado: Roman & Littlefield.

Wilkins, K. and Mody, B. (2001). 'Reshaping development communication: Developing communication and communicating development' *Communication Theory 11*(4), 385–396.

Williams, Raymond (1974). *Television: Technology and Cultural Form.* London: Routledge.

Chapter 8

Digital Technology and the Construction of "Glocal" Information Flows

Social Movements and Social Media in the Age of Sustainability

Emily Polk

An emerging body of research has focused on the influence of the Internet in coordinating and mobilizing social movements (della Porta and Mosca, 2005; Donnatella and Lorenzo, 2005; Fisher and Boekkoi, 2010; Garrett, 2006; Gurstein, 2011; Juris, 2005; Polk, 2013; Tufekci, 2011). Such research analyzes the potential for the Internet to serve as a participatory, nonhierarchical, and diversely networked space for social movement organization (Bennett et al., 2008). Wellman (2002), noting the multiple layers of networks made possible by the Internet, suggests that it provides "a ramp" onto the global information highway at the same time that it strengthens local links within neighborhoods and households. "For all its global access, the Internet reinforces stay at homes. Glocalization occurs, both because the Internet makes it easy to contact many neighbors, and because fixed, wired Internet connections tether users to home and office desks" (Wellman, 2002, p. 14).

This chapter seeks to explore the role of the Internet in facilitating the communication processes on a micro- and macrolevel of the Transition Network, a global social movement of more than 1,400 *Transition Towns* in more than thirty countries that have committed themselves to transitioning to greater sustainability and resiliency through a variety of actions that include the creation of alternative local economies, and food, education, and transportation systems. Building upon the premise that the Transition movement was able to achieve its rapid global growth largely because of its sophisticated online presence, this chapter explores first the ways in which the Transition movement used digital technology, specifically its Transition Network, the online hub of the movement and other social media such as Facebook and Twitter to accomplish the former and the latter. Second, it seeks to understand if, when, and how and in what ways Transition Amherst used the online Transition Network to support its own activities. Although research indicates that the Internet may be displacing geographically bounded participation in and obligation to singular, spacially based communities with a *networked individualism* (Wellman, 2002) whereby people are now participating in multiple networks with a responsibility only to him/herself, little research has been done to explore the paradox of how social movements like Transition—which are premised on encouraging others to return to the materiality of an embodied place, commit to the physically local, live in connection with and beholden to your neighbors—are using technology to do that.

Data was obtained primarily from ethnographic research conducted from November 2011 through February 2013 while I was an active participant in my own Transition town in Amherst, Massachusetts. It consists of participant observation, textual analysis of local and global Transition materials on and offline, and extensive interviews with members of Transition Amherst. Many of the names have been changed to protect the identities of the participants. At the time of this research, nearly a dozen community members actively participated in Transition Amherst, attending bimonthly meetings, as well as other community events hosted by the group. The initiative's activities included coordinating a Transition-themed book group and film series, as well as skill-sharing workshops on food preservation and bicycle repair. A Transition Amherst website was under construction.

My data suggests that the Transition movement supports glocalization processes by strategically using digital technology, specifically the Internet, for two primary reasons: (1) To *transmit information* (Transition methods, practices, guidelines, values, personal stories) on multiple levels to aid local Transition town initiatives, and (2) To *maintain control* of its branding and sustain its legitimacy and authority as a global movement toward community resiliency.

TRANSITION, OCCUPY, AND THE ARAB SPRING: MOBILIZING THROUGH DIGITAL TECHNOLOGY

Indeed, Transition builds upon and has developed alongside similar social movements that have been mobilized in part due to digital technology and social media, like the Occupy Wall Street movement and the Arab Spring, both of which used digital media to communicate and coordinate their actions. The emerging social movements share primary characteristics in relation to their communication online: (1) They have been and are mobilized and coordinated online, yet remain decentralized and participatory, (2) They suggest that they are not just trying to achieve new goals but rather shift the processes by which the goals are reached (namely by building support around ideas rather than individual leaders, something which is made possible by the non-hierarchical structure of the Internet), (3) They have cultivated numerous actor networks that recognize the necessary relationship between the physical public sphere and the virtual public sphere with regard to mobilizing action.

Yet Transition separates itself from other social movements by framing itself as *nonpolitical*. Thus while members of different factions of Occupy and the Arab Spring might use social media to coordinate large-scale protest demonstrations, the Transition movement is using the same digital communication infrastructures to publicize a blog about global activities on International Permaculture Day (April 29, 2012, Transition Network Facebook page). Both communications might be interpreted as a kind of "protest" of capitalism, and both invite local (embodied) participation, however, participating in the former connects the physical body to the physical protest of a system, while participating in the latter connects the physical body to creating an alternative to that system. Since the Transition movement's activities are not necessarily bounded to specific times and places (each initiative coordinates its own events and activities), its online communication tools can serve as a global resource and support structure, from which each initiative might pick and choose relative to its own needs, on its own time.

Pasek and Romer (2009) note that it's important to understand and differentiate not just the type of digital medium (social media, website, blog, etc.) but the purpose behind the message. They suggest six general purposes behind the use of information technology: informational (raise awareness, transmission of facts, information, and values), communicative (share stories, experiences, let people know how and when to mobilize), recreational, social (chat rooms, Facebook, Twitter), product consumption, and financial management.

The Transition movement is particularly adept in its use of information technology regarding the first two: *raising awareness* around peak oil and climate change and *communicating* the possibilities for positive social change when communities take small steps toward becoming more resilient.

CONNECTING THE GLOBAL TO THE LOCAL: DIGITAL COMMUNICATION IN THE TRANSITION MOVEMENT

The virtual home of Transition is its comprehensive and frequently updated website www.transitionnetwork.org. The site is visually pleasing and easy to navigate with a series of changing pictures and an information bar where users might access the most recent Transition "News" including the top stories (it is unclear who or how a "top" story is determined) and blogs from onsite bloggers, and, republished blogs from Transition Voice, This Low Carbon Life, Reslience.org's newsfeed and Transition US. Visitors can also click on the "Community" link where they can access initiatives, people, projects, events, a general forum where anybody can post a topic to discuss, social media, Transition by theme, and Transition "on one big map." Visitors who click on the "Support" link can find out what a transition initiative is, how to become official, how to build your local economy, how to work on inner transition (personal resilience), education (resources for schools), conflict advice, initiative website advice, and access researcher queries and articles. An additional "Resources" link provides more information on the ingredients of Transition, translations of Transition texts, and finally, a branding section that aids initiatives with graphic designs of Transition labels. The website's "Training" link offers information about the movement's trainers and courses that support Transition initiatives in adopting the model and methods. Currently the movement conducts trainings in more than twenty countries. Finally, the "Books and films" link offers visitors access to the dozens of books, movies, and videos written by and for participants in Transition initiatives.

The rest of the homepage highlights "Features of the Day"; for example, on May 17, 2013, Rob Hopkin's new book *The Power of Just Doing Stuff* was featured in the top news story on the home page. The rest of the site contains informational links: "What is Transition," "Why do Transition," "How to do Transition."

The site is a dynamic hub and its primary purpose is information transmission. The Forum section under the community link does allow for people to post and comment on topics by category. The most popular ones included the "General" category, "Researchers," "Web and Comms" and the "Themes": the most popular of which were energy and food. However, the main purpose of the site is to explain the Transition movement and provide visitors to the site with an opportunity to search through various initiatives around the world and access the resources and support for starting one's own Transition town initiative.

As noted earlier, there is no other environmental movement whose informational hub is as comprehensive, organized, and multilayered, especially in the context of digital media. In addition to a frequently updated website, Transition maintains an active presence on social media. Its Twitter account, has 15,200 followers and anywhere from two to half a dozen tweets a day. At the time of this writing, more than 160 Transition initiatives have their own Twitter accounts. The Transition Network's Facebook page has 12,476 *likes* and is updated with new posts once every few days. The Transition Network has a channel on YouTube, a group on Ning, a Flickr group, and a LinkedIn group. Although social media holds the promise of more horizontal interactivity that is participatory and nonhierarchical (Castells, 1997), it is worth emphasizing that for the Transition movement, social media is primarily used for *one-way information transmission*—(updates on initiatives, climate change studies, recently published books, job opportunities, etc.). Although a comprehensive study was not possible, a basic qualitative analysis of the social media sites suggests that updates, tweets, etc. come only from The Transition Network with very little sustained interaction via comments, *retweets*, etc. Although blogs from other initiatives are published on the Network, it is still at the discretion of staff of the Transition Network, a group of around ten people who work out of a small office in Totnes, England. Another example: At a Question and Answer session with Rob Hopkins at a screening for Transition's second documentary "In Transition 2.0," Ben Brangwyn (co-founder of Transition Network), shared some of the tweets from other initiatives who were also screening the documentary on the same night. Ben read tweets from England and New Zealand, each praising the "global Transition movement" (https://www.youtube.com/watch?v=reiCcRnYSFs). By publicly reading Twitter feeds at an event which is then posted on YouTube, the Transition movement is accomplishing several notable feats: (1) Establishing itself as competent with and adept at using social media and technology, (2) Using social media to project Transition's global reach and impact, (3) Achieving the former and the latter while maintaining control over the Transition brand by having a person who works for Transition select which tweets should be made public.

This generally *top-down approach to communication* maintains a unified voice for the movement, and controls the message and the branding even as proponents suggest that it is at its core participatory and grassroots. It's important to note that use of potentially democratizing tools like social media does not necessarily democratize all parts of a movement, but rather may in fact do just the opposite, when used strategically as a way to brand itself under a unified voice. It is also important to note that by pointing this out, I do not mean to imply or suggest that this is necessarily a bad idea. Indeed, it might be argued that Transition has been able to grow so rapidly because the movement has at its helm people who are intent on controlling the brand and unifying the voice of the values of the movement, if not the applied practice of it.

BUILDING COMMUNICATION INFRASTRUCTURES TO SUPPORT GLOCAL INFORMATION FLOWS

The Transition movement used its digital communication primarily to transmit values and serve as a support structure for cultural and behavior changes that support community resiliency. On macro- (*Transition Network*) and micro- (*Transition Amherst*) levels, the websites and social media were used as the communication medium for posting information about upcoming events and activities, raising awareness, and publishing the stories and experiences of people involved in the movement.

The *Transition Companion* suggests that short digital films and online photo galleries such as Flickr and Picasa, have been among the biggest successes in terms of raising awareness about what an initiative is doing, making a record of its activities, and "getting it out to lots of people" (p. 208).

Hector, a member of Transition Amherst, noted:

> You could see that it (the movement) is *organized* online but really it is more like a support structure that happens. So for example, Transition US doesn't coordinate Transition Amherst; it gives us support to do our own thing better. You know the coordination is not the coordination of the whole movement, it's providing the support and the inspiration and examples from other initiatives so that I don't have to do it in the dark, and also providing some of the support for publishing the information. (Interview, December 10, 2012)

The Transition movement supports local Transition towns with the communication infrastructures necessary to connect a local initiative with the global hub in various ways on its website. For example, its "social media" link under the Community tab section provides a way for local initiatives to hyperlink a "Transition Stories" badge to its website so that local initiatives can access Transition "stories" from all over the United Kingdom; local initiatives can also embed "widgets" onto their sites that reflect the latest news, events and projects of the Transition movement. The live widgets on the local website show the latest items in the topical section chosen as the widget and the item titles link back to pages on the Transition Network website. For initiatives that are just starting out, the Transition Network offers a comprehensive checklist for how to begin to create an online presence beginning with a basic web presence, blogs, and newsletter, and offer suggestions for different web services, including whether or not a *techie* is needed. The *Transition Companion* notes:

> Websites should be light and flexible, not big burly cumbersome beasts that scare everyone, including the beleaguered techie who volunteered early on and then found him or herself struggling to deliver something far too complicated that no one really understands. Your site must be understandable and usable by more than one person, preferably not a techie too. (p. 212)

The Transition Network negotiates the tension between advocating for an online presence with the movement's central call to cultivate and nurture neighbor-to-neighbor communication with a mix of humor (illustrated in the above quote) and a clear explanation of how social media can support face-to-face relationships, not supplant it.

> Facebook, Twitter and Ning (and Youtube and Flickr for videos and photos) have exploded in popularity in the last few years . . . They are very powerful tools, and great fun if you like that sort of thing (many don't). Transition Initiatives have used them to grow their memberships, coordinate events, connect with other initiatives, share ideas and make friends and much more. They use social networks in a mix, along with other media including mailing lists, their own websites, YouTube, Flickr and much more. The best way to see if you enjoy using them, and get value from their use, is to give them a try. It can't hurt. Transition Network has a presence in most of the social networks and we like to do that social networking thing, but deep down we feel that the best place to do "social networking" is with your neighbors. Doing it online is good, but shouldn't take priority over doing it locally (transitionnetwork.org/web-options-transition-initiatives).

TRANSITION AMHERST AND DIGITAL TECHNOLOGY: A LOCAL CONTEXT

The most frequently employed method of digital technology for Transition Amherst was email. Transition Amherst used two email addresses, one was for the "steering group" members only, and the other was a general email that went out to anybody who had ever written their name on a sign-up sheet at a Transition Amherst event, over 300 people. A variety of emails were sent by all members to both addresses on a weekly basis. Emails to the general listserv advertised upcoming Transition-related events, articles, books, and rarely received a reply. Emails to the smaller group were used primarily to share minute notes from the meeting, and to finalize decisions such as the wording of the mission statement and/or who would bring which supplies to an event. The latter often involved numerous emails from many members. The construction and development of a website was a priority for the Amherst group through the time of this research. "Updates on the website" was an item on the agenda of every single meeting even when the majority of the group's efforts and attention were focused on planning the "Great Unleashing," a town-wide celebration hosted by a Transition Town as a way to simultaneously share the work that the initiative has done and make a public call for greater participation from the community. Hector, a computer programmer, volunteered to develop the website for Transition Amherst using the WordPress blog template. Jarod and Jasmine, a couple who had recently moved to the area, and myself volunteered to be on a subcommittee that would support Hector in his effort via design and editorial suggestions. The website subcommittee met once and we each discussed the designs of other Transition town websites and elements that we could adapt for our own design. Although feedback on the website was always requested at Transition Amherst meetings, there were never any in-depth comments or suggestions, aside from general support for its development. Hector did spend a considerable amount of time teaching members how to upload material to the site. Transitionamherst.org went live in the spring of 2012 with a homepage that promotes the initiative's collectively written mission statement: *Our purpose is to foster vibrant and resilient community—in the face of rising energy-prices, climate change, and economic instability—by empowering one another to share our skills and gifts, and create a better life for all*, in addition to more specific practical accomplishments of the group (such as the reskilling workshops and the film series). Additional links offer website visitors the opportunity to find out more about events (past and upcoming) and blogs written about the projects (which are not regularly updated), local resources, and working groups. As noted earlier, all of the group members had access to the website and could post if they wished, however, Hector remained the primary person who manages the content and posts the materials.

That Hector, who is known in his neighborhood as a "survivalist" because he lives almost entirely off the grid, with no car, a rainwater recycling system for his water, solar panels for heat, and a garden where he grows much of his food, is also the "technology development" person for Transition Amherst, is worth noting. He understands his position at the intersection of prehistoric *survival* and *modern technology* as somebody who has the capacity to use online communication to educate others about how to "live the best that they can." Thus, he is perhaps a living metaphor for the larger movement's relationship with technology. Hector said:

> I enjoy the technology but I'm also a responsible homemaker in a rustic way so I do that as best as I can. The website is really useful for announcements so that people know it even if they don't have an email or remember where it was. I can really use it for reference so when I do something I can document it there, and there are all these links that people can use who are looking for more information. Also if somebody asks me "[Hector] can you publish this so that people know about it in our community," that's a way for people to know. (Interview, December 10, 2012)

Hector did emphasize, however, that while computers may be useful for information transmission and facilitating face-to-face connections, they should never replace or supersede human interpersonal communication because they are ultimately material symbols of human growth, and not sustainable.

> Hector: I never gave up my computer job. But I know that computers will not play a role in averting the collapse. They will just go away because they are a symbol of human growth. They killed off what was there and what was there in my opinion was better which was the stories of our grandfathers told to our fathers or in a home around a woodstove. That's really a multiage human connection. (Interview, December 10, 2012)

Hector also noted that in the context of online information transmission, there would always be a gap between watching it on a screen and the embodied experience of actually doing it. Information on the Internet might educate a person about how to, for example, survive by raising and killing chickens, but the act of actually doing it is not something that can be experienced physically by watching a video about it online. Thus while somebody might read about a Transition initiative's effort to create a local food economy, there is still a gap that the Internet cannot bridge between the intellectual knowledge gained from watching and the embodied experience of actually doing it.

> Hector: It's very different to know that we have to produce our food locally from killing a chicken. Being there and extinguishing a life, for me it's anguish, it's an emotional upheaval. When you read up on the Internet about local food production, you think great, somebody else will do it and then I get the chicken and then I get the vegetables but when you have to do it yourself it's different and that's not something you can get from the Internet. (Interview, December 10, 2012)

Although members of Transition Amherst prioritized the development of the website, it was unclear how often and/or for what reason people actually visited the site. Part of the reason it was unclear is that during the course of this research members would visit the site at Hector's request for feedback. It's not known if and when people would go of their own volition. The majority of members did not find out about the Transition movement or Transition Amherst from online communication, but rather via word-of-mouth from their friends. This may also be a testament to the average age of the group, the majority of whom did not grow up with the Internet.

TECHNOLOGY USE AND YOUTH: A GLOBAL APPEAL

Rosa, an Amherst College student who interned with Transition Amherst for three months, noted that her biggest contribution to the group was helping to build a social media platform. Rosa created the Transition Amherst Facebook page because she said it would make the group more accessible to young people. She said:

> My generation's grown up in a world characterized by technology and rapid transfers of information, thus I feel like we as a collective are best suited for utilizing platforms like the Internet, multimedia and social networking, which are essential for spreading awareness, and particularly as college students we tend to have extremely broad social networks spanning states and even countries that should be utilized. In this regard I think that the Facebook page was definitely the biggest contribution Ophelia [the other Amherst College intern] and I made to the group. (Email, February 5, 2013)

The Transition Amherst Facebook page has 159 *likes* and was last updated on July 19, 2013.

Rhea Harris a student at Hampshire College who helped to form Transition Hampshire, a college-based Transition initiative also noted the value of technology in connecting her efforts to a global community of people working on the same kinds of projects.

> I think there is definitely an inherent value to the technology. I mean we live in a place where we can go and see things people are creating on the other side of the world but we don't know what their values are or what they're feeling or how they fit into the global system of things and Transition enables us to see that. It allows us to see "Oh there are those people over there who care about the same things I do and they are doing what's right for their community and I'm doing right by mine and we can make that basic connection. (Interview, February 7, 2013)

Rhea also noted that while technology has contributed to the emergence of several recent global social movements, technology also enables people to see how these movements are different from one another. She said that many of her friends are very involved in Occupy and don't understand why she is so committed to Transition. She said:

> I think I definitely feel like I've had so many conversation with friends who say "What are you doing with Transition, come down to New York with me to Occupy and protest," and yeah I'd love to but I want to see everyone included not just people who are immediately drawn to things like Occupy. With Transition, you can also just grab so many people who might not outright agree with your politics but who agree with your values. There's a need for people to come together not just in one place and at one time but we could have all these different people doing things toward one purpose. (Interview, February 7, 2013)

OFFLINE/ONLINE A FALSE BINARY?

It is important to note that the success of these emergent social movements do not necessarily depend on their capacity to be active online. This is why it does not necessarily matter how many Twitter followers the Transition or Occupy movements have. I argue and have argued before (Polk, 2013) that what matters in terms of organization and effectiveness is how well the movements have been able to facilitate the relationship between online and offline activities: in other words, have these online communication infrastructures made it easier for people to meet up in person and coordinate successful activities and events that lead to greater community resiliency on the ground? Certainly for Occupy, the movement initially garnered media attention as a result of the visual spectacle of the tent camps set up in parks and the numbers of people demonstrating on Wall Street. That these physical actions were coordinated online is only as relevant as the numbers of physical bodies who actually show up to participate.

Although Occupy is mobilized online, Kevin Smart, a member of Transition Amherst who was in his early seventies, noted that social capital in the movement is achieved only by being physically present at the demonstrations. He said:

> I would go to the Occupy meetings but I didn't like those either. The problem with those is they espouse more violence than I again have the cultural value for. You have to go out there to have any credibility in the Occupy movement. You had to go to the demonstrations, be out there on the front lines. I don't have the resources to do that so I would have no credibility in the movement. (Interview, November 27, 2012)

For Transition Amherst, the online communication infrastructures have been less significant with regard to mobilizing action in the physical public sphere. On a macrolevel, as Hector noted, the Transition Network has provided access to support for different projects, including the building of a website. However, on a microlevel, members said they did not frequent the Transition Network frequently, and still seemed to prefer the materiality of the books and first-hand experience of people as a resource. For example, one of the more popular Transition Amherst activities was a book club and potluck that focused on the *Transition Companion*. The group met for about an hour before each bimonthly meeting to participate in an informal discussion facilitated by Hector. The meetings were an opportunity for members to bond, share experiences and ask questions. Each person had a book with them, suggesting perhaps that the materiality of the pages, the physical presence of the book was still meaningful to members, and a way for people to connect more deeply, even though all of the content was accessible online.

Jarod Paul noted that in this age of digital communication, one must have a relational view of the digital public sphere and the physical public sphere in the context of how well it facilitates connections. He said:

> You can't have one without the other. You can talk about which is more effective, which one has more variance, which tool is more effective in which context. But you need a contribution of all. An aspiration which I see is making the connections. How do we do that effectively? I think our culture is poor in this and it's sort of set up as a problem and an aspiration without a prescriptive solution. (Interview, December 12, 2012)

In addition to making connections, George Heart, a former member of Transition Amherst and sustainability professor at UMass, suggested that the Transition movement must also be about cultivating a sense of place. He found a way to negotiate the relationship between the offline and online spheres by asking his students to find their *place* in Amherst. He asked them to pick a physical location and "celebrate it." He then made a web page for his students and set up a blog with pictures so that the students could put it up online and share it with others.

> Even though they are only here for four years, students can still have a sense of place. You know, they were all over the town. It was impressive that every student had a place. And it wasn't Antonio's, you know. It was a tree they liked to go sit under. It was a place they could be alone and just be. It was always out doors. Four years your gonna be here, you've still gotta have a sense of place. (Interview, February 6, 2013)

ON LEADERSHIP, TECHNOLOGY, AND SOCIAL MOVEMENTS

Zeynep Tufekci (2011) points out that many commentators have related the diffuse, "leaderless" nature of recent social movements like the Arab Spring with the prominent role social-media-enabled peer-to-peer networks played in these movements. Such peer-to-peer networks are focused on actions and causes, not on specific leaders, and allow for an anonymity that makes it potentially easier to organize subversive political action. Similarly, Occupy participants mark this lack of leadership as a political value, a part of its identity, and it's ultimate goal of having greater distribution of wealth, opportunities and resources. Indeed the very first sentence which describes the movement on its website, occupywallstreet.org reads: "Occupy Wall Street is a *leaderless* resistance movement with people of many colors, genders and political persuasions." Such distribution requires a flattening of hierarchies and a dismantling of the notion of "leader" as a lone hierarchical figure of authority. This flattening of hierarchies is made possible with technology that can mobilize masses and coordinate diffuse networks not dependent upon a singular leader. Tufekci (2011) notes: "How it operates and makes decisions are synonymous with its very identity and survival. Occupy is the change that its members seek." Thus decisions are (supposed to be) made collectively, using consensus-based processes and handmade posters hang from makeshift stages at Occupy events that read "No Leaders!" (Kuttner, 2011).

I suggest that the Transition movement shares the same desires for a more participatory democracy, a flattening of traditional hierarchies, and seeks, like Occupy, to be the change its participants seek, not necessarily by demanding change from institutional structures, but rather through its collective actions toward community resiliency. I also argue however, that the valorizing of the leaderlessness of these movements is ultimately to its detriment. The Transition movement's comments on leadership are rare and confusing: simultaneously resisting the word "leader," while acknowledging the need for one(s). For example on January 27, 2013, The Transition Network posted the following status update on its facebook page:

> I'm quite allergic to the term "leader" but I'll make an exception in the case of this paragraph from a book called "Resilience: why things bounce back." Kinda encapsulates how we're trying to work in Transition.
>
> When we found a resilient community or organisation, we almost always also found a very particular species of leader at, or near its core. Whether old or young, male or female, these translational leaders play a critical role, frequently behind the scenes, connecting constituencies, and weaving various networks, perspectives, knowledge systems, and agendas into a coherent whole. In the process, these leaders promote adaptive governance—the ability of a constellation of formal institutions and informal networks to collaborate in response to a crisis.
>
> Those mycorrhiza can teach us a thing or two, eh?"

By posting the above as a status update on the movement's preeminent social media site, the movement is shaping a particular actor/network discourse around "leadership" while also suggesting perhaps that the word is to generally be avoided as should all potential allergens. The excerpt that follows, while offering an interesting metaphor for promoting adaptive governance in the pursuit of collaboration, suggests no specific or clarifying guidelines for how one might actually pursue the task as a "leader." Nine comments followed the status update, one of which requested an alternative to the word "leader," to which the Transition Network responded that they were "working on it."

Marshall Ganz (2010) defines "Leadership" as accepting responsibility to create conditions that enable others to achieve shared purpose in the face of uncertainty. Leaders accept responsibility not only for their individual "part" of the work, but also for the collective "whole." Leaders can create conditions interpersonally, structurally, and/or procedurally (p. 1). According to Ganz:

> Leading in social movements requires learning to manage the core tensions at the heart of what theologian Walter Brueggemann calls the "prophetic imagination": a combination of criticality (experience of the worlds pain) with hope (experience of the worlds possibility), avoiding being numbed by despair or deluded by optimism. A deep desire for change must be coupled with the capacity to make change. Structures must be created that create the space within which growth, creativity, and action can flourish, without slipping into the chaos of structureless-ness, and leaders must be recruited, trained, and developed on a scale required to build the relationships, sustain the motivation, do the strategizing, and carry out the action required to achieve success. (Ganz, 2010, p. 4)

The challenge, according to Ganz, is to cast a net wide enough to recruit others to do this work, create the capacity to train them, and offer the coaching to support their development (p. 7). Recent social movements like the Arab Spring and Occupy suggest that digital technology has the potential to facilitate the casting of the net and the necessary recruitment for global social movements like Transition, however, initiatives must prioritize the creation of the structures within which leaders might be developed and encouraged to build the interpersonal face-to-face relationships on the ground. This does not mean that an initiative must conform to a prescriptive definition of an authoritarian figure as leader. On the contrary, I suggest that an initiative determine the leadership needs within the context of its goals and then prioritize the cultivation and development of the people who can help lead the group in accomplishing them. Until this is a priority, global social movements facilitated online, run the risk of suffering from a lack of organization, poor internal and external communication, and somebody to hold others accountable for their actions.

CONCLUSION

A textual analysis of the Transition Movement's online activities including its website and social media outlets suggests that the role of digital technology (specifically the Internet) as an actor in the Transition Network, serves two primary purposes: Information transmission and global branding. Regarding the former, the Transition Network website and its various social media outlets provide a comprehensive and multilayered source of information for local Transition initiatives who want to learn more about Transition methods, practices, guidelines, values, personal stories. Regarding the latter, a global digital hub controlled by a small group of people employed by the Transition Network in Totnes England, allows the movement to maintain control of its branding and sustain its legitimacy and authority as a relatively "unified" global movement toward community resiliency.

Although Transition builds upon and has developed alongside similar social movements that have been mobilized in part due to digital technology and social media, like the Occupy Wall Street movement and the Arab Spring, Transition distinguishes itself from the others by framing itself as nonpolitical. Since the Transition movement's activities are not necessarily bounded to specific times and places (each initiative coordinates its own events and activities), its online communication tools can serve as a global resource and support structure, from which each initiative might pick and choose relative to its own needs.

Although the construction of a website was a priority for the Transition Amherst group, it is unclear how often members visited the site. All members had access to the password to upload content to the site. However, Hector remained the primary person who designed the site and posted content. The majority of members said they found out about Transition Amherst from somebody they knew, not from an online source. However, everybody had visited the Transition Network site at some point. Transition activity online was not referred to or discussed at meetings, which, as noted earlier may be a reflection of the group's age range, none of whom grew up with the Internet. Indeed, Rosa Pratt, the Amherst College student who DID grow up with the Internet said her greatest contribution was creating the Facebook page for Transition Amherst, which she felt was a priority for helping the initiative to communicate better with young people.

Finally, I suggest that there is an important (albeit tenuous) connection to be made between the digital facilitation of social movements like Occupy, the Arab Spring and Transition and the *leaderless* ethos that inform an important part of their identities. Online networks can be used to facilitate communication around actions and causes, not on specific leaders, and allow for an anonymity that makes it potentially easier to organize subversive political action. Networks provide the opportunity for an initial flattening of hierarchies and a dismantling of the notion of "leader" as a lone hierarchical figure of authority, a shared goal of movements that seek a more participatory, equitable and just democracy.

I also argue however, that strong leadership development is crucial to sustaining a long-term movement, and the valorizing of the leaderlessness of these initiatives is ultimately to its detriment. This is not to suggest that traditional hierarchical figures of authority need be in place for an initiative to work, on the contrary, it should be a priority to create the space and the structures necessary to select and train collaborative and skilled leader(s) who can nurture the strengths of the group, aid in successful facilitation, hold others accountable, and continue to cultivate relationships online and offline that will ultimately strengthen and sustain the movement.

REFERENCES

Bennett, Lance. W. (2003). Communicating Global Activism. Strengths and vulnerabilities of networked politics. *Information, Communication & Society*, 6(2): 143–168.

Castells, M. (2007). Communication, power and counter-power in the Network Society. *International Journal of Communication*, 1: 238–266.

Castells, M. (2004). "Informationalism, Networks, and the network society: a theoretical blueprint." In Castells M (ed.) *The network society: a cross-cultural perspective.* Northampton, MA: Edward Elgar.

Cohen, A. P. (1989). "The symbolic construction of community." London: Routledge.

DeLuca, K. and Peeples, J. (2002). From public sphere to public screen: democracy, activism, and the 'violence' of Seattle, *Critical Studies in Media Communication*, 19(2): 125–151.

Donnatella, D. and Lorenzo, M. (2005). Global-net for Global Movements? A Network of Networks for a Movement of Movements. *Journal Public Policy*, 25, 165–190.

Fisher, D. R. and M. Boekkoi. (2010). "Mobilizing Friends and Strangers: Understanding the role of the Internet in the Step It Up day of action." *Information, Communication & Society*, 13(2):193–208.

Ganz, M. (2010). "Leading Change: Leadership, Organization, and Social Movements." In *Handbook of Leadership Theory and Practice: A Harvard Business School Centennial Colloquium.* Nitin Nohria and Rakesh Khurana (eds.) Boston: Harvard Business Press.

Garrett, R. (2006). Protest in an information society: A review of literature on social movements and new ICTs. *Information, Communication & Society* 9(2): 202–224.

Gurstein, M. (2011). Up from Facebook: #Occupy—(Re)Building and Empowering Communities. Retrieved on November 29, 2011 from http://gurstein.wordpress.com/2011/10/22/up-from-facebook-occupy%E2%80%94rebuilding-and-empowering-communities/

Gurstein, M. (2011). Egypt: From the Iron Rule of Tyranny to the Iron Law of Oligarchy: Can ICT Change the Rules? *Gurstein's Community Informatics*, Retrieved on March 2, 2011 from http://gurstein.wordpress.com/2011/02/23/applying-the-ict-lessons-of-revolt-to-the-institutional-challenges-of-reconstruction-they-overthrew-hosni-mubarek-now-can-they-overthrow-robert-michels/

Hopkins, R. and Lipman, P. (2009). Who we are and what we do, *Transition Network.org*, 1–24.

Hopkins, R. (2011). "The Transition Companion: Making your community more resilient in uncertain times." UK: Green Books.

Hopkins, R. (2008). "The Transition Handbook: From Oil Dependency to Local Resilience." UK: Green Books.

Juris, J. (2005). "Global movements for global justice." In *Network Society—A Cross-Cultural Perspective*, M. Castells (ed.) Tennessee: Ingram.

Kreiss, D and Tufekci, Z. (2012). "Occupying the Political: Occupy Wall Street, Collective Action, and the Rediscovery of Pragmatic Politics." *Cultural Studies - Critical Methodologies*, 13:3.

Kuttner, P. (2011). "No Leaders? No Way. Leadership During Occupation." Retrieved on June 1, 2013 from http://culturalorganizing.org/?p=295

Occupy Wall Street: The Revolution Continues Worldwide. Retrieved March 20, 2012 from http://occupywallst.org/about/

Oliver and Marwell. (1992). "Mobilizing Technologies for Collective Action." *Frontiers in Social Movement Theory*, Morris, A. and Mueller, C. eds., New Haven: Yale University Press, 251–272.

Pasek, J., More, E., and Romer, D. (2009). "Realizing the Social Internet? Online Social Networking Meets Offline Civic Engagement." *Journal of Information Technology and Politics*, 6(3/4): 197–215.

Polk, E. (2013). "Revolutions, Social Media, and the Digitization of Dissent: Communicating Social Change in Egypt." In *Sustainability, Participation, and Culture in Communication*, 137–152. Servaes, J. (ed.) Chicago: Intellect.

Rosaldo, R. (1989). *Culture and truth: The remaking of social analysis*. Boston: Beacon.

Smith, A. (2011). The Transition Town Network: A Review of Current Evolutions and Renaissance. *Social Movement Studies*, 10(1): 99–105.

Transition US Governance Toolbox. Transition United States, 1–11. Retrieved on Feb 1, 2013 from http://transitionus.org/sites/default/files/TransitionUS_HowToGovernanceToolbox_v2.0.pdf

Tufekci, Z. (2011). "Can 'Leaderless Revolutions' Stay Leaderless: Preferential Attachment, Iron Laws and Networks." *Technosociology*. Retrieved on May 13, 2013 from http://technosociology.org/?p=366

Wellman, B., Quan-Haase, A., Boase, J., Wenhong, C., Hampton, K., Diaz, I., and Miyata, K. (2003). The Social Affordances of the Internet for Networked Individualism. *Journal of Computer-Mediated Communication*, 8(3). Retrieved on January, 12, 2012 http://jcmc.indiana.edu/vol8/issue3/wellman.html

Wellman, B., Boase, J., and Wenhong, C. (2002). The Networked Nature of Community: Online and Offline. *IT & Society*, 1(1): 151–165, www.ITandSociety.org

Wellman, B. (2002). "Little Boxes, Glocalization, and Networked Individualism" Revised Papers from the Second Kyoto Workshop on Digital Cities II, Computational and Sociological Approaches.

Chapter 9

A Trojan Horse in the City of Stories?

Storytelling and the Creation of the Polity

John Hartley

The international digital storytelling movement uses facilitated workshops to enable ordinary people to tell their own stories using digital means of production, editing, archiving and distribution (Lambert, 2006; Meadows et al., 2006; Hartley and McWilliam, 2009; Lundby, 2009). Digital storytelling has taken hold in various educational, therapeutic and screen-culture contexts, although early experiments designed to integrate it with broadcasting have not continued (Meadows and Kidd, 2009).

My concern is with the use of *storytelling* more generally for what may be called "*user-created citizenship*"; not simply for self-expression but also for the creation of a "we"-group (or deme) around activism of some kind. This leverages a much broader discussion around the role of stories in creating human polities. Like any pivot, this one allows me to swing around in quite a wide arc . . . from 12,000 years ago up to Rupert Murdoch, two years hence. I call for digital storytelling to *challenge incumbent stories* (which may occupy the sort of position in our civic headspace as did a certain horse for the Trojans), and hence to challenge the very constitution of polities based on aggressive parochialism or "universal adversarialism."

THE CREATION OF THE SELF

There is a growing body of scholarship on digital storytelling and self-representation (Thumim, 2012; Chouliaraki, 2012; Couldry et al., in press), including a long-running conference series, which has been held in Wales, Australia, Norway and elsewhere. The most recent conference was held in Ankara, Turkey, in 2013, hosted by a former PhD graduate of mine, Burcu Simsek. She introduced digital storytelling into Turkey as a part of her work with women's groups. It was designed to raise questions for the digital storytelling movement, about how its stories can win attention within the competitive context of digital media more generally, and how self-mediation can contribute to wider aims than simple self-expression—for instance, activism among "minority" identities, citizen science or "citizen documentary," and "do-it-yourself citizenship."

Digital storytelling promotes self-expression, and the digital storytelling movement is largely organized around the identity, authenticity and experience of the teller. It often sets its own work in opposition to the stories told in "mainstream" media.

Storytelling is also political, not in the current sense of party-politics and a professionalized political process, but in the older civic sense of "creating a polity" (what Anderson [1991] calls an "imagined community") among those who form its social "story circle"—the audience. The causal chain that I propose is that social groups of larger scale than kin require a mechanism for constituting themselves as one group (a deme or demos), and storytelling is one such mechanism. Thus "the audience" comes before "the citizen."

The storytelling I have in mind here is not devoted to winning votes or deciding the issues of the day, but to answering the more fundamental questions of how we know we are a "we"; and what "we" are like. [1] Stories create meaningful identities for a given "we"-community by setting the social world into a diegetic story-world, rendering social values into character, action and plot. "We" and "they" identities are personified into heroes and adversaries. These are tested in action, and at the end are either confirmed (comedy/marriage) or modified (tragedy/death). Semiotic and social structures are mutually constituted, which means that stories are necessarily about society as well as self; the two linked together are a "polity," a social group with a semiotic identity, organized for action and survival under conditions of uncertainty or adversity. The "we"-community or "deme" may be as small as a family, workgroup ("danwei") or "hunting party," or it may be of societal or global scale. I am borrowing the term "deme" from bioscience, where it refers to an interbreeding subpopulation of a given species. The word itself derives from the Greek "demos," where a deme was a population group within Attica, upon which citizenship was based (Hornblower and Spawforth, 2005; and see Hartley and Potts, forthcoming).

Mediatized or storytelling demes and polities are not coterminous with political governments. The "institution" of storytelling can place protagonists and their deme into much smaller units than that of city, province or nation, or much larger ones, right up to species, planet or cosmos. Digital storytelling activities to date seem to be clustered around the small-scale end of this gradient. Broadcast media compete in the middle- to large-scale, typically at the level of the nation, now expanding to global networks. The "outer limits" are explored by sci-fi, fantasy, and utopian/dystopian imagination. Like stories themselves, "new media" communities range from one extreme (the bedroom) to the other (the cosmos) (Hartley, Burgess and Bruns, 2013), such that "polities" or "demes" ("we-communities") can form around many kinds of affiliation other than traditional ethno-territorial co-presence. This kind of "long-tail" informal or user-created polity is the semiotic space where digital storytelling thrives.

Storytelling as an Evolved Institution of Language

In the field of evolutionary bioscience, "costly signalling" is a hot topic. The theory goes that for the purposes of sexual reproduction, competing males can assure sceptical females of the underlying quality of their genes if they are able to sustain some outward sign—antlers, showy plumage, a Rolex—that clearly costs them dearly in energy and risk, without possibility of deceit. Among humans, some evolutionary social scientists think a prime function of language (i.e., an explanation for its evolution) is for monitoring the *honesty of others' utterances*, in order to sustain cooperative groups. This is what brings storytelling (language) and citizenship (large groups) together. E.A. Smith (2010, 232) explains the underlying theory thus:

> Language facilitates complex coordination and is essential for establishing norms governing production efforts and distribution of collective goods that motivate people to cooperate voluntarily in large groups. Language also significantly lowers the cost of detecting and punishing "free riders," thus greatly enhancing the scope and power of standard conditional reciprocity. In addition, symbolic communication encourages new forms of collectively beneficial displays and reputation management.

Like Smith, Herbert Gintis emphasizes the importance of punishment to guarantee large-scale group cooperation (Bowles and Gintis, 2011). But language is itself a duplicitous medium. Anything that can be used to tell the truth can also be used to lie (Eco, 1976). *Humans have to learn how to tell the difference between truth and "free-riders" in speech as well as in genes*, and have evolved codes in language and "institutions of language" (e.g., text conventions, genres, registers, jargon; and also institutional "languages," e.g., the law, sciences, romance, with specialist functions and features) to automate the process. Thus, language-in-use, as a social system and not just as an abstract set of rules, must develop institutions to "improve the efficiency and accuracy of communication" (Smith, 2010, 242).

I would argue that one such institution is storytelling. *Storytelling shifts "costly signalling" from the individual level to that of the group or deme.* Groups that can build mutual knowledge and trust for solving "group action problems" will prosper compared with others. As Eric Beinhocker has argued, storytelling is how humans think, using inductive reasoning.

> Stories are vital to us because the primary way we process information is through induction. Induction is essentially reasoning by pattern recognition. . . . We like stories because they feed our induction thinking machine, they give us material to find patterns in—stories are a way in which we learn. (Beinhocker 2006: 126–7)

But note that this kind of "thinking machine" is "external" and social, requiring at least two parties. Stories and storytelling forms are how humans store and distribute acquired knowledge—they are a kind of oral/aural library system, promoting social learning in the act of using the format and its store (Ong, 2012; Lord, 1960). Selective and competitive pressures would tend to improve the means by which humans determine the trustworthiness of others (non-kin), how they determine causal sequence in phenomena (inductive reasoning; also plot), and how they preserve knowledge across time, distance, generations and even languages (social learning). Such improvements in the efficiency and accuracy of communication, such that communicative duplicity can be detected and punished for the benefit of group action, explain both the ubiquity and the formalization of storytelling, and its appeal. It is an "institution of language" not an invention of individuals, and even as we tell stories, stories tell us.

To the extent that a good story, well told, may reveal the underlying qualities of protagonists (personifications of groups) as well as the teller, storytelling itself is a form of costly signaling, and may be evolutionarily advantageous:

> We show that honest signaling of underlying quality by providing a benefit to group members can be evolutionarily stable, and may proliferate when rare as long as high-quality individuals are neither too common nor too rare, and the cost of signaling is sufficiently greater for low than for high-quality players. (Gintis et al., 2001)

As Gintis and his colleagues assert, not everyone is "high quality" (some signals are lying); and not every story is trustworthy. To sort the wheat from the chaff, stories carry *within them* signs of "honest signalling," and such signs are competitively advantageous. Among the most important is that of "being there"—the eyewitness or participant who can relate the story from first experience, the more scarred (mentally or physically) the better (scars being evidence of honesty that is hard to fake).

Because both tellers and the told can attend only to one narrative at a time, individual stories are also mutually competitive. This too results in an evolutionary arms race to improve storytelling as a form, to attract attention to specific tales, and to codify tried and tested fomulae. A chief test of truth is authenticity (but authenticity is signaled using conventional signs and codes). The authenticity of the observing and narrating self is crucial, unless this is a Trickster's tale (Hyde, 2008), in which case the story itself, through genre, poetics and meta-discursive signs, will signal that it is to be "read" differently. Hearers can discern what it cost a teller, and a protagonist, to bring a particular story to their attention, and humans seem well attuned to the difference between glib assertion and hard-won ("costly") authentic experience. This means that all stories are also "about" *storytelling*, carrying with them signs (or "metadata") that speak to *their* believability, which is presumably why so many plots are about defeating duplicity, confronting liars, and revealing hidden truths. This is why storytelling is a competitive arms race. Cutting through to attract people's attention is not a simple matter; knowledge of how to do it convincingly is encoded in narrative forms and conventions, and as the saying goes, even crazy listeners "know a hawk from a handsaw,"[2] so scepticism and inattention are built in.

Power Law Distribution

As far as a story's reach and impact go, storytelling is distributed along a power law curve, from a "winner-takes-all" head to the "long tail" (Anderson, 2006). There are a few stories that catch everyone's attention; and many stories that catch a few. Clustered around the head are stories about the polity (how our group came to be and why it matters), and stories about religion (outsourcing group-identity to the supernatural, to punish deceivers and free-riders). Clustered at this extreme too are high-investment social institutions and corporations, especially the media, schooling, and the law, all of which legitimate themselves with foundation stories that seem to coincide with those of the nation, polity or deme. They produce the small number of stories that *everyone* knows or sees (sacred texts; blockbusters).

At the other extreme of the "long tail," stories tend to be about the self (identity in conflict), about locality (our place; peasants' tales), or about travel (the world; sailors' tales) (Buonanno 2005). And here too is where we usually find digital storytelling.

This power law distribution is worthy of mention because stories told by global media corporations are not opposed, adversarially, to stories told by persons; they are on a gradient (in which players can change places). The difference between them is not antagonistic, because different types of story serve different functions for different scales of group. In short, digital storytelling is not *opposed to* mainstream media narrative; it is on a continuum with it. Thus, it is important not to dismiss storytelling emanating from global corporations, even if it seems to outcompete local or individual efforts in terms of investment, airtime and popular attention. It is true that commercial media production is competition for digital storytelling, but that doesn't make it an adversary or "they" identity. The digital storytelling movement can compete, by mastering not only the trick of authenticity ("costly signaling"), but also the requirements of inductive reasoning and social learning, for the solution of a significant group action problem, among a specific "we"-group.

Inevitably, even in advocacy or educational work that may reject commercial or marketing values, *self-expression* is not enough to achieve *communication* with others. To create a "we"-community, digital storytelling needs to be political and competitive, and open to the logics and potentiality of social networks and "social network markets" (Potts et al., 2008). Digital storytellers need to know enough about the "costly signalling" game to be able to use their messages for the "creation of a polity" around their advocacy.

This is not necessarily welcome news for activist agencies, because the strong asymmetry between high-investment commercial media and community-based self-expression is exactly why such agencies are active in the first place—they want to take mediation "back" from industry and relocate it in the community. But the very community they invoke is suffused with the model of commercial media narration, which is a kind of *cultural technology* that contemporary citizens carry around in their heads. "Everyone knows" how stories work because their forms are reiterated countless times in media, and these models are unselfconsciously rehearsed in everyday talk, play and the like. The arrow of time cannot be reversed: *"we" are mediated selves*. "Self-expression" needs to compete with the media of which it is a differentiated part.

The problem remains, whether competing takes the spontaneous form of informal and untutored popular culture, for example, "selfies" (Nelson, 2013), where a few randomly lucky signals may "go viral" while most don't, or the more elaborately facilitated and edited forms of digital storytelling. Both storytellers and their audiences live within a semiotic environment where stories must compete to gain attention. Hence, those who wish to use stories and digital media for self-expression and community advocacy must go beyond the identity and authenticity of the maker, to embrace communication and outreach (a.k.a. marketing).

Stories need some distinctive quality, which facilitated workshops are there to teach, because although everyone is familiar with a good story, technique has to be learned. In a competitive environment, "authenticity" is another ploy, so stories have to outsmart scepticism. Daniel Meadows argues:

> If citizens are to make their own TV on the kitchen table—as it were—then it is imperative that Big Media provides them with forms which can be readily learned, elegant forms which allow for an articulate contribution. We should

> make good Digital Stories, not bad television. . . . Digital Stories are indeed multimedia sonnets from the people but let's not kid ourselves that they grow on trees. (Meadows et al., 2006, 3)

The "Oracle of Girl World"[3]

An example of someone who took self-mediation from the blog-in-the-bedroom to global media presence, very much by using "*elegant forms which allow for an articulate contribution*," is Tavi Gevinson. She commenced with a home-based fashion-fan blog at age 11 (*thestylerookie.com*). Owing to its appealing authenticity and identity, plus her editorial and fashion style, not to mention her very young age, the blog attracted millions of hits and thence attention within the fashion world, which she was able to absorb and rework into further ventures. She learned how to deal with her own appearance and appearances, fame and other celebrities, without falling into the pitfalls that can go with publicity, marketing, and fashion-centered media. As one Canadian style magazine observed:

> Gevinson is riding the woven wave of her cult fashion credentials to promote and popularize feminism and teen empowerment, but her public appearances and TV interviews are not arrogant, self-congratulatory reality stunts, far from them. (Herman, 2013, 47)

Gevinson went on to use her own story to develop a powerful voice for girls, At 14, she launched an online magazine (*Rookiemag.com*), which she edited after school (still using the domestic bedroom as office and factory floor, marshalling the efforts of more than 50 contributors). At 15, she gave a high-rating TEDxTeen Talk,[4] and at 16 undertook a national US tour that achieved intense levels of participation among her peers. At 17 she spoke at the Sydney Opera House and Melbourne Writer's Festival.[5]

Tavi Gevinson is not much discussed in the digital storytelling movement, although she excels at "multimedia sonnets" and is prominent in style magazines with extensive popular and commercial reach. So digital storytelling *as a movement* has much to learn from her ability to combine self-representation with scaled up digital communication in a good cause (see Gevinson, 2012, 2013). It seems to me that her success is more than random luck; it attends to someone whose talent is authentic, well managed and current, making use of the storytelling institutions and communicative technologies to hand, in exactly the same cause—feminism for girls—that many digital storytelling advocacy groups espouse. Here is where copying may be seen not as derivative but as part of the process of social learning; a prompt to innovation.

THE CREATION OF NATIONAL CHARACTER

I turn now from the *communication* element of storytelling to the *political* aspect—the role of storytelling in "creating the polity." I argue that this has been a prime function of storytelling since before there *were* polities; and that it remains an important element of any attempt to create a "we"-community in the digital era—a "digital deme." As mentioned, the stories that "constitute the polity" cluster at the head of a long-tail distribution curve of stories circulating in a given society. As a result, they're not the sort of stories that generally crop up in the digital storytelling movement, although as we will see later on there's a connection between "polity" and "personal" that digital storytelling does seek to encompass.

My starting point is that the most important political narrative is the one that *constitutes* the polity. These stories are common in foundation myths, from Adam and Eve in the Bible's *Genesis* to Romulus and Remus in Virgil's *Aeneid*; they supply a reason or cause for the origin of humans in general or of a particular city or nation. Despite their own ancient and mythological origins, *etiological (causation) stories* are commonplace still, ascribing reason and meaningfulness to events and places that then explain the character of a nation. Even the US Constitution has one, deriving "natural law" from the supposed natural state of humanity, prior to civilization and government (and therefore taking priority over it). Such stories are repeated continuously, in school, in journalism and in fiction, all the way from Homer and Virgil to the nightly news and movies, making the polity (or "imagined community") anew each day by recreating the story of who "we" are.

One prominent genre of etiological stories is associated with national day celebrations. These are staged by both media and public authorities, with high production values and showy symbolic content, especially in modern nation-states born out of popular revolutions (France) or settler-based social experiments (the United States, Australia). Settler nations don't have mythical origins, except among their Indigenous populations, who were typically excluded from their modern constitutions. All the more reason, therefore, to provide themselves with modern myths and legends, and to make these the subject of public display.

In Australia's case the official national day (January 26) commemorates the landing of the fleet that established the British penal colony at Botany Bay in 1788. As a foundation event, it is contested. It's called "Invasion Day" among Indigenous groups.[6] As a summertime public holiday, it is associated with beach culture rather than constitutional reverence. As the "Cronulla riots" of 2005 demonstrated, Aussie beach culture can suddenly erupt into political prominence as a stage for conflict about national identity, including racist and flag-flaunting versions that seek to exclude migrant cultures (Hartley and Green, 2006). It is widely ignored by non–English-speaking communities, among whom it has low salience. Constitutionally, moreover, Australia Day is on the "wrong" day in January. After referenda in the six colonies, the "Commonwealth of Australia" was founded in 1901, by Act of Parliament and Royal Proclamation, on January *1*, not 26.

Small wonder that Australians gain a rather confused and impoverished sense of national origin from Australia Day. In recent years, its significance in this respect is being eclipsed by Anzac Day (April 25). Here, a strong sense of a DIY ("do-it-yourself") myth of national origin has built up since the 1990s. One popular highlight is the annual "pilgrimage" to Gallipoli.[7] Their destination is a peninsula in Turkey that was unsuccessfully invaded by French and British Empire forces, including Australians, New Zealanders and Indians, as well as British troops, in April 1915. They withdrew eight months later with high casualties on all sides, defeated by Ottoman forces under the command of Mustafa Kemal, later Atatürk, founder of the modern Turkish republic.[8]

Over recent decades this unlikely setting has become the place of origin of Australian and New Zealand national consciousness, and Anzac Day the most important day of national memorialization. Just as it is eclipsing Australia Day, so it also eclipses Remembrance Day (Veterans Day in the United States; November 11) as a memorialization of national war service. Unlike Remembrance Day, Anzac Day is a public holiday, marked by dawn services at war memorials around the country and at Gallipoli. More popular now than when veterans still lived, these gatherings attract many thousands of attendees of all ages.

The trek to Gallipoli is especially popular among backpackers and other young travellers, for whom it seems to function as a place of passage in their own process of self-realization. It attracts a music-festival sized crowd, who camp nearby and gather, many flag-clad, for dawn ceremonies. One historian has criticized the event for resembling a "Big Day Out" (i.e., an annual summertime rock concert).[9] He complained that "official presentations [at the 90th anniversary in 2005] included pop music which inspired dancing and couples were seen canoodling near graves" (King, 2013). He dreads the "prospect of a memorial service packed with excitable fans instead of mourners" at the forthcoming 2015 centenary.

The only explanation for this popular behavior is political narrative—the story, internalized by many, of how a distinctive national consciousness was forged in the crucible of war, despite the facts that the campaign was a failed sideshow, that more Australians died at the Western Front,[10] and that nationhood was established in 1901. The narrative (about character) has trumped the realities (military, legal and political).[11]

Naturally, the Australian news media are eagerly on hand each year, at Gallipoli and elsewhere, disseminating the story to a wider public and celebrating the purported national character through the apparently self-staged rituals of ordinary people at a widely distributed but simultaneous national corroboree. *The Australian* plays a prominent role in promoting Anzac day activities at home and in Turkey. This may be connected with the fact that the proprietor of *The Australian*, Rupert Murdoch, has a family connection with the original events. His father, Keith Murdoch, a young journalist, reported the campaign's failings back to Australian and British government officials, despite the efforts of military censorship (Fewster, 1982). His story established crucial elements of the enduring legend: specifically, the charge of British incompetence (among the general staff) and Aussie heroism (among the diggers). As a recent story in the UK press put it:

> Keith Murdoch's role at Gallipoli is not without controversy, given that the information he smuggled out was mostly second-hand and provided to him by the *Daily Telegraph* journalist Ellis Ashmead-Bartlett, an arch critic of the campaign. When Murdoch's ruse to smuggle Ashmead-Bartlett's report from the theatre of war was betrayed, he sat down in the office of the Australian High Commissioner in London and dictated his recollection of the report's contents into an 8,000-word letter to Australian Prime Minister Andrew Fisher. "It was an amazing document, a mixture of error, fact, exaggeration, prejudice, and the most sentimental patriotism, which made highly damaging charges against the British general staff . . . many of them untrue," said the Australian writer Phillip Knightley. "But the basis of the charges—that the Gallipoli expedition was in danger of disaster—was correct." (Burrell, 2013; and see Fewster, 1982)

This version of Gallipoli—Murdoch Senior as brave whistleblowing war correspondent circumventing censorship to tell the truth—is an essential part of its meaning. The other major influence on the creation of the legend was another Australian, Charles Bean (Seal, 2004). It was he who first claimed—and through his later monumental histories established—that "the consciousness of Australian nationhood was born" on Anzac day 1915.[12] Bean was determined to report what he saw, but what he saw was coloured by what he was looking for. He posed this quest as a question:

> How did this nation, bred in complete peace, largely undisciplined except for a strongly British tradition and the self-discipline necessary for men who grapple with nature . . . react to what still has to be recognized as the supreme test for fitness to exist?[13]

The answer was already clear to Bean: "fitness to exist" was bestowed by "*character*." He wrote in 1918:

> . . . the big thing in the war for Australia was the discovery of the character of Australian men. It was character which rushed the hills at Gallipoli and held on there.[14]

Bean went on to found the Australian War Memorial, and is himself remembered as the first Australian War Correspondent at Gallipoli and later on at the Western Front. But it was not Bean who established the meaning of Gallipoli. That honour went to Ellis Ashmead Bartlett, the English journalist who first broke the story in the British and Australian press.[15] Bean is referring to Bartlett here:

> The war correspondent is responsible for most of the ideas of battle which the public possesses . . . I can't write that it occurred if I know that it did not, even if by painting it that way I can rouse the blood and make the pulse beat faster—and undoubtedly these men here deserve that people's pulses shall beat for them. But War Correspondents have so habitually exaggerated the heroism of battles that people don't realize that real actions are heroic.[16]

Bartlett, who worked for the London *Daily Telegraph*, was the doyen of the Allied press gallery covering the campaign (Fewster, 1982). As well as scooping the story, he also shot the only movie film taken of action in the campaign. Later in the war he toured the United States, Australia and Britain with his story and movie (Fewster, 1982). It seems that the Anzac legend was created by him, as has long been recognized by historians:

> Undoubtedly [the] more important influence that [Bartlett] exerted over the legend is that he probably more than any other person determined how the tale was told. The feat of securing a foothold on inhospitable ground against stern resistance from well prepared defenders was a most praiseworthy achievement. But the troops were by no means assured of a glorious immortality by their efforts alone. It required gifted observers to pass the story on to the outside world if the landing were to receive the recognition it so richly deserved. . . . The essence of any legend lies as much in the related story as in the events it purports to describe. Ellis Ashmead Bartlett, by his efforts with pen, picture and speech was instrumental in first shaping then institutionalizing a legend which has, and will continue to be passed on for generations. (Fewster, 1982, 30)

Australians gathering at Gallipoli have likely never heard of Bartlett, but it is his story they come to honor. They won't be the first, because even at the time his "bardic" role (Hartley, 2009) was recognized by the diggers themselves. One soldier was moved to write in the *Anzac Book* (edited by Bean and published in London in 1916):[17] "It is fairly certain that future historians will teach that Australia was discovered not by Captain Cook, explorer, but by Mr. Ashmead Bartlett, war correspondent."[18]

Australian national character may have been tested for fitness in 1915, and this may or may not have revealed its authentic character; but no-one would have known anything about it without a narrator. It's the story that carried the knowledge, not the deed; and the communicator who "creates the polity," not the acts of members of the deme, however authentic or courageous.

ON NOT HAVING A STORY . . .

It took me many decades to realize that I had a personal relationship with Gallipoli, because in my part of the world, and in my early years, it was known as "the Dardanelles." In the UK popular imagination it was connected less with the Anzacs than with the name of Winston Churchill (First Lord of the Admiralty), whose bold conjecture it had been, and who therefore took the political blame for its eventual failure.

My grandfather was there. Of course he wasn't an Anzac—he fought for the "other side," as it were, that is, the British. Before that fact forms a stereotypical picture in your mind, I'd better say at once that he wasn't the type of British that so upset the likes of Bartlett and Murdoch, that is, an incompetent upper-class officer. He was strictly "other ranks" (a Quartermaster Sergeant, although later promoted, in the tradition of the Army Service Corps, to Captain). He must have had some connection with the Anzacs, because he kept a 1916 copy of *The Anzac Book*, which I now have.

Josiah Arthur Barnes came from "the Borough" (Southwark) and worked at Mount Pleasant, the biggest postal sorting office in the Empire. Before the war he'd been in the Post Office Rifles (Territorial Army), and after it he returned to the Post Office, eventually to become Secretary of the E.C. Sorters Branch of the Union of Postal Workers, which was founded in 1919. He was, and remained, solidly working class, but of the aspirational type. When I became aware of him, in his seventies, he played lawn bowls (I have the cups he won), he still drank his tea out of the saucer not the cup, and blew his nose on a khaki handkerchief.

To say he "fought" may give the wrong impression, because he served in the Army Service Corps, which was responsible for transport, food supplies, logistics—and the mail. According to a *1914–18* website:

> At peak, the ASC numbered an incredible 10,547 officers and 315,334 men. In addition were tens of thousands of Indian, Egyptian, Chinese and other native labourers, carriers and stores men, under orders of the ASC. Yet this vast, sprawling organization—so vital to enabling the army to fight—merits just four mentions in the Official History of the war.[19]

It seems there is no "nation-forming" story attached to the coordination of armies; neither the military itself nor its historians remember the contribution of a third of a million of its soldiers.

Although armies can't fight without organization, transport, logistics, food and communications, those who provide them are available to take the blame for what goes wrong. My grandfather kept a copy of the *Final Report of the Dardanelles Commission* (1917–1919, 81–2), the official government inquiry into the debacle. I found it forgotten on top of a cupboard long after his death. I wondered why he kept it for so long. Would it reveal anything about him? Looking through it, I found a chapter on "Postal Arrangements."[20] After excoriating the "negligence or stupidity of some postal clerk" (was this him?), the report mentions that "there was a good deal of looting of parcels" sent via the Army Service Corps Parcel Transit Service (was this him?).

The *Report* relates various scandalous stories of mails damaged, not delivered, delayed or sent astray (81). Noting, however, that the "amount of letters is stated to have grown to 1,000,000 a week," and the number of parcels up to 90,000 a week, the *Report* finally concludes that both the Post Office and Army Service Corps Parcel Transit Service bore responsibility for the "miscarriages," but that "we are of the opinion that, on the whole, no blame attaches to those who had the organization and conduct of the service" (82) (was this him?).

Doubtless there are some amazing stories here, but the heroes who managed to deliver over 30 million items to troops so far from their various homes in Britain, France, Australia, New Zealand, Egypt and India, not to mention those serving the Turkish army, have remained unsung from that day to this . . . at least officially. Apparently, "[t]here is no memorial to the Army Service Corps."[21] But in *The Anzac Book* (1916) it is clear that the fighting men felt very differently:

> Mails, too, are an anodyne. Their arrival eclipses considerations of life and death—of fighting and the landing of rations. The mail-barge coming in somehow looms larger than a barge of supplies. Mails have been arriving weekly for six months, yet no one is callous to them.[22]

The writer singles out letters first, "they put a man at home for an hour." But then he mentions the local newspaper, sent from home: "Intimate associations hang about the reading of the local sheet—domestic and parochial associations almost as powerful as are brought by letters" (22). Finally, there are parcels, objects of an "intensity of gloating expectation." Most prized were tobacco and food, but clothes, toiletries, and writing paper were also mentioned.

Of course, not all the parcels got through, and many that did were damaged. A "digger's alphabet" gives us some idea of what the troops thought of quartermasters—they were shirkers and looters: "Q is for . . . the Quarter-bloke, dodging the line"; "R is for . . . the Rum that the Quarter-bloke pinched" (quoted in Seal, 2013, 250–1).

In fact, the only memorabilia I have of my Grandfather are "spoils of war." He kept—over the fireplace—a set of murderous looking shells (which I now have), allegedly captured from the Turks and "proving" that the enemy used "dum-dum" rounds (lead-tipped, designed to flatten on impact). He also sported a collection of "trench art," including shell-cases converted into cigarette boxes, powder-puff boxes, etc., some with British silver sixpences (the king's head) let into the base, overlaying the German or Turkish script that indicated whence the shells originated. One such is dated "APR 15"). There's also a Turkish artillery shell fuse, and, had he not handed it in to the British Government in the 1950s during an amnesty, there would have been a captured Turkish revolver too. Instead, I have the receipt for it.

These bizarre objects are all I have of my grandfather; a clutter of memorabilia-without-memories. He died while I was away at the orphanage. My mother wrote and told me to pray for him—a tricky task, for he was C of E and a Freemason, my mother was Catholic convert, and I was at a Protestant orphanage. Would praying for him in the wrong religion be OK? Evidently my mother didn't think it mattered, for she buried him in the same grave as my nonreligious father, both of them in the Catholic bit of Margate cemetery. As the only surviving male in the family I was always going to inherit the memorabilia of war . . . and his masonic apron. I don't like them much but can't chuck them out, so there they stay, souvenirs of an unknown life, transmitted through time without any story attached to them, and almost completely meaningless, certainly to my daughters, who will no doubt have to decide what to do with them one day.

I barely knew him as a person. I can't remember much affection on either side. The only present I remember having from him was a parcel that mysteriously but excitingly arrived at the orphanage one year, when I was about 10. It contained a pair of boxing gloves, which I immediately read as a rebuke. He was encouraging me to make a man of myself (a sure sign that I had failed in that endeavor).

Neither my mother nor my grandmother told stories about him after he died. We discovered decades later that he had two sisters. Who knew? What did he do in the war? No one knew, despite the pictures of him on a camel by the Great Pyramid and wearing his officer's uniform so proudly. Was he brave? Did he perform herculean service? Was he on the take? Was his a war of "negligence and stupidity"? We'll never know, although I don't think so, because he called every house he lived in to the end of his life "Rafa," referring to the city of that name in Palestine, where his war took him after the Dardanelles and Egypt. I don't know if he was at the Battle of Rafa in January 1917, which prominently featured the Anzacs and Chauvel's Light Horsemen. It was a difficult battle, plagued by ammunition shortages, but it marked the entry of the Empire Expeditionary Forces into Palestine. Maybe that was the time of his life. But none of his family knows, because there's no story.

I could make this into a digital story (that's its structural place in this chapter), but such an indeterminate tale would go against the grain of the genre. The story would be telling you that there is no story. My grandfather's tale may have expressed the *British* national character ("a nation of shopkeepers") but not in ways that follow the accepted script, which doesn't associate glory with complex systems and logistics, or making sure that letters from home got through.

There are many digital stories about Anzacs, and they generally do follow the script. Some are made by school students who are doing oral history about veterans from their local area. Others are made by veterans themselves, or their widows or descendants. None that I can find has been made by the backpacking pilgrims to Gallipoli's Dawn Service. It does seem from my researches[23] that when it comes to digital storytelling about Gallipoli, there is a *pre-scripted story* that people differentially populate with their own identities, experiences and families. The individual experience may be unique but the template is copied. I can't tell you my Granddad's Gallipoli story because it's the other way round; the individual was commonplace but there is no template for this story. On the evidence of Gallipoli stories, it's the copy that wins the day; people want their story to fit in with the meanings that have already been scripted; it's a kind of votary offering to the ancestors. What they actually did or what they thought about it may be of less significance than what they descendants need for them to mean.

Original stories, on the other hand, seem to require a higher level of narrative investment than digital storytelling typically commands, and so it is at the national and corporate end of the spectrum where new stories about Gallipoli are still actively created. This may intensify in the run-up to the forthcoming centenary (April, 2015). What will the new story be? It's a pretty safe bet that it won't include my Granddad; but there are indications, to which I'll return at the end, that a very particular ancestor will feature prominently.

THE CREATION OF THE POLITY

Unbeknownst to the thousands gathering annually at Gallipoli's Lone Pine memorial, there is another lonely hill in Turkey where a much deeper myth of human political origin is beginning to be undermined and reordered. The place is Göbekli Tepe (Potbelly Hill), site of impressive ancient stone monuments that were erected before human settlement and before farming, predating Stonehenge and the Egyptian pyramids by *seven thousand years*.

The deeper "myth" that its discovery challenges is the accepted story of the "Neolithic Revolution." That theory was first synthesized by an Australian archaeologist, V. Gordon Childe (1925, 1936). His theory states that human civilization "dawned" with the invention of farming, which required hunter-gatherers to settle and thus enabled the development of cities. In other words, after a rapid domesticating "revolution," the economy (agriculture) and politics (cities) determined culture (civilization).[24] It is a strongly Marxist political narrative, following the base/superstructure model (Williams, 1973), which is not surprising as Childe was a convinced Marxist, political activist, and lifelong supporter of Stalin.[25]

Vere Gordon Childe is widely forgotten in his native Australia, except by "Rampaging Roy Slaven," whose new play *Vere [Faith]* was inspired by playwright John Doyle's discovery of his existence.[26] But he ranks as one of the most important archaeologists of the twentieth century: if not Indiana Jones then certainly his teacher.[27] He was reputed to loathe actual digging, although he excavated Skara Brae in the Orkney Islands. His strength was synthesis. He performed for archaeology the "modern synthesis" that Julian Huxley achieved for the biosciences, at about the same time. He was able to gather piecemeal discoveries and sites across Eurasia into a coherent story: the story of the Neolithic Revolution.

The *story* of the Neolithic Revolution (if not the science) has held sway ever since. As Egyptologist John Romer has recently put it:

> The "Neolithic Revolution" . . . that most useful phrase, was concocted by the Australian archaeologist Vere Gordon Childe in the 1920s . . . specifically to combat the then current climate of ethnic stereotyping in European archaeology . . . Not surprisingly, perhaps, given the contemporary climate, Childe's newly invented Neolithic Revolution, a two-word adventure story in itself, soon became part of Western history. (Romer, 2012, 32)

Childe's story—that "material prosperity . . . brought social and artistic progress in its wake"—exerted its own powerful influence on scientific thought. The science may be "tricked out with fashionable neo-evolutionary economics," but as Romer points out, "the very language of the inquiry" determines what will be found: the story precedes and determines the evidence, which is largely a work of the imagination, ascribing causal sequence to "the relics of the past."

Thus, for many decades there has been no need to argue that economics was primary and culture was dragged along "in its wake" . . . until the discoveries at Göbekli Tepe. Now, it seems, we may need to reverse the flow of causation, because the monument-builders there were hunter-gathers, who neither farmed nor settled. It turns out that symbolic ritual, including gathering, dancing, feasting and possibly worship, was staged in massive, elaborate stone-built circles that have no economic or residential function. They may resemble ancestors, who may be gathered in a ceremonial (story) circle, perhaps linking the living and the dead.[28] They are constructions of the *polity*—an expression of meaningfulness for the groups that built them. More important, they offer intriguing evidence that our standard formula has it wrong; it is not the economic base that determines the cultural and political superstructure, but culture that determines the polity, and the expression of the resultant facts (in organization and logistics, as well as stone and story), precipitates economic change.

These feats of construction both organized and represented what can be described as the earliest known political narrative. They predated and likely precipitated both farming and settlement. Klaus Schmidt, lead excavator of the Göbekli Tepe site, has concluded:

> The evolution of modern humanity involved a fundamental change from small-scale, mobile hunter-gatherer bands to large, permanently co-resident communities. The factor that allowed the formation of large, permanent communities was the facility to use symbolic culture, a kind of pre-literate capacity for producing and "reading" symbolic material culture, that enabled communities to formulate their shared identities, and their cosmos. (Schmidt, 2010, 253–4)

What Schmidt calls "shared identity" was *performed* here: in the monuments themselves, in the work needed to make them, and in the attendant ceremonies. Göbekli Tepe was an improbably early site of *mediation*, and in terms of causal sequence, it suggests that culture produced the polity and economy, not the other way round.

Schmidt told *National Geographic*, which itself was excited by the revision of Childe's theory: "Twenty years ago everyone believed civilization was driven by ecological forces. I think what we are learning is that civilization is a product of the human mind." (quoted in Mann, 2011). The *National Geographic* (June 2011) wanted the story to be about the "birth of *religion*," calling Göbekli Tepe "the world's oldest temple." Other archaeological publications have been content to follow this lead (Scham, 2008). Klaus Schmidt himself is not so sure about that: "we can not say with certitude if concepts of god existed at this time" (2010, 254).

So I'm not following *National Geographic* in its own "political narrative" about Göbekli Tepe: "We used to think agriculture gave rise to cities and later to writing, art, and religion. Now the world's oldest temple suggests *the urge to worship sparked civilization*" (June, 2011). Göbekli Tepe certainly challenges the notion of the Neolithic Revolution, but "the urge to worship" is not needed as a causal mechanism, since "worship," "religion" and "temple" are such loaded terms, saying more about now than then. There's no need to impose such interpretations, but the very existence of such monuments, made before farming and settlement had been "invented," does call attention to the direction of causation:

- Culture and technology (shared desire, purpose and capacity to build)
- Politics (organization of a complex cooperative community)
- Economy (intensive organization of food supplies, materials and logistics).

Narrative Creates the Polity

That Göbekli Tepe represents a cultural (rather than economic or civic) story doesn't seem to be in doubt. It externalizes and performs the identity and knowledge of a "deme" at an unprecedented level of ambition and complexity, and it does seem reasonable to assume that the builders told each other a motivating story as to their intentions when they built it: they were working to a purposeful "script."

The monument itself may be regarded as a "media platform" or "institution of language"—the first ever "mass medium" if you like—for storytelling on the grand scale. So my proposition is that Göbekli Tepe is a relic of *political narrative*—a story that called together a deme to common purpose, and so constituted the polity that needed to support that giant enterprise by inventing agriculture, husbandry, and settlement.

What Schmidt calls "the human mind" that produces civilization is of course a network of externally linked minds, which connect with one another and interact with circumstances, natural and political (demic), via storytelling. The oldest recorded story in the world is *The Epic of Gilgamesh*, originating in the Fertile Crescent not far from where Göbekli Tepe stands. A King Gilgamesh may have existed, at around 2600 BCE (about 7,000 years after Göbekli Tepe). A celebrated version of his story is preserved on the so-called Flood Tablet, from the library of Assyrian king Ashurbanipal at Nineveh (Iraq), seventh century BCE (nearly 9,000 years after Göbekli Tepe). Its decipherment in the decade following Darwin's *Origin of Species* caused a sensation (MacGregor, 2011), because on one of his adventures Gilgamesh is told the story of a flood in terms that bear an uncanny resemblance to the Biblical (Noah's) flood story, thereby challenging the latter's divine provenance even as it confirmed its historical plausibility. It re-projected the story of human civilization back from the supernatural to the political: it wasn't divine intervention that caused the Flood; it was a traveller's tale.

As a political narrative the *Epic of Gilgamesh* still resonates, for its plot remains familiar, even in the news media. Christopher Booker (2004) has identified seven basic plots that are structural transformations of ancient tales, continuing in contemporary stories, in literature and on screen. They are: Overcoming the Monster; Rags to Riches; The Quest; Voyage and Return; Rebirth; Comedy; Tragedy. Booker includes Gilgamesh among stories with the most basic plot, "overcoming the monster," along with the story of Perseus (the Gorgon), Theseus (the Minotaur), Beowulf (Grendel), Little Red Riding Hood (the wolf); and, more recently, Dracula, HG Wells' *War of the Worlds*, *The Guns of Navarone*, *The Seven Samurai/The Magnificent Seven*, *Jaws*, *Alien*, *Dr. No* and *Star Wars–A New Hope*.

Applying this model to political communication, we can readily see the basic plots reiterated on the nightly news. A telling example was George W. Bush's "Mission Accomplished" speech, delivered aboard USS *Abraham Lincoln* on May 1, 2003, immediately after the initial open combat phase of the Iraq invasion. Instantly controversial, as victory in the so-called War on Terror was far from "accomplished," the speech was also notable for the biblical rhetoric that Bush invoked to celebrate how US forces had "overcome the monster" unleashed at 9/11. The President concluded:

> All of you—all in this generation of our military—have taken up the highest calling of history. You are defending your country, and protecting the innocent from harm. And wherever you go, you carry a message of hope—a message that is ancient, and ever new. In the words of the prophet Isaiah: "To the captives, Come out! and to those in darkness, Be free!."[29]

Using the full resources of the modern "warfare state" (Edgerton, 2006; Sparrow, 2011), and in the face of monstrous unseen adversaries, Bush invoked an ancient story to *reconstitute* the polity.

Evolution Creates the Narrative

Something along the lines of a story that brings a "deme" together to overcome unknown adversaries, uncertainty and death, may explain the findings at Göbekli Tepe. Gilgamesh, Ashurbanipal, Isaiah and Bush indicate that stories like that survive more or less unchanged in form and function over millennia, with successive political leaders using the same words for the same purposes. This *longue durée* suggests that it is time to consider whether (despite John Romer's scepticism) what we are looking at has something to do with the cultural-evolutionary adaptation of humans, explaining one of the mechanisms by which "*Homo narans*" (Fisher, 1984) maintains large, complex "demes" of non-kin in cooperative, albeit competitive, polities.

Here, recent work of evolutionary bioscientist Mark Pagel (2012a, 2012b) is instructive.[30] Like the archaeologist Klaus Schmidt, he too sees culture as primary, where knowledge, technology, meaningfulness, identity and sociality produce the economy, politics and civilization. He calls culture the "survival vehicle" for the survival of human groups or demes, much as the human body is the survival vehicle for genes. In both cases, individual persons and their physical bodies are but carriers for the mechanism of survival, which is information codes—genes (individuals) and culture (demes) respectively. As a survival vehicle, culture is characterized not so much by aesthetics as by allegiance of individuals to their deme. Pagel argues:

> The fact that cultural allegiance is most vividly expressed not in ethical behaviour but aggressive parochialism suggests it has been instrumental in protecting human beings throughout their evolution. (Pagel, 2012b)

THE BIG GUNS OF STORYTELLING . . . FALL SILENT?

When it comes to telling stories, that "aggressive parochialism" is what I have called a "universal-adversarial" stance in journalism (Hartley, 1992a, 1992b). The "we" group is taken to be universal (it includes everyone in the deme); "they" groups are taken to be adversarial (enemy, threat, deviant, dissident, deranged).[31] My research has found that universal-adversarialism is a chief characteristic of modern journalism: "we" (say, Americans), represent all humans; "they" are out to get us. There is, then, no better way to express who "we" are than in what Charles Bean called "the supreme test for fitness to exist": *warfare*. This structural characteristic of stories about cultural identity permeates many types of discourse across many domains of life. It is not confined to actual wars or international politics, but crops up wherever what Thorstein Veblen once called "invidious comparison" is called for. In short, we're not just "wired for culture," as Pagel puts it, but our stories are "wired" for universal-adversarialism.

Storytelling can be characterized as a carrier of information codes. As such they are designed for imitation, copying, sharing, emulation. They are a distribution mechanism for *how* to think and *what* to think. As mentioned, they are a resource for inductive reasoning. Stories also store lessons, allowing social learning to cross generational, language and geospatial boundaries, reproducing the sequence of inductive logic that teaches us not only what to fear but what to do about it: how to outwit duplicitous adversaries, how to test unknown characters for truthfulness, how to signal prowess to enemies and lovers, how to behave courageously . . . and so on. Culture is the "survival vehicle" for groups (demes); stories are the survival vehicle for culture.

Stories like *The Epic of Gilgamesh* reflect archetypally on personal fear of death. The plot brings a realization of death's inevitability for the hero, but hope for reproduction through family, followers and "our" strong city. Individuals may die, but their actions benefit the group, which thereby survives. So it was too with "Mission Accomplished." President Bush said: "Those we lost were last seen on duty. Their final act on this earth was to fight a great evil, and bring liberty to others." The dead made us free.

Culture and stories are mechanisms for transmitting cooperation and social learning (Thomas and Seely Brown, 2011), and for developing externalized forms of shared knowledge (e.g., language, customs, institutions, technologies, tools, etc.) that help to promote the survival of the group across time and place, even against the interests of individuals within the group, who die for unrelated genes, as it were. Cooperation overcomes "selfish" genes (Dawkins, 2006) by casting members of the same tribe or deme as "honorary relatives" (Pagel 2012), who look out for other members of the group, even though no genes are shared between them, and who, through acts of "costly signalling," seek to impress strangers, even though they'll never meet them (Gintis and Bowles, 2011; Miller, 2009).

This is how cooperative trustworthiness is *tested*, allowing onlookers to judge what a given claim has cost the speaker; hard-won experience scoring higher than braggadocio, and it explains why *truth, trust, troth, and truce* have the same etymological roots (Hartley, 1992b, 48). In this sense, all storytelling is political, constituting the "we"-community, seeking to create polities of trust, to expound the costs of cooperation for characters and deliver its symbolic rewards (Boyd, 2009). Culture demands high levels of altruism toward the group and high levels of trust for insiders. Concomitantly, it instils distrust for outsiders or strangers. In contemporary news media, universal-adversarial journalism (Hartley, 1992a) creates a "they" identity, not only for direct enemies (monsters) but also for Tricksters (Hyde, 2008), who may be masquerading as "honorary relatives" to gain advantage of our deme's knowledge systems and information codes—to steal our semiotic cattle. Evolved mechanisms to counteract this kind of theft may include *different languages* (the "tower of Babel"), an early form of intellectual property protection, perhaps; and secret, arcane, cabbalistic or hermeneutic knowledge, including secret men's and women's business among Indigenous demes. Modern nation-state genres of we/they exclusionary tactics include the differentiation of "our" publicity from "their" propaganda (Hartley, 2006).

Storytelling seems to be universal among humans, but stories themselves have evolved only within specific "we"-communities, often quite small or tribal demes. It is only in the past century or so that communications media, economic development and social network markets have expanded sufficiently to reach global scale. With organized or coordinated global media and networks comes the possibility that the differentiation of "we" from "they"—friends, family and lovers from strangers and enemies—may not be so easy any more, as stories themselves become universal in scope. Movies, music, and publishing all aspire to global audiences and readerships; stories do well that appeal across previously impermeable demographic boundaries (viz. JK Rowling). "We" identities become much more abstract and distributed across complex networks.

How has storytelling kept pace with these changes? The most important change is that informal "polities" can now be self-created, using the long-tail characteristics of large-scale social networks, where like-minded affinity or identity groups gather from among otherwise heterogeneous populations and communities of interest cocreate their own political narrative, inaugurating an era of user-created citizenship.

When "We" Become "They"

With the emergence of global communication networks with billions of users, the universal-adversarial formula that has been a primary feature of stories that "produce" a polity for "we"-communities in national media and journalism, now gets in the way. How can an adversarial sense of "we," of trust for us but fear of "they" identities, be shared across the community when that group begins to approximate to the entire species? In mainstream politics, narratives are emerging where humanity is both "we" and "they" at once—stories about climate change and environmental sustainability, for instance, or those about "man's inhumanity to man" in war, displacement and refugee migration. Humanity at large is seen as the causal agent of those problems, and thus "our" own adversary. We have literally become our own worst enemies. The monster our heroes must overcome is . . . ourselves.

Human culture and technology seem to have evolved rather faster than human storytelling formats, so the universal-adversarial pattern no longer fits the facts. Demes are no longer "tribal" or even nation-states. With contemporary digital media, we live in a semiosphere that is manifestly global and local at once. "Our" deme may be organized around quite different rules of association among strangers than the ones that govern national citizenship.

Young Australasians camping out on the hills surrounding Anzac Cove are associating themselves with a political narrative of national origins, but they are also members of many other networks, complexly interconnected and of global extent, intensely meaningful for those involved but not necessarily shared by the people in the next tent, such that the distinction between "we" and "they" is as meaningless as a distinction between Australasian and Turkish graves on the peninsula, or between heroes and postmen.[32]

However, there seem to be no digital stories about this form of consciousness; only dutiful prayers to the ancestors, following a script written generations ago by War Correspondents. So maybe digital storytelling isn't as radical and progressive as its commitment to self-expression for the ordinary person seems to suggest. Maybe it needs to copy more forward-facing models; Tavi Gevinson, perhaps.

"The Legacy of My Father"

The alternative is that it will end up copying mainstream media unwittingly, again, because the "big guns" are preparing, as we speak, for the 2015 centenary of Gallipoli. Not only are they gearing up, they are even ganging up, in an alliance of opposites that is as surprising as any. This one is between Rupert Murdoch and the BBC. That there is no love lost between these titans of storytelling is well known. However, the "star" of "one of the most important programmes" in the BBC's coverage of the centenary of the Great War has been announced: Rupert Murdoch himself, who is being interviewed for a BBC film that will "tell the tragedy of Gallipoli." *The Independent* had the story:

> In an hour long interview with the BBC, conducted at the New York headquarters of his global business News Corp, the media mogul has talked of his pride in his late father's actions and how they inspired him to begin a career in newspapers. . . . He said he was anxious to keep the memory of his father's achievements alive. "I have always kept in mind very much the, if you like, legacy of my father and the influence he had on me and I have his picture prominently on the wall of my study at home. We do feel—I feel—that's a family obligation."[33]

What was his father's legacy? According to this story, "Keith Arthur Murdoch was a young Australian war correspondent who changed the direction of the war by exposing the 1915 Gallipoli campaign as a disaster." The filmmaker, Denys Blakeway, told *The Independent* that "the 60-minute film, due to be screened on BBC2 in 2015, 'will hinge on Rupert Murdoch's father's action.'"

Clearly, the narrative die is already cast. Rupert Murdoch's father "changed the direction of the war," and *Gallipoli* (2015) will "hinge" on his action. Murdoch already has form in this respect. He bankrolled Peter Weir's 1981 movie *Gallipoli* (starring Mel Gibson).[34] That film, to Weir's later "regret," falsified the event for purposes of national pride:

> The charge at The Nek on August 7, 1915, which provides the film's climax, did take place but an Australian, rather than a British officer, ordered the final charge. The film gives the opposite impression, something Peter Weir has said he regrets. "The implication was that we were Pom bashing" he told David Stratton . . .[35]

Thus, through the heroics of Rupert Murdoch's father, Australians "overcame the monster"—*not the Turks but the Colonial Power*—and founded a nation based on white egalitarian mateship. As the Minister for Veterans Affairs and Minister Assisting the Prime Minister on the Centenary of ANZAC put it, in a speaking at the Dawn Service at Gallipoli in 2013:

> Although it was so dreadful, it has become central to our nation's story. A hallmark in defining our nationhood and what we see as important in terms of mateship, service, sacrifice, courage and commitment.[36]

Australian mateship being what it is, Anzac day is now sponsored by VB beer (CUB). Their "Raise a Glass" campaign is fronted by retired army general Peter Cosgrove, who says: "Wherever you are, whatever you're drinking, raise a glass to those who serve." VB will even arrange for General Cosgrove, AC, MC, to phone you with a wake-up call for the Dawn Service.[37] This seems to be the political narrative that has already been marshalled to "inspire a new generation to understand what happened," as the BBC's Director General put it at the launch of the broadcaster's plans for World War I centenary coverage.[38]

DIGITAL STORIES TO CONSTITUTE A NEW POLITY

So we should return finally to digital storytelling. Can it do better than this? The *form* of digital storytelling has been well established, and its *purpose*, to open up the storytelling capabilities of digital media to everyone, remains laudable. But the *content* of vernacular, unrehearsed stories unwittingly reproduce the political narrative that "constitutes the polity" by mythologizing "our" origins—unless conscious effort is made by digital storytelling activists and agencies to try something new. If a "new generation" needs to be inspired to "understand what happened," would it not be better to develop an alternative approach, based on the example of people from that generation, such as Tavi Gevinson, rather than relying on big-gun War Correspondents and mythmakers like the Murdochs, Weir, Bean, Bartlett and Cosgrove. If we truly want to honor our Granddads, and find new ways to constitute the globally networked polity, where foe-creation is self-destruction, we must find room for stories that do not go "over the top," but, rather, understand how the very concept of "we" is as much of a threat to ordinary people's understanding, well-being and peace as any monster. As someone ("eleni b" from Greece, wonderfully) wisely noted on Pinterest:

> The metaphor behind the Trojan Horse: The audience accepts the story because, for a human, a good story always seems like a gift. But the story is actually just a delivery system for the teller's agenda. A story is a trick for sneaking a message into the fortified citadel of the human mind.[39]

As the citizens of another legendary city in Turkey discovered, it's always wise to beware of Greeks bearing gifts. Is there now a Trojan Horse in the city of stories? This chapter has been driven by an interest in the potential for digital media and social networks to democratize storytelling, in the hope that non-professional people may tell a wider range of stories than the few that dominate national politics, the movies, journalism and education. The hope is that "user-created citizenship" will *revise* not *reproduce* our understanding of "who we are as a people" (as British PM David Cameron said about World War I commemorations).[40] But the indications are not altogether positive. Digital storytelling in the context of Gallipoli seems to have been captured by institutional agencies that use it to disseminate existing meanings, while the generative journalistic story goes unchallenged (even when aspects are known not to be true). If digital storytelling is to play a corrective role in the forthcoming centenary of World War I, it will face powerful competition from the pros,[41] so it needs to develop much sharper self-consciousness about the importance of storytelling as a whole, and a much more sophisticated understanding of the generative role that narrative plays in constituting who "we" are. Given that digital media and social networks have already made what constitutes "our" deme more risky, complex, open, uncertain and multivalent than ever before, it is urgent for progressive innovations like the digital storytelling movement to catch up.

NOTES

1. "Storytelling" is loosely defined—the term may incorporate elements of "show" as well as "tell"—ceremony, dance and song as well as character, plot and action; it is social as well as semiotic, and visual as well as aural/textual.
2. Shakespeare, *Hamlet* II, ii, 273–4: www.bartleby.com/70/4222.html
3. Michael Schulman (July 27, 2012) "The Oracle of Girl World." *New York Times*: www.nytimes.com/2012/07/29/fashion/tavi-gevinson-the-oracle-of-girl-world.html
4. "Still figuring it out" (April 2012), had over half a million views at last count: www.youtube.com/watch?v=6osiBvQ-RRg
5. Sources: www.drawnandquarterly.com/newsList.php?item=a5230864510be8; http://ideas.sydneyoperahouse.com/2013/tavi-gevinson-tavis-big-big-world-at-17/ and http://tickets.mwf.com.au/session2_mwf.asp?s=532

6. See for instance: www.australiaday.com.au/studentresources/indigenous.aspx; and Peter Gebhardt (2013) "A national day of shame." *Sydney Morning Herald*, January 24, www.smh.com.au/federal-politics/society-and-culture/a-national-day-of-shame-20130123-2d7b3.html

7. There was no such thing as an Australian citizen till 1949, only British subjects. Aboriginal people were not citizens until 1967.

8. See www.anzacsite.gov.au

9. See www.bigdayout.com

10. The Australian War Memorial at Villers-Bretonneux lists 10,000 Australian dead with no known grave, www.dva.gov.au/commems_oawg/OAWG/war_memorials/overseas_memorials/france/Pages/villers%20bretonneux.aspx

11. Former Prime Minister Paul Keating, speaking in 2008, criticized this trend, saying: "we still go on as though the nation was born again or even, redeemed there. An utter and complete nonsense." www.abc.net.au/news/2008-10-30/anzac-gallipoli-gatherings-misguided-keating-says/188086

12. *The Story of Anzac* (1921), cited in the *Australian Dictionary of National Biography* (1979) http://adb.anu.edu.au/biography/bean-charles-edwin-5166; and in the Australian government's official ANZAC website www.anzacsite.gov.au/1landing/beanbio.html

13. *The Story of Anzac* (1921), cited in the *Australian Dictionary of National Biography* (1979) http://adb.anu.edu.au/biography/bean-charles-edwin-5166

14. *In Your Hands, Australians* (London, 1918), cited in the *Australian Dictionary of National Biography* (1979) http://adb.anu.edu.au/biography/bean-charles-edwin-5166

15. Bartlett's story, as published in the *Sydney Morning Herald*, May 8, 1915, is reproduced in full in Seal (2013), 23–30.

16. Charles Bean, personal records, "Ashmead Bartlett and a crisis," item 892, 3DRL/6673, Australian War Memorial 38, cited in www.anzacsite.gov.au/1landing/beanbio.html

17. 1st edn, edited by C.E.W. Bean, published by Cassell, London, 1916; 2nd edn, published by Sun Books, Melbourne, 1975; 3rd edn, edited and published by the Australian War Memorial www.awm.gov.au/shop/item/9781742231341/#.UnGrZiRhNdg

18. "The Raid on London," by Private Pat Riot, 9th Battalion. *The Anzac Book*, 143–5.

19. The Long, Long Trail www.1914-1918.net/asc.htm

20. www.nla.gov.au/apps/cdview/?pi=nla.aus-vn2035864-s83-e and www.nla.gov.au/apps/cdview/?pi=nla.aus-vn2035864-s84-e

21. The Long, Long Trail: www.1914-1918.net/asc.htm.

22. "Glimpses of Anzac," by Hector Dinning, Aust. ASC. *The Anzac Book*, 17–23.

23. See, for instance http://museumvictoria.com.au/discoverycentre/websites/making-history/student-uploads/; http://www.dpc.vic.gov.au/index.php/veterans/victorian-veterans-virtual-museum/digital-stories-in-our-words; https://open.abc.net.au/openregions/qld-tropical-north-93gv1ap/posts/remembering-billy-26wn8wx; http://generator.acmi.net.au/gallery/category/conflict-australians-war; https://www.raiseaglass.com.au/ (click on "Your Stories"); www.au.dk/en/about/news/single/artikel/cultural-memory-more-relevant-than-ever/.

24. But, for a contrary view, see www.telegraph.co.uk/science/science-news/5604296/Is-farming-the-root-of-all-evil.html

25. The Wikipedia entry on Childe is a good place to get a flavor of the man, his times, his work and its impact, http://en.wikipedia.org/wiki/V._Gordon_Childe.

26. Story at www.sydneytheatre.com.au/magazine/posts/2012/september/feature-vere-gordon-childe.aspx

27. http://indianajones.wikia.com/wiki/Vere_Gordon_Childe

28. "Pillars at the temple of Göbekli Tepe—11,600 years old and up to 18 feet tall—may represent priestly dancers at a gathering." *National Geographic*: http://ngm.nationalgeographic.com/2011/06/gobekli-tepe/musi-photography

29. Text at www.cbsnews.com/stories/2003/05/01/iraq/main551946.shtml. Isaiah's "captives" were held by the Assyrian king Sennacherib. The next king of Assyria but one was Ashurbanipal, so the link to Gilgamesh is not so far fetched.

30. See also a useful review of Pagel, 2012a here, www.independent.co.uk/arts-entertainment/books/reviews/wired-for-culture-the-natural-history-of-human-cooperation-by-mark-pagel-7573966.html

31. Note that the names of many premodern nations, for example, Nyungar (Western Australia) or Cymru (Wales), simply mean "human being" or "compatriot." Conversely, many languages use pejorative terms for outsiders—all non-Ancient Greeks were "barbarians"; the Hebrew for "opposite," adversary, accuser is "satan"; Wales is derived from the Norman-French word for "foreigner." The Apaches evidently have a bet each way: "The word 'apache' comes from the Yuma word for 'fighting-men' and from the Zuni word meaning 'enemy'" (www.greatdreams.com/apache/apache-tribe.htm).

32. See his words about this when he visited Gallipoli in 1934 as President of Turkey: www.awm.gov.au/encyclopedia/ataturk.asp

33. This and related quotations from here, www.independent.co.uk/news/uk/home-news/bbc-unveils-the-star-of-its-first-world-war-anniversary-coverage--rupert-murdoch-8884028.html

34. See: www.mcc.murdoch.edu.au/ReadingRoom/film/dbase/2002/gallipoli2.htm, with details of Murdoch's funding (through Associated R&R).

35. Paul Byrnes, at: http://aso.gov.au/titles/features/gallipoli/notes/.

36. Warren Snowdon speaking on April 25, 2013, www.theaustralian.com.au/in-depth/anzac-day/gallipoli-calamity-defined-a-nation-warren-snowdon-told-dawn-service-at-anzac-cove/story-e6frgdaf-1226629387147#sthash.fhqgofsJ.dpuf

37. https://www.raiseaglass.com.au/. And see Pete Mitcham, aka Prof Pilsner (April 25, 2013) "Raise a glass—of scepticism," *Brewsnews*: www.brewsnews.com.au/2013/04/raise-a-glass-of-scepticism/. Wake-up call: https://www.victoriabitter.com.au/2013/04/this-anzac-day-theres-no-excuse/

38. www.radiotimes.com/news/2013-10-16/bbc-to-create-digital-cenotaph-for-ww1-anniversary-with-biggest-ever-broadcasting-season

39. www.pinterest.com/pin/254383078925276152/

40. Quoted here www.independent.co.uk/news/uk/home-news/jeremy-paxman-blasts-david-cameron-over-wwi-centenary-comments-8866630.html

41. As well as allowing Keith Murdoch pride of place in the Gallipoli story, the BBC plans to do exactly what digital storytelling is good at: "we will tell well-known

stories from fresh perspectives and original stories so far untold." (Adrian Van Klaveren, controller for the BBC's World War I Centenary): www.mirror.co.uk/tv/tv-news/rupert-murdoch-joins-bbc-digital-2459368

REFERENCES

Anderson, B. (1991) *Imagined Communities: Reflections on the Origins and Spread of Nationalism*, 2nd edn. London: Verso.

Anderson, C. (2008) *The Long Tail: Why the Future of Business is Selling Less of More.* Rev, edn. New York: Hyperion.

Baym, G. (2000) 'Constructing moral authority: "We" in the discourse of television news.' *Western Journal of Communication*, 64:1, 92–111.

Beinhocker, E. (2006) *The Origins of Wealth.* Cambridge MA: Harvard Business School.

Booker, C. (2004) *Seven Basic Plots: Why We Tell Stories.* London: Bloomsbury.

Bowles, S. & Gintis, H. (2011) *A Cooperative Species: Human Reciprocity and Its Evolution.* Princeton University Press.

Boyd, B. (2009) *On the Origin of Stories Evolution, Cognition, and Fiction.* Cambridge, MA: Harvard University Press.

Buonanno, M. (2005) 'The "Sailor" and the "Peasant": The Italian Police Series between Foreign and Domestic.' *Media International Australia*, 115, 48–59.

Burrell, I. (2013) 'BBC unveils the star of its First World War anniversary coverage—Rupert Murdoch.' *The Independent (UK)*, October 16. Online: www.independent.co.uk/news/uk/home-news/bbc-unveils-the-star-of-its-first-world-war-anniversary-coverage--rupert-murdoch-8884028.html

Childe, V. G. (1925) *The Dawn of European Civilization.* London: Kegan Paul, Trench, Trubner.

Childe, V. G. (1936; rev. edn 1951) *Man Makes Himself.* London: Watts.

Chouliaraki, L. (ed.) (2012) *Self-Mediation: New Media, Citizenship and Civil Selves.* London: Routledge.

Couldry, N., MacDonald, R., Stephansen, H., Clark, W., Dickens, L. & Fotopoulou, A. (in press) 'Constructing a digital storycircle: Digital infrastructure and mutual recognition.' *International Journal of Cultural Studies.*

Dardanelles Commission (1917–1919) *The Final Report of the Dardanelles Commission.* London, HMSO. Online: http://nla.gov.au/nla.aus-vn2035864.

Dawkins, R. (2006) *The Selfish Gene. 30th Anniversary Edition.* Oxford: Oxford University Press.

Eco, U. (1976) *A Theory of Semiotics.* Bloomington: Indiana University Press.

Edgerton, D. (2006) *Warfare State: Britain, 1920–1970.* Cambridge: Cambridge University Press.

Fewster, K. (1982) 'Ellis Ashmead Bartlett and the making of the Anzac legend.' *Journal of Australian Studies*, 6:10, 17–30. DOI: http://dx.doi.org/10.1080/14443058209386844

Fisher, W. (1984) 'Narration as a human communication paradigm: The case of public moral argument.' *Communication Monographs,* Volume 51, Issue 1.

Gevinson, T. (ed.) (2012) *Rookie Yearbook 1*. Montreal: Drawn and Quarterly.
Gevinson, T. (ed.) (2013) *Rookie Yearbook 2*. Montreal: Drawn and Quarterly.
Gintis, H., Smith, E.A. & Bowles, S. (2001) 'Costly Signaling and Cooperation.' *J. theor. Biol.* 213, 103–119 doi:10.1006/jtbi.2001.2406.
Gintis, H. & Bowles, S. (2011) *A Cooperative Species: Human Reciprocity and Its Evolution*. Princeton: Princeton University Press.
Jenkins, H., Ford, S. & Green, J. (2013) *Spreadable Media*. New York: New York University Press.
Hartley, J. (1992a) *The Politics of Pictures: The Creation of the Public in the Era of Popular Media*. London & New York: Routledge.
Hartley, J. (1992b) *Tele-ology: Studies in Television*. London & New York: Routledge.
Hartley, J. (2006) 'The best propaganda: Humphrey Jennings.' *The Silent Village* (1943).' In: McKee, A. ed. *Beautiful Things in Popular Culture*. Malden, MA & Oxford: Wiley-Blackwell.
Hartley, J. (2009) 'TV Stories: From Representation to Productivity.' In Hartley & McWilliam, eds.
Hartley, J., Burgess, J., & Bruns, A. (eds.) (2013) *A Companion to New Media Dynamics*. Malden, MA & Oxford: Wiley-Blackwell.
Hartley, J. & Green, J. (2006) 'The public sphere on the beach.' *European Journal of Cultural* Studies, 9:3, 341–62.
Hartley J. & McWilliam, K. (eds.) (2009) *Story Circle: Digital Storytelling Around the World*. Malden MA and Oxford: Wiley-Blackwell.
Hartley, J. & Potts, J. (forthcoming) *Cultural Science: The Evolution of Meaningfulness*. London: Bloomsbury.
Herman, S. (2013) 'Tavi Gevinson.' *The Lab Magazine* (Canada), 7, 44–51.
Hornblower, S. & Spawforth, A. (2005) *The Oxford Classical Dictionary*, 3rd edn. Oxford: Oxford University Press.
Hyde, L. (2008) *Trickster Makes This World: How Disruptive Imagination Creates Culture*. Edinburgh: Cannongate.
Lambert, J. (2006) *Digital Storytelling*. 2nd edn. Berkeley, CA: Digital Diner Press.
Lanham, R. (2006) *The Economics of Attention: Style and Substance in the Age of Information*. Chicago: Chicago University Press.
Lord, A. (1960) *The Singer of Tales*. Cambridge, MA: Harvard University Press.
King, J. (2013) 'It's Anzac Day—not the Big Day Out.' *Sydney Morning Herald*, April 20. Online: www.smh.com.au/national/its-anzac-day--not-the-big-day-out-20130419-2i5a4.html#ixzz2j9lR6hPO
Lundby, K. (ed.) (2008) *Digital Storytelling, Mediatized Stories*. New York: Peter Lang.
MacGregor, N. *A History of the World in 100 Objects*. London: Viking.
Mann, C. (2011) The Birth of Religion. National Geographic, June: http://ngm.nationalgeographic.com/print/2011/06/gobekli-tepe/mann-text
Meadows, D., Heledd, L., & Evans, C. (2006) 'How public broadcasting serves the public interest in the digital age'. *First Person: International Digital Storytelling Conference*, Australian Centre for the Moving Image, 5 February: www.acmi.net.au/global/docs/first_person_meadows.pdf
Meadows, D. & Kidd, J. (2009) '"Capture Wales." The BBC Digital Storytelling project.' In Hartley & McWilliam, (eds.), 91–117.

Miller, G. (2009) *Spent: Sex, Evolution, and Consumer Behavior*. USA: Penguin Books.

Nelson, O. (2013) 'Dark undercurrents of teenage girls' selfies.' *Sydney Morning Herald*, July 11. Online: www.smh.com.au/comment/dark-undercurrents-of-teenage-girls-selfies-20130710-2pqbl.html.

Ong, W. (2012) *Orality and Literacy: Technologizing the Word; 30th anniversary edition with additional chapters by John Hartley*. London: Routledge.

Pagel, M. (2012) *Wired for Culture: The Natural History of Human Cooperation*. London: Allen Lane.

Pagel, M. (2012b) 'The Culture Bandwagon.' *New Humanist*. Accessible at: www.eurozine.com/articles/2012-02-21-pagel-en.html

Potts, J., Cunningham, S., Hartley, J. & Ormerod, P. (2008) 'Social network markets: a new definition of the creative industries.' *Journal of Cultural Economics*, 32:3, 166–85.

Scham, S. (2008) 'The World's First Temple'. *Archaeology*, 61:6, http://archive.archaeology.org/0811/.

Schmidt, K. (2010) 'Göbekli Tepe—the Stone Age Sanctuaries. New results of ongoing excavations with a special focus on sculptures and high reliefs.' *Documenta Praehistorica*, XXXVII, 239–56: http://arheologija.ff.uni-lj.si/documenta/authors37/37_21.pdf

Seal, G. (2004) *Inventing Anzac: The Digger and National Mythology*. St Lucia: University of Queensland Press.

Seal, G. (2013) *Great Annzac Stories: The men and women who created the Digger legend*. Sydney: Allen & Unwin.

Smith, E.A. (2010) 'Communication and collective action: language and the evolution of human cooperation.' *Evolution and Human Behavior*, 31: 231–45.

Sparrow, J. (2011) *Warfare State: World War II Americans and the Age of Big Government*. Oxford: Oxford University Press.

Thomas, D. and Seely Brown, J. (2011) *A New Culture of Learning: Cultivating the Imagination for a World of Constant Change*. Online: CreateSpace Independent Publishing Platform.

Thumim, N. (2012) *Self-Representation and Digital Culture*. London: Palgrave.

Turner G. (1993) *National Fictions: Literature, Film and the Construction of Australian Narrative*, 2nd ed. Sydney: Allen & Unwin.

Williams, R. (1973) 'Base and Superstructure in Marxist Cultural Theory.' *New Left Review*, Series I, no. 82; reprinted in C. Mukerji and M. Schudson, (eds.) (1991) *Rethinking Popular Culture: Contemporary Perspectives in Cultural Studies*. Berkeley and LA, CA: California University Press, 407–23.

FUNDING

ARC Linkage Project: LP110100127: "Digital storytelling and co-creative media: the role of community arts and media in propagating and coordinating population-wide creative practice."

Part III

APPLICATIONS

Chapter 10

Communicating Neoliberal Development

A Critical Analysis of Grameen Bank Programs for Women

Karin Gwinn Wilkins

Communication for sustainable social change implies that development strategies explicitly promote long-term processes that benefit future generations. This approach extends beyond more immediate development goals that attempt more direct and short-term objectives within the duration of a defined project cycle, instead building structures that are meant to endure and facilitate strategic interventions to improve social welfare. Mediated technologies are incorporated in a variety of ways in strategic sustainable social change, contingent upon assumptions made about the roles of communication strategies in relation to substantive problems and proposed solutions. In this chapter, I explore how communication technologies are understood to benefit women in microenterprise programs funded through the Grameen Bank, a well-known institution in the development industry. This critical analysis is positioned within a conceptual framework that highlights global contexts in which neoliberal agendas dominate.

NEOLIBERALISM, CELL PHONES AND POOR WOMEN

Neoliberal approaches to development permeate discourse within the industry, structuring programs while linking interventions to global capitalism. Critical analyses of development highlight how rhetoric and practice reinforce neoliberal assumptions about the nature and process of social change, privileging individual consumption over normative shifts and policy change (Dutta, 2011; Richey and Ponte, 2011). This work builds from scholarship that explores communication about development, concerning constructions of appropriate beneficiaries, along with designated problems and strategic solutions. This research accentuates power dynamics as central to analyses, building from political-economic and feminist critiques of development, focusing on discourse concerning Grameen Bank programs.

Microenterprise programs finance small-scale interventions designed to build individual financial capital in market economies. Interventions may attempt to improve access to financial services and business information about marketing and distribution, or to encourage policies that support regulation in favor of individual financial transactions and small business banking. Communication technologies become prominent in these enterprises as products, both as hardware and software, as processing technologies, such as data entry and distance learning, and as support, through computer and other skills training. Women's participation can be as diversely understood within this rubric as ranging from ownership or management, to paid or volunteer labor, or as discussants of problems and solutions.

As a development strategy microenterprise works within a neoliberal framework privileging economic growth as a global development enterprise (Nederveen Pieterse, 2009; Sparks, 2012). Privatization of the development industry has emerged with the *glamorization* and visibility of private donors (Edwards, 2009). Many of these private-aid initiatives focus on individual empowerment goals, resonating with a neoliberal approach to social change that emphasizes entrepreneurship in market economies.

The Grameen Bank programs illustrate a highly visible development approach, not only in the industry but also in global public discourse. Muhammad Yunus, 2006 Nobel Peace Prize winner, becomes identified as the central character in this narrative, along with the cell phone and poor women. Before exploring this discourse it is worth considering how communication becomes conceptualized as mediated tools working toward development goals, as well as how development as an industry communicates assumptions about the sustainability of social change interventions.

The development industry has attended to women's and gender issues through quite disparate ways over time, after ignoring women's contributions as active agents in communities for decades prior to the United Nations *Decade for Women* beginning in the 1970s (Wilkins, 2005). Current attention to women's issues and gender concerns is integral in global discourse promoted by dominant development institutions, through their articulation of the global *Millennium Development Goals* (UN, 2011). Among other concerns, the Millennium Development Goals (MDGs) focus on reducing maternal mortality, achieving universal access to reproductive health services, and improving rates of female participation in formal education, paid employment, and elected political positions. While attention to women in development is valuable, concerns with gender highlight potential concerns with the ways in which programs articulate problematic assumptions about women and men in society.

Feminist critiques of mainstream development enterprises highlight the patriarchal nature of discourse that privileges men's roles over those of women (Wilkins, 2007). Part of the problem lies in development approaches that subjugate women as passive targets of strategic communication campaigns, constructed as participants when contributions are valued in the service of imperialism, patriarchy, and capitalism. Toward imperialism, women become prominently part of justification for military intervention in their role as victims in need of a savior; toward patriarchy, women become targeted in population and nutrition programs as conduits for reproducing and feeding new generations; and toward capitalism, women become relegated to small-scale entrepreneurial micro-lending schemes within the confines of the capitalist market. This final approach informs the justification for further analyses of the Grameen Bank model, although quite prominent in the field not adequately studied in terms of latent consequences.

This research explores how discourse of Grameen Bank programs implemented in Bangladesh has constructed the role of communication in social change, specifically in descriptions of telephones, and the role of women in development, in articulations of gender. This discourse is considered through published works on these programs by development professionals, as well as public discourse.[1]

THE PROMISES AND PROBLEMS OF GRAMEEN AS DEVELOPMENT

The narrative of microenterprise begins with the promise of 2006 Nobel Peace Prize winner Muhammad Yunus, proclaiming micro-lending as the solution to overwhelming poverty. His initiative through the Grameen Bank relied on private-sector banking services to offer loans to defined groups for small-scale entrepreneurship. In his 2007 book on the subject, Yunus describes having distributed about 2.6 million loans, mostly to women (95%; p. 235), toward generating income, supporting housing, and funding education for children of Grameen Bank borrowers. The microfinance industry grew into a halo of "saintly aura" (MacFarquhar, 2010), invoked by Kristof and WuDunn (2009) as a "revolution sweeping the developing world" (p. 187) with over $60 billion in assets by 2010 (McAnany, 2012).

Microcredit became quite well-known and popular within the development industry since the 1997 Summit, then attracting more than 3,000 people. By the 2002 Summit, Yunus (2007) estimated that 54 million families across the world had benefitted from microcredit (pp. 256–260). And four years later, Sinclair reports that professionals at a 2006 Global Microcredit Summit in Nova Scotia felt like "saviors" (2012, p. 129). Advocates and heretics (as Sinclair calls himself) agree that microfinance became institutionalized as a dominant development scheme (McAnany, 2012, p. 107).

The Grameen Bank structured itself as "a private-sector self-help bank" with a "social consciousness" (Yunus, 2007, p. 203). The social goals of the organization are at odds with the underlying structural foundation of the Bank. On the one hand Yunus recognizes that "(p)overty is not created by the poor. It is created by the structures of society and the policies pursued by society" (2007, p. 206). Critical of the World Bank, national governments, and other development agencies, Yunus declares his goals as working to reduce poverty and promote human rights. The prescription for this diagnosis though is to engage in a different style of banking, directed toward enabling loans to those with few resources, within the structure of a free-market economy. Along with this structural arrangement come resonating ideological themes, such as individual empowerment and self-reliance.

Self-reliance serves as an important refrain in the mantra of micro-lending, both at institutional and individual levels. Although early years in the Grameen Bank (1977–1988) were supported by philanthropic organizations such as the Ford Foundation and the MacArthur Foundation, Yunus exhibits pride in transitioning his work toward becoming an independent bank without external charitable support. Kristof and WuDunn agree that "markets and microlending are proving a powerful system to help people help themselves" (2009, p. 187). This implied autonomy becomes connected with conceptualizations of sustainability that encourage economic resilience by avoiding dependency on external agencies (Arun, Heeks, and Morgan, 2004).

Public discourse resonates with the positive reviews thus described in the field. Most of the news articles reviewed (88%) offer quite positive depictions of the Grameen Bank and its founder, frequently referring to Yunus as a Nobel Peace Prize winner, who "pioneered" microfinance. As an example, an often referenced essay by MacFarquhar (2010) characterizes Yunus as the "darling of the development world," who . . . "having pioneered the practice by lending small amounts to basket weavers in Bangladesh, was awarded a Nobel Peace Prize for it in 2006. The idea even got its very own United Nations year in 2005." Yunus is celebrated as both savior and entrepreneur, aiding in "the newly independent nation of Bangladesh (which) was suffering a famine," creating a bank that "succeeded by reversing traditional banking rules" (Hundley, 2006). This discourse celebrates the commercial success of the bank, by suggesting that the "world of finance is realizing Muhammad Yunus's idea is more than mere charity—it is good business" (Sappenfield and Trumbull, 2006). A few articles note Yunus' caution though that while "banks can make a profit . . . this is what loan sharks do" (Sappenfield and Trumbell, 2006). The profit angle comes through in constructions of both the success of the bank as well as the perceived sustainability of women's projects, "saving" "poor, self-employed women" from "dire poverty;" the *Washington Post* (1998) specifically states that "one third have escaped dire poverty" but no evidence is given.

Microfinance has grown in complexity, enhancing the potential for individual investors within these broader market schemes. As one illustration, the KIVA microfinance program (http://www.kiva.org/about) initiated in 2005 allows individuals to contribute money on their website to support loans to featured and photographed applicants. In its first four years KIVA raised one hundred million dollars through this strategy (McAnany, 2012). Individual donors might be attracted to this scheme that suggests loans are avoiding costly intermediaries and serving people more directly, but clear and accurate information on the actual processes of distribution, and even actual interest rates charged to clients, does not feature on these sites (Rosati, 2012; Sinclair, 2012). This concern brings us to a consideration of how limitations of this type of intervention are articulated.

LIMITATIONS OF MICROFINANCE AS DEVELOPMENT

David Korten's introduction to Hugh Sinclair's recent (2012) critique of microfinance explains how difficult it can be to criticize this $70 billion industry, given the dominant if naïve celebrations, hailing "microfinances as an almost sacred solution to global poverty" (p. 216). Although Sinclair does not resist the idealism of microenterprise, he builds a case against the public relations effort that has established this as a dominant model of development with little documented evidence of impact or consequences. Although there are many trumpeted anecdotal success stories of "female empowerment" and charts of loans repaid to banks, the industry operates with little regulation. Instead of offering a more substantial critique of the system, Sinclair worries that a potentially useful development model has been "hijacked," given the greed of profit-driven management and the neglect of industry regulation (2012, p. 218).

Sinclair does raise concerns with implementation, given the physical distance and technological challenges of monitoring loans (2012, p. 11). John Hatch is referenced in this volume as well (p. 78) making a critical but not well-known point that 90% of microloans finance consumption of goods such as televisions, rather than actual enterprise schemes. These small-scale businesses are also hard to sustain over time, as bigger corporations such as Walmart emerge and dominate local commerce.

Few news articles (fewer than one out of five) allude to some of the limitations with the Grameen banking model or program. These concerns include the high interest rates charged (24–36% according to Ved, 2001); the lack of mobility out of poverty for those taking loans; and the lack of replicability in other countries with different economies of scale.

Yet outside of this public discourse the Grameen Bank model has faced criticism both within and outside of Bangladesh. Yunus was forced to leave his position with the Bank in 2011 (Sinclair, 2012). A year later, the editorial board of the *New York Times* (2013) published a glowing review of Yunus and Grameen, accusing the Bangladeshi government of having little justification for "nationalizing and breaking up widely admired Grameen Bank, which pioneered the business of lending small amounts of money to poor women who want to start and grow businesses. Lawmakers should reject these destructive ideas and stop meddling in the affairs of this important financial institution, which serves 8.4 million rural women."

According to *New York Times* journalist Neil MacFarquhar (2010), while the global averages for interest rates on these types of loans have been about 37%, loans in Mexico averaged about 70%, with Nigeria and others also being exceedingly high. And while interest rates for KIVA in the United States may be under 10%, interest rates in southern Sudan, as an illustration, are as high as 88% (Sinclair, 2012, p. 174). The appropriate boundary for interest rates on loans has become a serious point of contention, with attempts to balance the competing goals of helping the poor with intending to profit.

Moreover, actual interest rates are not reported to individual investors on websites such as KIVA and Microplace (MacFarquhar, 2010), meaning that people with good intentions toward contributing to these efforts would not know that their investments might be aiding exploitive practices. This concern leads to another identified problem in the industry: *lack of transparency*. Independent agencies can conduct valid assessments, but without regulation and accurate data their work can be challenged.

More importantly than these serious issues of lack of transparency and of accountability, the articulated goal to resolve poverty has not been met. Most of the microfinance clients are not among the extremely poor (Sinclair, 2012, p. 7). While a few individuals may improve their circumstances, there is no good evidence to suggest that broader conditions of poverty or of inequality have changed (p. 201).

Concerns with the sustainability of microfinance highlight the importance of context. Individual projects may not be able to create sustainable social change if the macroeconomic conditions privilege larger corporations over small businesses. Without supportive governmental policies, these development programs have limited potential for sustainability.

Although not all of these loans have been devoted to entrepreneurship (some given for purchasing goods or supporting housing or education costs), microenterprise has commanded most of the popular attention in development discourse on microfinance. Moreover, not all entrepreneurial loans employ communication technologies. However, their projected function in social change processes suggests an importance worth exploring.

ROLE OF MEDIATED COMMUNICATION IN DEVELOPMENT

Communication processes as well as technologies are integral to social change, both in terms of strategic intervention and academic observation. Mediated communications have been articulated as integral to development in academic and institutional discourse on social change, for their potential to support economic gain, social connections, political transparency, and cultural identities (McAnany, 2012). In programs designed to alleviate poverty, ICTs have played a central role in development discourse as instruments to promote commerce and citizenship.

Although ICTs take on many technological manifestations in projects, and the Grameen Bank has distributed loans for many purposes and media tools, this analysis focuses on the construction of the cell phone's role in development specifically, given the prominence of this medium within descriptions of loans distributed to women through this program. I do not intend to suggest that most of the funding has gone toward entrepreneurial projects that involve telephones, but that these are the projects highlighted in global development discourse, by development industries as well as public discussions.

Although digital and other media may contribute to many types of social change interventions, the ICT literature privileges their economic and bureaucratic uses. The microenterprise literature applauds access to information more broadly and to ICTs more specifically as "*empowering*." Yunus believes that selling telephone access not only gives women a way to earn income, but also forwards their "modernization" (2007, p. 228). Expanding this view of ICTs toward democratizing potential, he suggests that:

> Information and communication technology gives us reason to hope that we are approaching a world free of power brokers and knowledge brokers. Individuals will be in command. There will be no screening authority on center state. This is particularly exciting for all disadvantaged groups, voiceless groups, and minority groups. (2007, p. 253)

Television also becomes part of this glorified equation through which mediated communication multiplies effects of strategic intervention toward development goals, whether through explicitly designed content, as in social marketing or entertainment education, or through the structure of its industry. Despite a lack of empirical evidence (see critiques noted by Hornik and McAnany, 2001), enthusiasts attribute lower fertility rates in Brazil and in India to the introduction of expanded television options in remote and rural communities (cited in Kristof and WuDunn, 2009, p. 245). Arguing a potential link between television options and birth rates, these advocates echo earlier hopes by Lerner (1958) that "changes occurred because TV brought new ideas into isolated villages that tended to be very conservative and traditional," more specifically such that "rural viewers came to recognize that the "modern" way is for women to be treated as human beings" (Kristof and WuDunn, 2009, p. 245).

Given the banking framework structuring this approach to development, not surprisingly economic incentives dominate, in terms of individual profit as well as national economic growth. At the individual level, cellphone adoption is applauded in "low-income and poor communities, where the low cost of phones and the availability of cell networks even in remote areas has fueled the rapid growth"; the problem, in this framework, is not that people may not have use for phones, which is assumed, but that "(p)olitical instability and dictatorships make it hard to work with telecom service providers, and some central banks are reluctant to cooperate with companies that could take away their control over their citizens' finances" (Kang, 2010). Within the national model of development, this assumption of instability and corruption with poor country governments inhibits individuals' abilities to take advantage of telephones. A more specific construction attaches the category of "poor" to that of "women."

Accentuating the entrepreneurial potential of telephone sales in rural markets, the Grameen model justifies its value in terms of "having demolished the long-dominant stereotype of the poor as not creditworthy . . . Grameen has set its sights on the misguided view of the rural poor as inevitably isolated from the global economy" (Washington Post, 1998). *Those who focus on economic growth as a central goal, appreciate this connection to capitalism.* In some visions this model is referred to as "*inclusive capitalism,*" which encourages "bottom-up economic development, a way of empowering individuals by encouraging entrepreneurship as opposed to more traditional top-down approaches in which aid money must filter through a bureaucratic chain before reaching its beneficiaries, who by virtue of the process are rendered passive recipients . . . For this reason, the cellphone has become a darling of the microfinance movement" (Corbett, 2008).

The cellphone not only features in small-scale entrepreneurship, but also as a product that unites national with corporate enterprise. The arm of Grameen that focuses on the phone "teamed with Norwegian telecom Telenor to create Bangladesh's largest mobile network, and to bring solar-powered phones to remote villages" (Sappenfield and Trumbull, 2006). Grameen and Telenor together were reported as having invested almost US$200 million, both to sell phones directly to urban residents as well as sponsoring service through loans to women in rural areas (Reed, 2002), resulting in "annual revenues of about $1 billion" (Corbett, 2008). This business model has been touted as having "revolutionized communications in Bangladesh" (Corbett, 2008), creating a "profitable business" (Newseek, 1997) as well as a "microcredit movement" (Page, 2000).

Within this rubric of profitable enterprise, strategic programs are designed to alleviate poverty by offering loans to proposed entrepreneurial plans. The group most frequently identified as a target and benefactor of these loans is women.

THE ROLE OF WOMEN IN SUSTAINABLE DEVELOPMENT

The Grameen model focuses on loans to women, with the assumption that economic and social change happens more quickly when credit is extended directly to women rather than to men (Yunus, 2007, p. 72). Poverty, according to Kristof and WuDunn, results more from men's poor judgments in spending than from women's decisions, believed to be more in the interests of their families (2009, p. 192). Without detailed evidence, Yunus asserts that women who borrowed financing from Grameen were more likely to use contraception and ultimately have lower birth rates (2007, p. 134). The Grameen model became known for focusing on poor women living in villages purchasing telephones (Gumucio Dagron, 2001).

Although their intentions may be to privilege women in these funding schemes, monitoring done to demonstrate actual percentages given to female-only groups are rare (Sinclair, 2012, p. 142). Moreover, it is difficult to document situations in which husbands send their wives to secure loans that men then use themselves. These programs do little to shift gender norms, when men profit from their wives' successful management of these loans but do not alter their attitudes toward gendered roles or take on more domestic responsibilities. Some argue that family dynamics can be troubled by a shift in economic control, while others see microfinance strategies as better able "to bolster the status of women, and to protect them from abuse, than any laws could accommodate. Capitalism, it turns out, can achieve what charity and good intentions sometimes cannot" (Kristof and WuDunn, 2009, p. 187). Sinclair concludes that the "savings of some of the poorest women of Africa were being stolen" (2012, p. 157), yet their image helps justify the work of these agencies, even when repeating the same photograph of African women across various funding sources (p. 160).

Grameen Bank programs target and fund women intentionally, with the assumption that women act responsibly on behalf of their families and require direct strategic attention. Described as "phone ladies" by Yunus and repeated in public press (in about one-third of these articles), women "buy specially designed cellphone kits costing about $150, each equipped with a long-lasting battery. They then set up shop as their village phone operator, charging a small commission for people to make and receive calls" (Corbett, 2008). These women are able to purchase telephones by taking Grameen loans, which are repaid with interest. As Yunus explains, "(n)ow we have over 200,000 telephone ladies all over Bangladesh," he said of the women, who soon could add Internet service to their entrepreneurial repertoire. "Those who argued these women would not know how to use the technology have been proven wrong" (Hundley, 2006). According to the *Washington Post* (1998), through this system one "telephone lady in each village buys a phone set on credit and pays for it gradually by charging other villagers small sums for each use. Early experience has shown a sizable demand—to communicate with relatives working overseas, to announce family weddings, to check whether a strike in town has ended before bringing produce to market." Women contribute then to development through their participation in local market economies.

While women may be portrayed in more active roles in this field, their agency tends to be defined in terms of borrowing money (about one-third of the articles) or running small businesses (less prominent theme but present). The phrase repeated across narratives is that "most borrowers" are "women." Only two of these sixteen articulated women's roles as moving beyond borrowing money to managing small businesses, such as "starting cellphone businesses in poor villages in Bangladesh and elsewhere" (Hopkins, 2006). Sheikh Hasina, Bangladeshi Prime Minister, describes being "so charmed when they told me that they will give cell phones to the rural poor women for doing business, and they will thus come out of extreme poverty . . . (but that she did) not see any phones in the hands of the poor women, (through) which they are supposed to earn money" (BBC, 2010). Not articulated as subjects but as objects of investments, some articles applauded banks that chose to "invest" in "girls and women" (UNDP, 2012).

Less prominent (19%) than agency being defined in terms of small business entrepreneurship was attention to women's roles in society, "leading the digital world and social movements" (UNDP, 2012). Women's empowerment, implying agency to enact social change, tends to be defined instead in terms of familial roles, such as "send(ing) children to school" (Sappenfield and Trumbull, 2006) or improving "the living conditions for their children" (Hundley, 2006).

Emphasizing more of a victim status, about a third of the articles described women as "poor," and therefore vulnerable. Ved (2011) suggests that "poor women in rural areas . . . would not dare to even enter a regular bank." Using the dichotomy of tradition vs. modernity overly emphasized in development discourse, women's problems are situated as stemming from "religious principles" (Washington Post, 1998). While this is a common refrain in development discourse more broadly (Wilkins, 1997; Wilkins, 2005), women's roles are not seen as merely passive in this development model.

However, while women are constructed as active agents in their local economies, this role is limited to that of small-scale entrepreneurship and to free-market participation. Understanding sustainability as "long-term productivity" (Arun, Heeks, and Morgan, 2004), focusing development on short-term business plans loses sight of changing economic conditions. Moreover, contexts of access and regulation of communication technologies, particularly in terms of telephone and Internet services, might limit entrepreneurial potential. And attributing value to women in terms of their familial roles draws attention away from their economic ambitions. Sustainability of these development strategies can be quite limited when focusing on loans to women for small business enterprise.

COMMUNICATING NEOLIBERAL DEVELOPMENT

This case illustrates ways in which this discourse contributes to neoliberal triumphs at the expense of sustainability, in that through Grameen Bank development, capitalist systems are justified, with women's roles reduced to that of bank borrowers and small-scale entrepreneurs who contribute to free-market systems. As a dominant framework, development concerns with poverty have shifted from a rhetoric of modernization to microcredit, sharing an interest in free-market economies. Rosati explains that the "post-WWII transnational capitalist project of development . . . has unified the world—so-called developed and developing nations alike—in a total subservience to the market, to capital, to the products of our own work" (2012, p. 97). While this discourse benefits global industry, there are serious latent implications ignored by the celebration of the Bank's success.

Given that microfinance has been judged "exceptionally successful in parts of Asia," moreso than in Africa (Kristof and WuDunn, 2009, p. 191), focusing this review of programs in Bangladesh allows us to consider other features of this regional context. Within the region, scholars have raised concerns over the privatization of industries for the public good, such as public sector software (Thomas, 2012) and information technologies (IT; Chakravartty, 2012). In explaining her critique of neoliberal economic reforms to the IT industry in India, Chakravartty explains that as "access to and competence in new information and communication technologies has become a defining feature of modern citizenship in much of the global South, it becomes pressing to understand the complex and ambivalent discourse around public private partnerships and digital inclusion" (2012, p. 65).

In the quest for poverty alleviation, not only does this goal remain elusive, but unintended consequences of these interventions may also cause more harm than good for children, women, and men. Unregulated systems of production offer more opportunities for child labor and exploitation. Women, as the majority group working in garment factories, are killed and injured when avoidable fires and building collapses trap them (AP, 2013). Bangladesh has been the site of several recent tragic fires killing over one thousand people in one incident alone, despite rhetorical proclamations by global clothing companies that they would improve working conditions in the country's US$20 billion per year garment industry exporting mostly to European and US markets (AP, 2013). Women's safety, either in the family or in the workplace, does not feature in this glamorization of small-scale entrepreneurship.

In these projects women's roles become articulated as consumers (purchasing phones) and entrepreneurs (selling and marketing products; securing loans). At a manifest level, an initial review of evaluations suggests that some women have been able to secure income through these programs, and attain what they call empowerment, operationalized here as individually perceived self-esteem; problems in the system though, such as a competitive private market place and insecure economic environment result in many women not being paid full-time, not having benefits, and being paid late if at all (Arun, Heeks, and Morgan, 2004). Moreover, at a latent level, these programs need to be understood within increasing private investment in the region in communications technologies, and in private foundations (such as Gates and Dell) with wealth based in the global communications industries (Wilkins and Enghel, 2013).

The development industry is faced with increasing attention to accountability, toward rigorous assessment of programs (McAnany, 2012). Critical scholarship can offer more comprehensive frameworks contributing to an accountability that positions research in terms of understanding how best to solve social problems, and not just assessments of whether donor interests are being met. Situating discourse analysis within political-economic conditions of global capitalism informs our understanding of development in ways that should contribute to improved praxis. Attention to sustainability demands that we consider accountability over time, in our quest to address critical concerns with global poverty.

NOTE

1. To consider public discourse, this review is limited to global sources identified from a Lexis-Nexus search of digital news archives from 1997 until 2012. Sixteen

articles specifically addressing the "Grameen" Bank work in "Bangladesh," that included attention to "women" and "ICT" or "phone" were found from *Africa News, New Strait Times (Malaysia), BBC South Asia, New York Times (4), Washington Post (2), the New Yorker, International Herald Tribune (2), USA Today, Christian Science Monitor, St. Petersburg Times,* and *Newsweek.* Clearly this does not represent all news published on the topic, but only that is available in English through these archives. Along with published industry discussions (noted in reference list), these sources allow analyses of dominant and challenging themes in this discourse.

REFERENCES

Arun, S., Heeks, R., & Morgan, S. 2004. *Researching ICT-Based Enterprise for Women in Development Countries: A Livelihoods Perspective.* University of Manchester, UK: IDPM.

Associated Press (AP). 2013. Bangladesh Garment Factory Fire Kills 10 People. http://abcnews.go.com/International/wireStory/bangladesh-garment-factory-fire-kills-20506935. Accessed October 8, 2013.

British Broadcasting Corporation (BBC). 2010. Bangladesh PM pledges probe into alleged graft by Grameen Bank. *BBC Monitoring South Asia*, December 6.

Chakravartty, P. 2012. Rebranding Development Communications in Emergent India. Nordicom 33. Special Issue. Communication, Media and Development: Problems and Perspectives, pp. 65–76.

Corbett, S. 2008. Can the Cellphone Help end Global Poverty? *New York Times.* April 13. MM, p. 34.

Dutta, M. 2011. *Communicating Social Change: Structure, Culture, Agency.* New York: Routledge.

Edwards, M. 2010. Why 'Philanthrocapitalism' is Not the Answer: Private Initiatives and International Development. In Kremer M, van Lieshout P, Went, R (eds.) *Doing Good or Doing Better: Development Policies in a Globalizing World.* Amsterdam: Amsterdam University Press, pp. 237–254.

Gumucio Dagron, A. 2001. *Making Waves: Stories of Participatory Communication for Social Change.* New York: Rockefeller Foundation.

Hopkins, J. 2006. Giving your business what it needs to fly; three successful entrepreneurs share what they learned that put their firms on path to prosperity. *USA Today.* October 26. p. 1E.

Hornik, R. & McAnany E. 2001. Theories and Evidence: Mass Media Effects and Fertility. *Communication Theory Special Issue,* 11(4), pp. 208–227.

Hundley, K. 2006. A success story. *St. Petersburg Times.* May 5. p. 1D.

Kang, C. 2010. For the Poor, cellphones can offer lifeline. *Washington Post.* September 8. p. A16.

Kristof, N. D. &WuDnn, S. 2009. *Half the Sky: Turning Oppression into Opportunity for Women Worldwide.* New York: Vintage Books.

Lerner, D. 1958. *The passing of traditional society.* Glencoe: Free Press.

MacFarquhar, N. 2010. Banks Making Big Profits from Tiny Loans. *New York Times*. April 13. http://www.nytimes.com/2010/04/14/world/14microfinance.html?pagewanted=all&_r=0. Accessed October 8, 2013.

MacFarquhar, N. 2010. High Rates beginning to fray lifeline for the poor; Micro-credit's backers see some banks exploiting the needy for large profits. *International Herald Tribune*. April 15. p. 2.

McAnany, E. 2012. *Saving the world: A brief history of Communication for Development and Social Change*. Urbana: University of Illinois Press.

Nederveen Pieterse J. 2009. *Development Theory: Deconstructions/Reconstructions*. London: Sage Publications Ltd.

Newsweek. 1997. Big Bang for a small buck. *Newsweek*. February 17. p. 58.

New York Times Editorial Board. 2013. Bangladesh Takes Aim at Grameen Bank. *New York Times*. August 6. http://www.nytimes.com/2013/08/07/opinion/bangladesh-takes-aim-at-grameen-bank.html?ref=muhammadyunus. Accessed October 8, 2013.

Page, S. 2000. Clinton praises bank's battle against poverty Bangladeshi villagers bused to meet president following security threat. *New York Times*. March 21. p. 6A.

Reed, S. 2002. Private Sector; Helping the Poor, Phone by Phone. *New York Times*. May 26. p. 2.

Richey, L. & Ponte, S. 2011. *Brand Aid: Shopping Well to Save the World*. Minneapolis: University of Minnesota Press.

Rosati, C. 2012. Media and the Democratisation of Privation: Towards New Communicative Geographies of Anti-Poverty. In S. Melkote, (ed). *Development Communication in Directed Social Change: A Reappraisal of Theory and Practice*. Singapore, AMIC, pp. 95–126.

Sappenfield, M. & Trumbull, M. 2006. Big banks find little loans a Nobel winner, too. *Christian Science Monitor*. October 16. p. 1.

Sinclair, H. 2012. *Confessions of a Microfinance Heretic: How Microlending Lost its Way and Betrayed the Poor*. San Francisco: Berrett-Koehler Publishers, Inc.

Sparks, C. 2007. *Globalization, Development, and the Mass Media*. London: Sage.

Thomas, P. N. September 2012. Public Sector Software, Participatory Communications and Social Change. *Nordicom* 33. Special Issue, pp. 77–90.

United Nations Development Programme (UNDP). 2012. ICT and Telecom: Summit Sparks Record-Setting Global Conversation on Power of Tech, Social Media. *Africa News*. September 25th.

Ved, M. 2011. Why Grameen Bank needs a closer look. *New Straits Times* (Malaysia). April 11. p. 19.

The Washington Post. 1998. A different Asian bank. Editorial. *The Washington Post*. February 6. p. A24.

Wilkins, K. & Mody, B. 2001. Reshaping Development Communication: Developing Communication and Communicating Development. *Communication Theory Special Issue*, 11(4), pp. 1–11.

Wilkins, K. 2007. Confronting the Missionary Position: The Mission of Development/The Position of Women. *Communication for Development and Social Change: A Global Journal*, 1(2), pp. 111–125.

Wilkins, K. 2005. Out of Focus: Gender Visibilities in Development. in O. Hemer & T.Tufte (eds.). *Media and Glocal Change - Rethinking Communication for Development.* University of Göteborg: NORDICOM, pp. 261–270.

Wilkins, K. & Enghel, F. 2013. The Privatization of Development through Global Communication Industries: Living Proof? *Media, Culture and Society.* 35(2), pp. 165–181.

Yunus, M. 2007. *Banker to the poor: micro-lending and the battle against world poverty.* New York: Public Affairs.

Chapter 11

Visual Technology, Youth Interventions, and Participation

Two Cases from the Netherlands

Ellen Hommel, Rico Lie, and Anneke Smelik

In this chapter, we explore the impact of visual technology on youth participation. Our research addresses the benefits of the use of visual technology in education and the implications for contemporary adult educators. We empirically ground our research by looking at two specific projects, which use participatory video as an intervention method with youth in order to strengthen participation. In the first part of this chapter, we sketch the characteristics of multimedia with the fast development of new media technologies and the vast impact of media in contemporary culture. In the second part we discuss the impact of technological convergence on youth culture. We will argue that youth participation, defined as active citizens' involvement, is a prerequisite for learning and sustainable change. In the third part we discuss in more detail participatory practices that are embedded and implemented in everyday life in more detail. We take a close look at two cases from the Netherlands, exploring the conditions of participatory video for strengthening youth participation. We focus on the added value of visual media, building on the work by Pat Thompson (2008) who developed criteria that can be used for assessing the value of visuals in research. We also discuss the role that adult professionals—as filmmakers, facilitators and mediators—play in these projects, using the theoretical work of John Dewey (1938) and Paolo Freire (1970, 2005).

THE CONTEMPORARY AGE OF VISUAL MEDIA

It almost goes without saying that visual images are abundant in everyday life (Mirzoeff, 1999; Sturken and Cartwright, 2009). Technological development is the main reason for the visualization of everyday life (Mitchell, 2005).

Visual culture is to a large extent informed by what Walter Benjamin (1968 [1935]) has called *mechanical reproduction*, the technological possibility of infinite reproduction of images. The impact of visual technology started with the invention of the *camera obscura* and later the photo camera, before exploding into the multimedia of today (Crary, 1990). Over the past years we have grown accustomed to a visual overload (Smelik, 2011a). From huge screens in movie theaters to small screens that we can put in our pocket—visuals are everywhere: in the bus, metro or train, on buildings, in our kitchen or bedroom, in the office, on our phones. Even when we visit the hospital we are confronted with images probing our bodies in the form of X-rays, echography, MRI scans, to name just a few possibilities of medical imaging (Smelik and Lykke, 2008). All of these images flow to us day and night from all possible angles. Not only do we look at pictures and watch films, but we also make connections to the world, with words, images and sounds, anytime, anywhere, and to anyone.

In our discussion of the impact of new media technologies, we first define media by its actual hardware, like film, television, the Internet or newspapers, although this definition obviously simplifies the current media landscape (Smith, 2011). However, the hardware is equally made up by its software, by what McLuhan (1967) has famously called "the medium is the me(a)ssage." McLuhan argued that the invention of a medium always brings about fundamental changes in the consciousness of a society. The history of media technology demonstrates that a new medium not only brings new opportunities, but also influences the functioning of the already existing ones (Bolter and Grusin, 1999). The invention of film changed the function of photography from reproduction to capturing the "right" moment, while "the comparison between film and television suggests that television's distinguishing characteristic is its ability to broadcast live, potentially uniting a country or the entire world as we watch the same images together" (Smith, 2011, p. 121).

Smith further explains that with the arrival of a new medium, it is not immediately clear what that medium will eventually achieve in society and what its impact will be on its users. This also works the other way around. A technological invention only becomes a "real" invention when it is widely accepted by society. As Smith (2011) argues, "Although designers created the technology with a clear need in mind, the technology's purpose is never clear till it is placed in a social context" (p. 125). The introduction of a new medium and its possible success or failure, will therefore always involve a negotiation between technological development and social integration.

Cinema and television series have been the two leading *visual storytelling formats* for the most part of last century. They have determined the technological developments and have been driven by the quest for visual fidelity, or what is often called realism. It is important to understand the paradoxes of realism. In a world that is saturated by images, pictures have become complex, ambiguous and contradictory. Hardly anyone adheres to any idea of a naïve mimesis, that is, the idea that the image is a simple copy or mirror of reality. Mitchell argues that in visual culture the image has become increasingly complex, shot through with power, discourses, institutions and technologies (Mitchell, 1994, p. 16). Yet digitalization techniques have been a driving force in a push for realism, paradoxically in genres that have never been realistic in the first place; just think of fantasy, starting from *Jurassic Park* to *Harry Potter* or *The Lord of the Rings*, or computer games ranging from *Lara Croft* to *War of the Worlds*. Not only do fantastic genres and media demand screen realism, spectators and users equally yearn for authenticity.

We need media theory to understand the contemporary desire for authenticity (Smelik, 2011a, 2011b). Both in postmodern theory and in media studies the idea of a "society of the spectacle" has become widely accepted. Though the phrase was coined by the French Marxist Guy Debord in the 1960s to initially condemn the mass media, the term has a much wider meaning in today's society. It has become one of the organizing principles of our economy, society, and everyday life, according to Douglas Kellner (2005). While the spectacle was first related to the realm of fiction, fashion, theme parks and the like, in the last decade the spectacle has become part of the media covering reality. This means that the real on television, whether be it news or reality shows, is often spiced up for easier consumption. Examples are the sensationalist coverage of disasters or the fictionalization of reality shows. Geoff King (2005) introduced the notion of the "spectacle of the real" to refer to the conjunction of spectacle and reality. Reality or performance, true or untrue, original or copy, street fashion or fashion show: the different strands of fact and fiction become entangled in a Gordian knot.

When "the real is no longer what it used to be, nostalgia assumes its full meaning," as the French philosopher of postmodern society, Jean Baudrillard (1983, p. 12), asserts. Nostalgia for the real, or the authentic, is the result of modern mass media turning everything, including reality, into a spectacle. Viewers simply yearn for what is lost: the real. The quest for authenticity can be understood as a resistance to regimes of representation that turn each image into a spectacle or performance. But there is a paradox at work here. As Gilmore and Pine (2007) have argued, people crave real experiences; they want the authentic thing. In their book *Authenticity* they illustrate the contradictions of the experience economy: in an increasingly unreal world, consumers desire something real, original, genuine, sincere—in a word, the authentic—but they have to pay a lot of money to have it organized or produced for them. The public may seek what Gilmore and Pine call the "really real," but—as Walter Benjamin already predicted in 1935—in a media culture of the spectacle, the real and the authentic are lost objects never to be retrieved from the "lost and found" department.

Returning to the question addressed in this chapter—the impact of visual technology on youth participation as active citizens in society—we need to take into account the desire of people today for that lost quality of authenticity and a sense of the real. We argue that one of the important ways of achieving this is by active participation. New media such as video, the Internet, and social media have provided consumers with mobile and accessible media. Media scholars have elucidated on the process of "remediation" (Bolter and Grusin, 1999), the way in which new and old media remediate one another. Others have pointed to the high convergence between forms of the media, for example, Henry Jenkins (2006a). For Smith (2011, p. 127) convergence occurs when a medium becomes a mix of components adapted from other media. Remediation and convergence are not in itself new phenomena, but both have been increased and accelerated by digitalization.

Video first brought moving images into the hands of consumers. When the digital video camera appeared around 1995, and digital editing became accessible at the turn of the century, consumers could become producers and distributors of their own media products. Today, consumers can make their own movie on their mobile handheld devices, combine different apps, edit the material and share it immediately without any loss of quality. With these interfaces, users have access to the world in their hands.

Interface is the significant notion here, implying a technological shift from recording fidelity to interactive communication. This development involves a shift from spectator to user. Whereas films are still characterized by visual spectacle, narrative structures, and psychological credibility, games offer interactive participation (Raessens, 2005). Social media remediate older media such as photography, radio, cinema, and games (Bolter and Grusin, 1999), allowing consumers to create any media experience and share it immediately through the same interface that created it. In today's media landscape participation is almost self-evident (Jenkins, 2006b). The hierarchical relation between the producer of information and its consumers has changed forever. Today, we refer to consumers more as users who can produce, distribute and consume high-quality media messages in large quantities. Access and interaction are two conditions that are applicable for almost all citizens in Western countries. However, for Carpentier (2011) these two conditions are not necessarily equated with participation. Access and interaction do not in themselves guarantee participation or influence. Some scholars have been disappointed in the Internet as a factor of changing society as Curran et al. (2010) explain:

> While it is true that social media provide a pleasurable means of self-expression and social connection, enable people to answer back to citadels of media power and in certain situation (. . .) may support the creation of a radical counter public

> (. . .). Social media are more often about individual than collective emancipation, about presenting self (. . .) rather than changing society, about entertainment and leisure rather than political communication (. . .) and about social agendas shaped by elites and corporate power rather than radical alternative. (p. 180)

The technology of the Internet cannot then be separated from social structures and contexts. The Internet is constituted by the way it is organized, designed, and controlled; indeed, it is vested in powerful corporations, supported by software and hardware restrictions, and regulated by governments (Dijck, 2013).

In this section we have argued that the advent of new media such as the Internet and later of social media, have instigated a cultural shift that has deep implications for everyday life (Boomen et al., 2009). The interface enables the construction of new realities as well as new forms of collaboration and participation. In the context of this chapter it is important to note that young people—at least in the affluent West—actively participate in a landscape of visual and social media. The youth of today is media-savvy (Smelik and Versteeg, 2013), which may entail a changing role for adult educators. In order to assess such changing roles we focus in the following section on the level of participation, where the involvement of citizens is not restricted to institutionalized politics, but where participatory practice is embedded in everyday life (Carpentier, 2011).

YOUTH CULTURE

A long tradition of youth research exists within different disciplines such as sociology, psychology, pedagogy, criminology, and cultural studies. The study of youth is ambiguous. Multiple perspectives exist and interdisciplinary studies are rare. For example, in the study of youth cultures is a "false binary" (Furlong, 2013) between those who study culture as a representation and those who focus on socialization. But nowadays in spite of different approaches they find each other in a focus on antisocial behaviour, the vulnerable and the risks of growing up. As Furlong (2013) argues:

> While early work within the "cultural tradition" (. . .) had a core concern with understanding the dynamics of processes of social reproduction through cultural resistance by young people, all too often contemporary work on cultural dimensions of young people's lives has focused on "spectacular," deviant or risky expressions while sidestepping core sociological concerns relating to the ways in which inequalities are reproduced across generation. (p. 146)

The different disciplines of youth studies recognize that the age of youth can be characterized as a period where young people are trying to escape the pressure of adults to build an independent life. Furlong (2013) describes this period as semi-independent; a state of being in-between that is constructed differently across time and society. *Identity* is a core concept in sociological studies of youth. Different social science disciplines share some core assumptions about the contemporary concept of identity. In relation to the previously mentioned digital and visual culture, one of them is: ". . . that the protraction of the youth phase and the increased complexity of socio-economic contexts have implications for the development of identity" (Furlong, 2013, p. 125). In Western countries, identity is a strongly individualized lifetime project. Du Bois-Reymond (2009) and Arnett (2004) describe this period as a time to experiment with lifestyles, relations, and jobs, a time of optimism, and a time to chase dreams. On the other hand, the extended youth phase makes young people more dependent on adults and for a longer period of time (Cote, 2009). The strongly individualized nature of this phase requires youngsters to make substantial choices. The relevant question is therefore: are young people competent enough to make these choices? Both the negative and positive aspects of the extended youth phase require attention. Furlong underlines this: "Young people may not feel prepared to make choices or may find choices blocked or constrained at times when they want to take action" (Furlong, 2013, p. 10). Despite her positive view, Du Bois-Reymond (2009) also makes it clear that the idea of a "choice biography" does not mean a total freedom to choose whatever you want.

The study of youth can also be seen as an opportunity to analyze the contemporary complex society through cultural resistance or—contrariwise—how young people conform to new or different conditions (*cultural adaptation*). It is young people who are the first to adopt new technologies.

Because of the extended youth phase, young people have time, space, and energy to experiment with the possibilities of the technological convergence. It is therefore interesting to find out how young people integrate the possibilities of the digital and visual culture in their daily lives. Most of the research available seems to be motivated by advertisers for reaching their target group or by institutes who are concerned and suspicious about young people's behavior. Longitudinal research in the Netherlands is conducted by the Institute for Addiction Research and focuses, for example, on Compulsive Internet Use (CIU) as a new disorder. To trace Internet addiction, this organization conducts a monitor about the Internet use of youth. In 2012, young people still used laptops, desktops, and the game console the most. In 2012, 56% had a smartphone and almost 33% a tablet (Rooij and Schoenmakers, 2013). YouTube was the site used most on the Internet and mobile phone. Every two years Stichting Promotie Televisiereclame (SPOT) conducts market-driven research in the Netherlands into the time people invest in consuming different media. In 2010, young people between 13–19 years were identified by SPOT as heavy users of online videos.

We look at this generation who grew up with technology convergence as composed of experts in using technological possibilities. In this chapter we prefer to use the perspective of these young people as (political) citizens who have demands and priorities that differ from those expressed by older citizens. It is therefore important to stimulate involved and responsible citizenship grounded in trust and expertise rather than suspicion. Different forms of video technology have become integrated into all kinds of daily life activities. Young people are, as we have seen, heavy users of online videos, yet we are not sure to what extent young people are aware of aspects of media technology in the broader context of society and the influence of visual culture on representation and issues of realism and authenticity. The question then arises: how can adults support them in becoming active citizens by using visual and social media. We have specified this question by exploring the conditions of participatory video for strengthening youth participation in everyday life.

EDUCATION, PARTICIPATION AND VIDEO

Education, Participation and Visual Literacy

We postulate that participation defined as *active involvement* is a prerequisite for learning and sustainable change. This means that participation and education are inseparable. This is underpinned by two important well-known authors: John Dewey (1859–1952) and Paolo Freire (1921–1997). As we have seen, the classic hierarchical relation between the producer of the information and the user of that information has changed through the convergence of technology. As a consequence, the relation between adult educators and young people has changed too. Greater scope for the co-creation of video productions and participation of young people in them has been allocated in the learning processes and classic critical pedagogical advances in this.

Dewey is one of the most important educational reformers of the twentieth century. Experience, interaction and participation are the kernel of his ideas. In the contemporary age of visual media his ideas about the significance of participatory democracy get new meaning. Young people learn, and learn better, when they are actively involved. Learning is not seen as a direct and passive transmission of knowledge. Education is a continuous process of reconstructing experiences. "Participation is the concept by which Dewey clarifies how humans proceed from individual to social meanings.

Participation is the connecting element between the psychological and the social factors in education" (Berding, 1999, p. 4).

Democracy is more than politics. Democracy is a way of living together. It is about exchanging experiences, interaction and communication. Education can contribute to democratization by being democratic in itself. This idea gives a point of reference for contemporary participatory democracy. For Dewey (1938), knowledge is temporary because a society is always changing and developing warranted assertibility. Participatory processes have a pedagogical and a political intention.

One of the main critiques of Dewey relates to the role of educators in the process of growing up. Sometimes Dewey is misinterpreted about the meaning educators can have. Berding (2011), for example, explains how to transform the ideas of Dewey in today's education where the (adult) educator plays an important part. The adult educator is a mediator, a facilitator with his or her own knowledge who initiates and supports learning processes.

Freire, a philosopher and influential theorist of critical pedagogy, teaches us that all education is politics. For Freire an important condition is to speak the language of the community. Conscientization is his core concept, which is about the development of critical consciousness through action and reflection within a community (Freire, 1970). Literacy within the community is a basic condition. Participants are experts in their own everyday life and this is conditional for the development of a critical consciousness. Experience is therefore central in the process, as it is with Dewey. The educator needs to connect with the language of the participants. The methods used to identify and codify themes, problems and actions need to fit into their everyday lives. Literacy is political and conditional for democracy. We will elaborate on these critical pedagogical concepts of democratization and education in the second part of this chapter.

Literacy in the contemporary age of visual media has a new meaning. It is not only important that citizens can read and write, but it is also important to raise consciousness about the use of new media. New literacy, cultural literacy, media literacy, or visual literacy are new concepts for explaining the complexity of understanding today's society (Buckingham, 2000). Walter Benjamin ends his essay *A Short History of Photography* (1931) with the note: "It has been said that "not he who is ignorant of writing but ignorant of photography will be the illiterate of the future." But isn't a photographer who can't read his own picture worth less than an illiterate?"

We focus on a specific aspect of new media literacy, namely *visual literacy*. Visual interventions that aim to strengthen participatory processes need to strengthen visual literacy as well. Visual interventions made by adults to stimulate participatory processes and the development of young people have increased in the last two decades. Pink (2009) describes through a series of case studies how applied visual anthropology offers methods of research and representation to projects of intervention. In a similar study, Thompson (2008) describes in a similar study how "doing visual research with children and young people" offers diverse communication opportunities, namely:

> "Images communicate in different ways than words. They quickly elicit aesthetic and emotional responses as well as intellectual"
>
> "Some researchers argue that image-based research will particularly allow those children and young people who have difficulty with words an alternative means of expression"
>
> "Other researchers suggest that through the creation of images young people are more ready to express their beliefs and emotions"
>
> ". . . they (children) seem to take pleasure in the process, suggest that they are "getting something" out of their participation . . ." (p. 11)

We will use these opportunities formulated by Thompson in the second part of the chapter to address the added value of a visual medium—in our case video—in participatory youth interventions. But first we will have a look at the history of participatory video as this is the field of the two cases that will be discussed later.

The History of Participatory Video

Conducting visual research with youth often takes the form of participatory video projects. The various motives that are given for using the participatory video can basically be divided into two categories with the following objectives: (1) to collect academic knowledge and (2) to empower people and communities. The first is called "participatory video research" (PVR), the second "participatory video" (PV).

The first-known scientists who introduced the notion of participatory cinema are Judith and David MacDougall (MacDougall, 1975). David MacDougall is an anthropologist, who produced ethnographic films as a negotiable property that lies within a conceptual triangle formed by the (film) subject, the filmmaker and the audience (Banks, 2001). David MacDougall entered actively into the world of the subject, with his camera tied up on his shoulder even when he was not filming, so people got used to this man and his camera. MacDougall renunciated the idea of the observational objective camera and used the camera and his film to provoke dialogue. The researchers showed their film material to their subjects and gave them the possibility to correct and reflect upon it. Through such an exchange, a film can not only provide information the filmmaker or researcher perceives the world, but moreover about how the subjects perceive theirs. In this way film is not to be seen as physical evidence of an event or occurrence, but is a way of communication.

The first-known PVR project, where the researcher hands over the camera to the subject is *Through Navajo Eyes: An Exploration in Film Communication and Anthropology* (Worth and Adair, 1972). This is one of the first research projects that "gave the other a voice." The project set out to research the hypothesis that language creates the everyday reality of a culture. The researchers hoped to create new perspectives on this hypothesis. They asked themselves:

> What would happen if someone with a culture that makes and uses motion pictures taught people who had never made or used motion pictures to do so for the first time? Would they use the cameras and editing equipment at all? If they did, what would they make movies about and how would they go about it? (Worth and Adair, 1972, p. 3)

The detailed descriptions give an insight of the complexity of this process. Their overall conclusion was:

> We feel that this method may offer the fields of anthropology, communication, cognitive psychology, and the humanities a new research technique, another method for getting at the way people structure their own humanness. Our investigations seem to confirm that this method does help to reveal culture as determined and organized by the people within that culture. (Worth and Adair, 1972, p. 253)

The first-known participatory video (PV) project for empowerment is "The Fogo Process" in 1967 (see, for instance, White, 2003). Globalization creates distance between decision makers and the everyday experience of local people. Lack of information can create isolated communities, especially when these communities are already geographical isolated, like the Fogo islands in the North of Canada. The researchers enabled isolated communities to tell their stories and reflect on their problems through film. The locally produced footage is used to communicate with decision makers and other stakeholders, to create dialogue and interaction and promote social change.

"'The Fogo Process' . . . , provides real evidence of how people who have been marginalized by the economic and political structure of the world system can renew and empower their local communities and transform conditions of uneven development" (White, 2003, p. 123). Another project worth mentioning here is the "SkyRiver" project, in which Tim Kennedy participated for almost 30 years with the Alaskan Natives in the Greater Anchorage Area (Kennedy, 2008).

The study of participatory video is multidisciplinary. An example is Kindon's (2003), social geographic research about "Maori's and 'the relationships between place, identity and social cohesion in communities.'"

She uses three different ways of filmmaking: first, the Maori community is trained to produce films by themselves, second, the researchers filmed semi-structured interviews, and, third, the researchers made their own film, based on oral history. The process of filmmaking has the aim to collect academic knowledge as well as to build capacity. In capacity-building activities, active involvement is a prerequisite for learning and sustainable change. For almost thirty years, the organization Educational Video Centre in New York has produced documentary films in school classes with youth at risk. In this period of time, they experienced that producing documentaries with young people has different efforts, such as actively exploring their everyday lives, gathering information and making sense of it, developing multiple literacy skills, creating real work for real audiences to make a difference in their community by going out into it (Goodman, 2003).

PV and PVR have in the tradition of Participatory Action Research (PAR) and is mainly elaborated in the academic fields of development studies and pedagogy. The PV and PVR are diverse and multileveled. They give the unheard a voice, support community building, empower individuals, and can be used to collect data for producing academic knowledge (White, 2003; Thompson, 2008; Pink, 2009). However, the study of participatory video is still fragmented. It is important to frame the multilevel approach of participatory video and take advantage of the opportunities it provides in this contemporary age of visual media. Developing multiple literacy skills and especially visual literacy is an advantage, which is becoming more and more important today, as we have seen in the first part of this chapter. In the second part of this chapter we focus on two organizations who developed PV in the time that video became portable and available for the broader public. With the turn of the century digital editing became accessible too and consumers became producers and distributors of their own video products. It is in this period that the two organizations we studied started to develop their work with PV.

PARTICIPATORY VIDEO IN PRACTICE: TWO CASE STUDIES

Participatory video often refers to handing over the camera, but participation can also be accomplished in the process of design and (decision) making (see, for instance, Lie and Mandler, 2009). The case studies we describe in this section address both forms of participation. The studies focus on two organizations, All About Us Film Factory (AAUFF) and Bosch film. Both organizations conduct exemplary participatory visual projects in the Netherlands. We reflect on these projects to better understand how we can use visual practices in the contemporary age of visual media. We look at which methods are used by the organizations and explore and discuss the added value of the visual medium, as well as the role that adult professionals play in the interventions.

All About Us Film Factory (AAUFF) is a nonprofit organization that in the last ten years has developed a method that stimulates young people to film their personal stories. The approach was developed by doing. AAUFF wants the young people to become active (media) citizens who "influence the media as much as the media influences them" (www.allaboutusfilmfactory.com). The method aims to provide technical skills, which are conditional for the three main objectives: individual empowerment, media literacy and giving young people a voice. In this case study we focus on their first project *Zeedrift* (2003) in the Netherlands. This project can be regarded as forming the basis for the method. For comparative reasons we also look at the project Hey Mum, Tell Me . . . ! (2006).

Bosch film is a profit-making company that produces film projects with social engagement. It produces films and TV programs about current topics, preferably about controversial topics or topics that divide. The directors of Bosch film focus on the human point of view in all their projects. Therefore, it is necessary to collect the stories of the subjects involved. They use professionals for directing, editing, and producing, so they can guarantee the quality that is demanded by broadcasting institutions. They also cocreated implementation with the organizations involved for educational purposes. Their first participatory project was *The Street is Ours*, a directors' reaction to the somewhat negative one-sided way of looking at youth problems by adult professionals. The film was used to stimulate dialogue between decision makers and urban youth. In this chapter we also focus on their project *Find Out!*, an entertainment and educational intervention for young people. It is a television program about sex, drugs, and alcohol. The program is based on the principles of the Users-As-Designers concept. For every episode a group of young people was selected and became responsible for the production of that episode. They also performed in that specific episode.

The Added Value of Using Video

As we have seen in *Doing Visual Research with Children and Young People* (2008), Thompson listed four opportunities for using images with children. If we apply these opportunities to our cases, we can make the following analysis.

"Images communicate in different ways than words. They quickly elicit aesthetic and emotional responses as well as intellectual responses" (Thompson, 2008, p. 11).

In our cases we have seen that some of the films were used to stimulate dialogue among different stakeholders or used in educational settings.

The Street is Ours is a low-budget film, which was shot in two weeks. The preproduction, including the initial free dinners, the script writing, and the acting classes took about a full year. The reactions were strongly divided. Some professional youth workers claimed that this film would not stimulate the empowerment of ethnic minorities, because it deepened clichés. In almost every city in the Netherlands, Bosch film organized a meeting between decision makers and street youth.

AAUFF has also integrated this aspect of bonding and binding in the making process. In their method it is important to create a feeling of belonging and make the participants feel connected. Using video images is a way to make this happen and enable people to get to know each other in a very short time period. Every meeting stimulates self-expression and the young people learn to give honest, critical feedback about the products of others.

"Some researchers argue that image-based research will particularly allow those children and young people who have difficulty with words an alternative means of expression" (Thompson, 2008, p. 11).

We cannot identify this aspect clearly in our cases, because we have no information about the young people who participated in the projects. Moreover we do not know if they had difficulty with words before they started participating in the projects.

"Other researchers suggest that through the creation of images young people are more ready to express their beliefs and emotions" (Thompson, 2008, p. 11).

In our cases the technological opportunities were limited. Nevertheless, by analyzing the films we can see that this medium gives young people the chance to raise their voice, to tell their own stories, and show their emotions. This is especially evident in the individual films which are part of *Zeedrift*, where one of the participants shares his deepest thoughts about loneliness and belonging. In *Hey Mum, Tell Me . . . !* the young girls give us insight into their beliefs and emotions. In the moments where they brainstorm about the concept of motherhood, we can see and feel the importance of motherhood.

". . . they (children) seem to take pleasure in the process, suggest that they are 'getting something' out of their participation . . ." (Thompson, 2008, p. 11).

The aspect of *pleasure* is dominant. To create a film together with professionals is fun and provides directly noticeable results. In the *Find Out!* project the young people indicated that they learned more about how television is made than about the subject itself, but they were all satisfied with their film and proud of being part of the process.

Applying Thompson's opportunities to our two cases showed us that video adds value to youth interventions. First of all, using video enables young people to get to know each other in a very short period of time and at the same time increases the feeling of belonging and being connected. Second, video empowers young people to raise their voices, tell their own story and show their emotions. Finally, using video is fun and this has a positive influence on the quality of the participation of the youth.

Process, Product and Levels of Participation

The two organizations of our cases, AAUFF and Bosch film, conduct participatory projects with *different aims*. These aims are related to the process of making a film and the film as the final product.

AAUFF wants young people to become active (media) citizens. Individual empowerment is based on personal expression of their own life and daily experience. Personal development creates self-confidence. In this process, media literacy is an important aspect. The aim of AAUFF's projects is to raise consciousness about mainstream media. The young people experience in practice the differences between authentic stories and manipulated stories in the mass media. Most of the projects aim to create real stories about young people that differ from the one-sided vision of young people in mainstream media. By this way of working they focus on the individual and group level. The style of the films does not conform to mainstream film formats, but makes use of unsteady and sometimes out of focus shots, rustling sound and jumpy editing. As an immediate effect of this rather rough style, the stories come across as true-hearted, honest and raw. It fits with the authenticity that the makers were looking for; as a young viewer said: "this is really for real." This quote illustrates altered perceptions of reality that relate to the use of visual technology and lies at the heart of the discussion on the changed sense of the real which we provided in the first part of this chapter.

Bosch film produces films and TV programs about current issues, preferably about controversial topics or topics that divide. In all their projects, the directors of Bosch film focus on the human point of view and collect the real everyday life stories and experiences before shooting. From a genuine interest they confront the viewers, reflect on contemporary society and make the unheard heard. They use film professionals to guarantee that the end product can be shown to the mainstream public. Bosch film is more product-orientated than AAUFF. Bosch film's main objective is to produce films and videos, but only when the subjects are really involved. Participation is a must to tell the story lived.

The outcomes of the participatory video processes can be situated at *different levels*. On the microlevel they empower young people and make them literate. They give participants self-confidence and self-respect. On the meso level they stimulate dialogue; horizontally (with peers) as well as vertically (with different multileveled stakeholders). They create community awareness and stimulate processes of bonding and bridging. On the macrolevel the outcome is mainly about representation in broadcast media and giving young people a voice and a face.

Professionals in Education

As we argued above, in contemporary society young people are more familiar with the possibilities of new media than many adults. We also stated that we can look at young people as experts on their own life. The question then arises what these statements mean for the work of professionals in PV processes. The cases in this chapter date back to 2006, but are still illustrative for processes of youth participation. Although the kind of visual technology that young people use has changed throughout the years, the principles of participation and education still have relevance as a point of anchoring for today. Notwithstanding, things have changed rapidly and maybe only a few years ago access to high quality visual technology was only possible for professionals. Interventions by adult professionals were unavoidable for young people who wanted to join PV processes.

Professionals as Facilitators

The adults in the projects are all professionals in filmmaking. In our cases the adult professionals were involved in directing, editing, and producing to guarantee broadcast quality. The increasing possibilities of portable filming and editing were the inspiration for conducting participatory video projects at AAUFF. AAUFF facilitates the use of professional and semi-professional portable equipment. Nowadays high-quality portable equipment is available for almost everyone and this has changed the role of the facilitator.

Filmmaking is a process in which many choices must be made on a technical, ethical, and aesthetical level. The ethical part of the process is an important one, because professional filmmakers are aware of the consequences and the impact that visual technology can have. The professionals who we interviewed all pointed out that it is their responsibility to protect the young people from any possible negative consequences of showing their everyday life experiences and emotions. In one of the *Zeedrift* films, titled *Being Bad is Beautiful* this is most visible when criminal aspects occur in relation to Muslim fundamentalism. Professional integrity is a matter of finding a balance between the realism of the story, the privacy of the participants, and the impact on the viewer. Bosch film's projects are always designed by professionals working in the interests of the young people. In the editing stage, all aspects of the final film have to be in balance, including images, sound, music, titles, etc. They all make (and break) the story. Images are layered and tell a story in different ways. Bosch film's view is that the accountability lies with the professionals.

Professionals as Participatory Educators

In this chapter we briefly discussed the ideas of two important founding fathers of participatory processes, Freire and Dewey. Dewey sees young people, and Freire the oppressed, as experts in their own lives. Conscientization and literacy are at the core of Freire's ideas. Literacy is a prerequisite for equal development and dialogue. In his method he developed a way of consciousness-raising by using the language and daily experience of his participants. We can presume that in this age of contemporary visual media, visual literacy is a prerequisite. In his method he starts with collecting and selecting words and expressions of the participants. Subsequently, he begins the process of codification, whereby the dialogue leads to consciousness-raising. We also saw this at AAUFF and in a less intensive form at Bosch film. Both organizations gather themes from the daily experience of young people and address them as experts of their own life. AAUFF expresses this as follows: "Social discipline and acquiring an open, yet critical understanding of the media leads to a series of highly individual films that truly represent the world of young people, their environment, and youth culture at large" (www.allaboutusfilmfactory.com).

Dewey argues that education is a continuing reconstruction of experience. The adult educator is the mediator and facilitator of the learning processes. In every meeting, AAUFF stimulates self-expression, where the continuing reconstruction of experience is focused on an individual level. The young people create their own content and form. The preproduction stage of *The Street is Ours* and *Find Out!* can also be seen as a continuing reconstruction of experience. In this case it is situated on the level of youth and street youth as a community. During the shooting period the director remains connected to the everyday life of the young people instead of working with fully scripted dialogues. By doing so, he/she creates room for improvisation. In the follow-up manifestations, the dialogue between the community and the stakeholders can also be seen as the reconstruction of experience.

In our view *dialogue is participation* when this dialogue is authentic and when all the participants, young people and adults, are able to reflect on and discuss their own assumptions. We saw that in the study of youth most of the professional meanings are based on suspicion instead of on trust and expertise. Suspicion undermines open dialogue and makes the reconstruction of experience impossible. In our cases, this is most visible in the production process of *Find Out!*. Three nonprofit health organizations were involved in its production. Bosch film's assignment was that the episodes needed to be interesting for young viewers by using peers as designers. It also had to be a health education program about sex, drugs, and alcohol. These conflicting aspirations of dealing with top-down information on the one hand and designing from the bottom-up on the other hand resulted in tension. Such was the case, for example, when the young people wanted to devote an episode to synthetic drugs like XTC. This episode was not in accordance with the policy of the organization and they decided not to make this episode. Another episode, "A Match with a Hangover," was pretested for any possible negative effects before it was allowed to be broadcast. When different stakeholders are involved, the adult professionals need to take heed of the danger that participation does not become a matter of tokenism.

Professionals as Mediators

Professionals stimulate, support and guide young people to construct authentic real life stories. The guiding principle of both organizations is a genuine interest in the experience of the young people. AAUFF uses this for creating the final film in a participatory setting. AAUFF works with a maximum of twelve young people in one group which is an important condition for this intensive participatory process. Bosch film concentrates on reaching mainstream media and participation is deployed in a functional way in this process. In *The Street is Ours* they started their research by inviting young people "from the streets of Amsterdam" for free dinners. Finally, they spoke with more than hundred youngsters. In the Users-As-Designers project *Find Out!*, 109 young people participated as designers. Bosch film reaches a great number of young people with its functional use of participation (see also Bouman and Draaisma, 2006).

In both organizations the professional is a mediator who supports the learning processes. These professionals are focused on the living experience of young people. They approach young people with trust and as experts. The adults use their professional skills and experience in the process of co-creation, to learn together and sometimes to protect the young people for negative consequences of the visual impact.

CONCLUSION

In this chapter we focused on the impact of visual media technologies on the participation of youth as active citizens in society. In the first part of this chapter we discussed how the contemporary age of visual media leads to convergence between the different media. New media such as video, the Internet, and social media have provided consumers with mobile and accessible media.

As a result of the new media technologies, today's consumers can produce, distribute and consume high-quality media messages, which has increased possibilities for participation and collaboration. While this creates huge opportunities for youth participation as active citizens in society, access and interactive communication are not enough to guarantee active participation. In the case of active participation through the use of new media, we argued that it is important to take into account the desire of youth to reconstruct a lost sense of the real and authenticity.

In the second part of the chapter, we argued that visual communication is an elementary component in youth culture today. Young people are the most intense users of interactive media. As they are thus experts in the new media, different forms of media technology have become integrated into all kinds of daily life activities. Yet, we have also argued that young people may not be as critically aware of certain aspects of media technology as we may hope. After all, media education is still a rather new phenomenon in most schools. It is therefore important to redefine the role of professionals in youth interventions that make use of videos or other media. Here, we were inspired by Dewey's ideas about education as an exchange of experiences and Freire's ideas on the necessity of literacy and consciousness-raising.

In the third part of the chapter, we argued that video (and similarly the social media of today) can be characterized by its mobility and accessibility. The added values of visual communication are increasing the feeling of belonging and being connected, empowerment by giving young people a voice and means to show emotions and expressions, and finally the fun factor. From the two cases that we discussed, AAUFF and Bosch film, we can conclude that the benefits of the use of video in education are threefold. On an individual level, participatory video can empower young people, make them more literate in visual communication, and give them more self-confidence and self-respect. On the level of communities, they can stimulate dialogue, both horizontally with peers and vertically with different multi-levelled stakeholders. Participation through the media thus creates community awareness and stimulates processes of bonding and bridging. Finally, on a macro level the product of participatory video gives young people a voice and a face when it is broadcast.

Returning to the main question of this chapter we conclude that using visual technology in education and especially in youth interventions not only has a positive influence on participation, but it also enables professionals to connect to the current age of visual media. In this new age we need to work with redefined notions of reality and authenticity and this has consequences for the work of the professionals. Consideration of this *role of professionals* in youth interventions has become crucial. Professionals add value as committed facilitators and mediators to stakeholders and decision makers. They can support, guide, stimulate, and protect young people in the process of co-creation. As professional facilitators they can build trust with youth and work in the interest of the young people. As mediators, they can serve as the bridging component between individuals, communities, and organizations. When adult professionals learn how to connect with this new generation of experts in visual culture and base their interventions on trust and professionalism, the young people will learn in a participatory way.

REFERENCES

Arnett, J. J. (2004). *Emerging adulthood: The winding road from the late teens through the twenties*. New York: Oxford University Press.

Banks, M. (2001). *Visual methods in social research*. London: Sage Publications.

Baudrillard, J. (1983). *Simulations*. New York: Semiotext(e).

Benjamin, W. (1968 [1938]). The work of art in the age of mechanical reproduction. In *Illuminations. Essays and Reflections*. (pp. 217–251). New York: Schocken.

Benjamin, W. (1931). *A short history of photography*. Retrieved from http://imagineallthepeople.info/Benjamin.pdf

Berding, J. (1999). *De participatiepedagogiek van John Dewey. Opvoeding, ervaring en curriculum*. Leiden: DSWO Press.

Berding, J. (2011). *John Dewey over opvoeding, onderwijs en burgerschap; Een keuze uit zijn werk*. Amsterdam: Uitgerverij SWP.

Bolter, J., & Grusin, R. (1999). *Remediation. Understanding new media*. Cambridge: MIT Press.

Boomen, M. van den, Raessens, J., Lehmann, A. S., Lammes, S., & Schäfer, M. (Eds.) (2009). *Digital material: Tracing new media in everyday life and technology*. Amsterdam: Amsterdam University Press.

Bouman, M. & Draaisma, E. (2006). Find Out!; Een televisie-interventie volgens de user-as-designerbenadering. *Tijdschrift voor Communicatiewetenschap*, *34*(3), 307–318.

Buckingham, D. (2000). *After the death of childhood: Growing up in the age of electronic media*. Cambridge: Polity Press.

Carpentier, N. (2011). *Media and participation; A site of ideological-democratic struggle*. Bristol: Intellect.

Cote, J. (2009). Youth-identity studies: History, controversies and future directions. In A. Furlong (Ed.), *Handbook of youth and young adulthood*. Abingdon: Routledge.

Crary, J. (1990). *Techniques of the observer. On vision and modernity in the nineteenth century*. Cambridge: MIT Press.

Curran, J., Fenton, N., and Freedman, D. (2012). Misunderstanding the Internet. London and New York: Routledge.

Debord, G. (1970). *Society of the spectacle*. Detroit: Black and Red.

Dewey, J. (1938). *Experience and education*. New York: Collier Books.

Dijck, J. van (2013). *The culture of connectivity. A critical history of social media*. Oxford: University Press.

Du Bois-Reymond, M. (2009). Models of navigation and life management. In A. Furlong (Ed.), *Handbook of youth and young adulthood*. Abingdon: Routledge.

Freire, P. (1970). *Pedagogy of the oppressed*. New York: Herder and Herder.

Freire, P. (2005). *Teachers as cultural workers: Letters to those who dare teach (expanded edition)*. Boulder: Westview Press.

Furlong, A. (2013). *Youth studies; An introduction*. Abingdon: Routledge.

Gilmore, J., & Pine, J. (2007). *Authenticity. What consumers really want*. Boston: Harvard Business School Press.

Goodman, S. (2003). *Teaching youth media; A critical guide to literacy, video production and social change*. New York: Teachers College Press.

Jenkins, H. (2006a). *Convergence culture: Where old and new media collide*. New York: New York University Press.

Jenkins, H. (2006b). *Fans, bloggers, and gamers: Exploring participatory culture*. New York: New York University Press.

Kellner, D. (2005). Media culture and the triumph of the spectacle. In G. King (Ed.), *The spectacle of the real. From Hollywood to 'reality' TV and beyond* (pp. 23–36). Bristol: Intellect.

Kennedy, T. (2008). *Where the Rivers Meet the Sky. A collaborative approach to participatory development*. Penang: Southbound.

King, G. (2005). Just like a movie?: 9/11 and Hollywood spectacle. In G. King (Ed.), *The spectacle of the real. From Hollywood to 'reality' TV and beyond* (pp. 47–57). Bristol: Intellect.

Lie, R., & Mandler, A. (2009). *Video in development: Filming for rural change*. Wageningen/Rome, CTA/FAO.

MacDougall, D. (1975). Beyond observational cinema. In P. Hockings (Ed.), *Principles of visual anthropology* (pp. 109–124). The Hague/Paris: Mouton de Gruyter.

McLuhan, M. (1967). *The medium is the massage: an inventory of effects*. USA: Bantam Books.

Mirzoeff, M. (1999). *An introduction to visual culture*. Abingdon: Routledge.

Mitchell, W. J. T. (1994). *Picture theory. Essays on verbal and visual representation*. Chicago: University of Chicago Press.

Mitchell, W. J. T. (2007). *What do pictures want?: The lives and loves of images*. Chicago: University of Chicago Press.

Pink, S. (Ed.) (2009). *Visual interventions; Applied visual anthropology*. New York: Berghan books.

Raessens, J. F. F. (2005). Computer games as participatory media culture. In J. F. F. Raessens & J. Goldstein (Eds.), *Handbook of computer game studies* (pp. 373–389). Cambridge, MA: The Mit Press.

Rooij, A. J. van, & Schoenmakers, T. M. (2013). *Monitor internet en jongeren 2010–2012. Het (mobiele) gebruik van sociale media en games door jongeren (The (mobile) use of social media and games by adolescents)*. Rotterdam: Center for Behavioral Internet Science & IVO.

Smelik, A. (2011a). Mediating memories: The ethics of post-9/11 spectatorship. *Arcadia. International Journal for Literary Studies*, *45*(2), 307–325.

Smelik, A. (2011b). The performance of authenticity. *Address. Journal for Fashion Writing and Criticism, 1*(1), 76–82.

Smelik, A., & Lykke, N. (Eds.) (2008). *Bits of life. Feminism at the intersections of media, bioscience, and technology*. Seattle: University of Washington Press.

Smelik, A., & Versteeg, S. (2013). *Sociale media in de klas. (Social media in the classroom. A survey of the use of social media in secondary schools)*. Nijmegen: Radboud University Nijmegen. Retrieved from www.annekesmelik.nl (under Books).

Smith, G. M. (2011). *What media classes really want to discuss; A student guide*. London and New York: Routledge.

SPOT (2010). *Tijdsbestedingsonderzoek 2010*. Retrieved from http://www.spot.nl/onderzoek/tijdbestedingsonderzoek

Sturken, M., & Cartwright, L. (2009). *Practices of looking. An introduction to visual culture* (2nd ed.). Oxford: Oxford University Press.

Thompson, P. (2008). *Doing visual research with children and young people*. London and New York: Routledge.

White, S. (2003). *Participatory video: Images that transform and empower*. London: Sage Publications.

Worth, C., & Adair, J. (1972). *Through Navajo eyes; An exploration in film communication and anthropology*. Bloomington: Indiana University Press. Retrieved from isc.temple.edu/TNE/introduction.htm

Chapter 12

Analytical Antidotes to Technological Determinism

Learning from (Digital) Citizenship and Participation in Medellín, Colombia

Melissa Brough

Many see the Internet as providing new spaces for civic and political engagement, public dialogue and deliberation, and the tools for the participatory shaping of these very spaces (e.g., Benkler, 2006; Bollier, 2008; Chadwick, 2006; Kelty, 2008; Shirky, 2008). For some, the digital media landscape has "expanded the options and lowered the barriers of several forms of participatory activities . . . transforming the relationships between the political, social and cultural realms" (Kahne, Middaugh, and Allen, 2014, 8). The interactive, networked, horizontal, and in some cases, open architectures of Internet-based communication platforms seem to far exceed the potential of older forms of media to engage citizens of all ages in discourse and action in the public sphere, fueling various practices of "digital citizenship."

Like the concept of citizenship itself, *digital citizenship* is socially constructed and has been variously conceptualized and applied. The term has been used to describe simply, "the norms of behavior with regard to technology use" (Ribble, Bailey, and Ross, 2004, 7), reducing digital citizenship to online etiquette and detaching the concept from notions of citizenship that encompass political or civic participation. A more substantive definition is offered by Mossberger, Tolbert, and McNeal, who argue that digital citizenship is "the ability to participate in society online . . . [the] capacity, belonging, and the potential for political and economic engagement in society in the information age" (2008, 1, 2). Such scholarship links digital citizenship with public participation to analyze the ways in which the Internet can be understood as an expansion of contemporary society, hosting digital public spheres (e.g., Bennett, 2003; Byrne, 2008; Mossberger, Tolbert, and McNeal, 2008).

While understanding the connections between online and offline civic/political engagement is increasingly necessary, it is easy to slip into sweeping, technocentric generalizations that conflate digital participation with meaningful participation in public life, and gloss over important qualifiers such as who participates and how, who sets the parameters of this participation, and how these spaces of participation are structured. Understanding the social implications of using digital communication technologies for social change requires case-based studies and contextual specificity that take these qualifiers into account. In this chapter, I juxtapose two such case studies of the use of digital communication technologies for citizen participation in Medellín, Colombia. I propose four key dimensions for analysis to avoid technological determinism in the analysis of digital communication for social change: horizontality, dialogical communication, openness, and autonomy. These are informed by scholarship from Latin America on participatory communication and from the United States and Europe on participatory culture and digital media and learning, thus bridging bodies of literature that have developed largely in isolation from one another. I also critically analyze the limits of digital participation, and new media literacies that are relevant to communication for social change in the contemporary moment.

DIGITAL CULTURE AND PARTICIPATION IN MEDELLÍN, 2006–2012

Since the mayoral administration of Sergio Fajardo (2004–2007), the city of Medellín, Colombia, has received global attention for its innovations in urban planning, its emphasis on citizen participation through activities like participatory budgeting,[1] and a rapidly growing digital sector. Under Fajardo and his successor, Alonso Salazar (2008–2011)—both of the independent Compromiso Ciudadano party[2]—Medellín's reputation as a hub of narotrafficking and armed violence was eclipsed by an optimistic narrative of transformation and pro-social development fueled by citizen engagement. During this time, the municipal government of Medellín invested heavily in cultivating "digital citizens" as a way to increase social inclusion and public participation, and as part of its broader development agenda to become a "digital city."[3] This was outlined in the government's 2004–2007 Plan de Desarollo (Development Plan) to encourage:

> the construction of a "*Governable and Participatory Medellín*" . . . promoting citizen participation to help improve living conditions; a "*Socially Inclusive Medellín,*" contributing to the quality of education and social inclusion through

> access to technology; a "*Productive, [and] Competitive . . . Medellín*" . . . [to meet] the competitive demands of the global economy; and a "*Medellín Integrated with the Region and the World,*" allowing Medellín to interact with the world through ICTs. (Escobar Arango, 2008, 34)[4]

Major corporate actors in the ICT sector, like Microsoft and Hewlett-Packard, took notice of Medellín's transformation and brought some of their business to what is now being called Latin America's largest technology district (Agencia EFE, 2013). Some even see the city as an emerging "Silicon Valley of Latin America" (Ha, 2012).

Since 2004, the municipal government's development agenda with regard to ICTs has taken form in various ways. Most significantly, through Medellín Digital—renamed Medellín Ciudad Inteligente (Medellín, A Smart City) in 2013—it aimed to integrate digital communication tools and practices into all of the city's public schools and into some of its processes of governance, as a way to increase efficiency and transparency. The government also hoped this would promote digital innovation and entrepreneurship in the new technology sector.

On the other end of the spectrum, grassroots actors, such as citizen bloggers, online community radio, and other alternative media, have also shaped the relationship between digital communication and participation in the city. One example is the Corporación para la Comunicación Ciudad Comuna (referred to as "Ciudad Comuna"), a local, youth-driven, citizens' media project that has used a variety of new and traditional media to meet its social change goals. Ciudad Comuna reflects both Latin America's history of popular communication, as well as innovations in the use of digital media to promote citizen engagement.

Drawing on the two cases of Ciudad Comuna and Medellín Digital, this analysis considers what kinds of (digital) communicative practices help to promote citizen participation and social change in the contemporary moment, and what spaces or structures of communication support these. The case studies focus on the first five years of the initiatives, spanning approximately the time from July 2006 to December 2012; this period correlates with the Fajardo and Salazar administrations, when citizen participation was a key public agenda.[5] The vast differences between these grassroots and institutional initiatives reflect divergent conceptions of digital citizenship and participation, resulting in distinct communicative architectures and practices, all of which have fundamental implications for how we understand social change in the digital age. While the two cases are not comparable in scale, I contrast them in order to highlight these fundamental differences and to problematize the assumption that digital citizenship is inherently (more) participatory.

BRIDGING PARTICIPATORY COMMUNICATION AND DIGITAL MEDIA STUDIES:

Dialogue, Horizontality, Openness, and Autonomy as Dimensions of Analysis

The theorizing of participatory communication that has developed over decades, particularly in Latin America, offers insights that help mitigate against the tendency to simplistically conflate the use of digital communication technologies with participation in public life. At the same time, digital media studies offer insights about the changing dynamics of communication in the digital age that are useful for "updating" some of the thinking on participatory communication. Together, these bodies of work suggest several key dimensions of analysis.

First, participation as it has been theorized across many fields—from participatory communication, to urban planning and development, to media studies[6]—almost always implies or presupposes two-way or multidirectional dialogue (as opposed to one-way transmission of information) as one of its central dynamics. Well before the proliferation of the Internet, proponents of participatory communication advocated horizontal, dialogical modes of communication to promote democratic practices, more locally relevant solutions to development concerns, and greater social equity and inclusion (Bordenave, 1994; Freire, 2003; Singhal and Devi, 2003; Servaes, Jacobson, and White, 1996; Barranquero, 2009; Carpentier, 2011; Gumucio Dagron, 2001; Huesca, 2002; Servaes, 1999).[7] In "Participative Communication as Part of Building the Participative Society," Juan Diaz Bordenave argued that, in Latin America, "We have witnessed the gradual movement from an almost exclusively *diffusion approach* to a *group communication* approach and, more recently, to a *participative approach*, not in the commercial media, but in what is called *popular communication*" (1994, 43, emphasis in the original). What makes it participatory is an emphasis on the active involvement of grassroots and marginalized groups whose voices are not typically heard or valued, helping to amplify these in local decision making and development processes. It usually entails placing the means of media production—or what is referred to in digital media studies as content creation—and distribution in the hands of such groups.

Citizens' media is a form of participatory communication; coined by Clemencia Rodríguez, it refers to "communication spaces where citizens can learn to manipulate their own languages, codes, signs, and symbols, empowering them to name the world in their own terms. Citizens' media . . . purposely cultivate processes of transformation and empowerment in their producers and audiences" (2011, 24, 25; see also Rodríguez, 2001). Rodríguez describes citizens' media as "highly participatory by providing access and space for people to participate in all phases of media production" (Rodríguez, 2001).

In participatory communication, this "access and space" is specifically characterized as relatively horizontal and autonomous.[8] In the contemporary digital context, it might also be characterized as relatively "open," and openness can be analyzed at three levels: content, communicative architecture (including the technologies used), and management. Openness in terms of content may mean being open to public participation in the production and editorial processes, as well as openness to a variety of thematic content and a diversity of viewpoints. In terms of management, openness may mean public participation in key decision making for a communication initiative. Open communicative architecture may mean public administration of a website or the use of open source software for decentralized design, coding, and management, or a communication tool such as theater that has relatively low barriers to entry and relatively high transparency in terms of how it functions.[9] These four dimensions of horizontality, dialogue, autonomy, and openness (see Figure 12.1) can inform critical analysis of efforts to promote participation and social change through communication, whether digital or not.

The capacities of digital communication systems appear highly amenable to more open, autonomous, horizontal, and dialogical exchanges within and across societies. In relation to questions of communication, power, and participation, Manuel Castells (2009) argued that user-generated content (UGC) platforms, which are interconnected with mass communication systems, increase the opportunities for the participation of individuals in spaces of public deliberation and may help to amplify citizen voices. He calls this *mass self-communication*:

> It is mass communication because it can potentially reach a global audience, as in the posting of a video on YouTube, a blog with RSS links to a number of web sources, or a message to a massive e-mail list. At the same time, it is

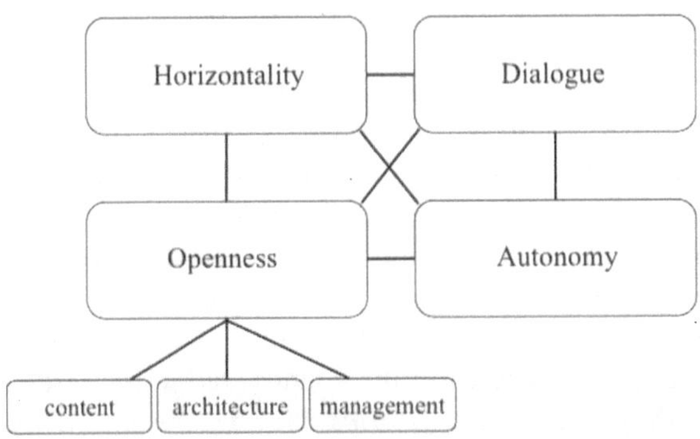

Figure 12.1 Four Dimensions for Analysis of Participation through Digital Communication.

> self-communication because the production of the message is self-generated, the definition of the potential receiver(s) is self-directed, and the retrieval of specific messages or content from the World Wide Web and electronic networks is self-selected. The three forms of communication (interpersonal, mass communication, and mass self-communication) coexist, interact, and complement each other rather than substituting for one another. (Ibid., 55)

Similarly, in the digital media landscape Nick Couldry sees the potential amplification of more/new voices (particularly if they travel across UGC to mainstream media platforms; see Brough and Li, 2013); greater mutual awareness of these new voices; and "new intensities of listening. . . . Governments cannot any longer say they don't hear" (Couldry, 2010, 140–141). These affordances have been amenable to the growth of *participatory cultures*, which Henry Jenkins has characterized as having low barriers to cultural production, artistic expression, and civic engagement, and in which participants are both consumers and producers of the culture (Jenkins, 2006; Jenkins et al., 2006).

Digital technologies themselves do not determine participatory practices, but it is clear that the structural attributes that enable mass self-communication *may* facilitate more horizontal, dialogical, open, and autonomous communication, thus enhancing practices of participatory communication as well as participatory culture.[10] However, expanding the technological capacities for civic and political practices to take place in the digital sphere does not inherently promote a more participatory public culture—that is, one that is characterized by low barriers to participation, dialogic communication, and relatively horizontal decision making about issues of public consequence, and in which the voices of non-hegemonic groups are valued (Brough, 2014). As the following cases illustrate, the particular configurations of (digital) communication platforms for citizen participation vary widely and these differences have significant analytical and practical implications.

CIUDAD COMUNA

The citizens' media initiative known as Ciudad Comuna is based in Comuna 8, one of Medellín's 16 *comunas* (districts), situated at the eastern edge of the city. This collection of mostly poor and working class neighborhoods is best known for the violence it witnessed during the urbanization of Colombia's armed conflict in the late 1990s and early 2000s; and dynamics of violence continue to plague the district. Comuna 8 is comprised of 18 or 34 neighborhoods, depending on whom you ask, some of which are informal settlements of people forcibly displaced from the countryside in the last 15 years. Such areas are known in Medellín as *barrios periféricos* (poor neighborhoods on the periphery of the city), marginalized from the center of the city geographically as well as politically and economically. Of Comuna 8's roughly 150,000 residents, nearly 40% are categorized as living in the lowest of Colombia's six socioeconomic strata, and another 40% in the second-lowest strata. In 2011, approximately 20% had no formal education; 35% had completed primary school, 40% secondary school, and less than 3% had a college degree (SISBÉN, 2011).

Ciudad Comuna described itself as a communication collective and a "*referente* [example] of development and youth participation that supports the construction of social fabric in Comuna 8 and in the city more broadly" (Ciudad Comuna 2010a, 2). They published a community newspaper, *Visión 8*, and website; produced video and other multimedia investigative journalism on a number of locally relevant themes; offered a communication "school" consisting of popular education[11] workshops in photography, audio and video production, and journalism; and held public film screenings and events.

Ciudad Comuna was created in 2006 with the support of a local community-based organization (CBO), and was legally constituted as an independent organization in early 2009. Its four founders ranged in age from 18 to 26 at the time. The eldest and most experienced of these served as an informal leader, though the collective operated primarily through decision-by-consensus. The project grew out of a budget allocation from the 2006 local participatory budgeting process, in which residents of Comuna 8 earmarked certain funds for the development of community media. With support from the CBO, the group made a successful bid to be the recipient of these funds and launched Ciudad Comuna as an independent *colectivo de comunicación* (communication collective).

The content Ciudad Comuna published was decided upon and contributed by local citizens, with an emphasis on human rights, local planning, and development. The group developed a methodology for offering training in multimedia journalism to local community members to broaden the range of contributors. They saw these popular education-style trainings as crucial to reaching their broader goals of promoting critical, proactive civic engagement, particularly among youth:

> Training is fundamental, because even though [in the workshops] we're thinking about media production, it's almost like that's the pretext; because really what we're doing is a process of educating so that youth of the *comuna* find a space of participation, a space in which to propose [ideas/solutions], to speak out, to criticize, to develop their ideas. . . . If someone is able to see themselves reflected in [the] media they will be much more conscious of their context, of their media, of their reality, and from there can begin to be a protagonist in processes of participation that generate transformation.[12]

By 2013, more than 100 local youth had participated in these trainings (Ciudad Comuna n.d.).

The community newspaper *Visión 8* was circulated online and offline; at the time of this research in 2011, 10,000 hard copies of each issue were distributed throughout the *comuna* (see Figure 12.2).[13] It was intended to:

> serve as a tool for the development of . . . *convivencia* [peaceful coexistence], inclusion, integration and dialogue among the communities [of] Comuna 8 in the city of Medellín, and to be the principal resource for the diffusion of [information about] the Development Plan of Comuna 8 in Medellín, which . . . emerged from community initiatives, and in which there is a collective vision from all social sectors in terms of what the future development of the *comuna* should be. (Ciudad Comuna, 2010a, 8)

In addition, *Visión 8* aimed to serve as an archive of the history and collective memory of the *comuna*, and to provide informed critique of local social and political issues—particularly relating to human rights—to prompt civic mobilization and the development of community-based solutions.

Although the newspaper *Visión 8* was their centerpiece, Ciudad Comuna described itself first and foremost as a communication collective with a transmedia approach that bridged multiple on- and offline platforms, determined by their social, rather than technological, goals. In addition to circulating a digital version of the newspaper, the collective also published content on web-based video streaming sites like YouTube and Blip.tv (with the numbers of views ranging up to 2,700 per video[14]), and online photo galleries like Flickr. In 2011, the collective launched an online community radio station called *Voces de la 8* (Voices of Comuna 8). They further disseminated new content on Facebook, the primary online social network used by the majority of their online readers.

In 2011, the collective had nine dedicated members (primarily between the ages of 16 and 22), 12 other occasional participants, and three adult advisors who periodically provided logistical, administrative, and editing support. During my field research, I also observed approximately 35 workshop participants, the majority under the age of 25, with a roughly equal split between males and females. Other community members of various ages also contributed content for publication. Finally, Ciudad Comuna had an editorial board that was open to the public and included volunteers (both youth and adult) from the community who represented a variety of local organizations, associations, community leaders, and informal groups. In 2011, approximately 30 people participated; in 2012, after multimedia efforts to recruit more residents, approximately 130 participated, 50 of whom then contributed content to Ciudad Comuna's subsequent publications.

Figure 12.2 Cover of Visión 8, December 2011. Image courtesy of the Corporación para la Comunicación Ciudad Comuna.

CHALLENGING RELATIONS OF POWER THROUGH HORIZONTAL, DIALOGICAL COMMUNICATIVE PRACTICES

The establishment of Ciudad Comuna challenged existing relations of power within the *comuna* from its inception. While the earliest edition of the newspaper included space for traditional local leaders[15] to publish their own content, Ciudad Comuna rejected the perception shared by many of these power holders that *Visión 8* should serve their information and communication needs as public figures. They argued instead that it should serve the interests of the citizens of the *comuna* and offer in-depth, investigative coverage of issues of local concern determined by a broader range of perspectives. In other words, Ciudad Comuna wanted to make the process of creating the newspaper—a form of media traditionally associated with top-down, vertical information dissemination—into a more horizontal one in which the concerns and voices of a broad array of citizens and community groups could be amplified. Some of the local leaders who had been on the editorial board of the newspaper's first edition stepped down, refusing to participate. However, by reaching out to community-based organizations and other nontraditional leaders in the community, Ciudad Comuna reestablished its community editorial board, broadened the range of participants and content contributors, and strengthened their reputation in the *comuna* as a citizens' media project.

The collective also used horizontality and dialogic communication practices in its organizational decision making. The founding four members of Ciudad Comuna comprised the Junta Directiva (board of directors), which took legal responsibility for the association and made some of its strategic planning decisions. Yet many decisions about programming and other operations were made by consensus in the Ciudad Comuna Assembly, composed of all of the collective's approximately 24 participants, primarily youth under age 25.

WORKING TOWARD OPEN AND AUTONOMOUS SPACES OF COMMUNICATION

Ciudad Comuna's approach included a significant degree of openness across all three levels of content, communicative architecture, and management. Decisions about Ciudad Comuna's thematic priorities for its content were made in the open editorial board meetings to ensure direct input from a variety of community members outside of the collective. Topics covered a wide range of local and more universal concerns, including information on the local development plan and participatory budgeting process; coverage of protests and other mobilizations; coverage of violence and testimonials from victims of violence; and content focusing on human rights, education, health and nutrition, gender equity, and the merits of open/free software.

While Ciudad Comuna received contracts from the municipal government and was funded in part by the city's annual budget through the participatory budgeting process,[16] the collective carefully and strategically maintained their autonomy, and much of the content they published was highly critical of the administration:

> Despite functioning with [some] resources from the participatory budget . . . we've maintained a very critical posture and we've had a lot of autonomy to speak out . . . on the programs that the administration brings to the Comuna 8. . . . In the majority of cases we have more criticisms than points to celebrate. . . . The media we create become tools through which people can denounce situations of exclusion, abandonment [by the state], marginalization. . . . This process of communication may be the only tool that the people of Comuna 8 can use to refute information that comes from the municipal administration or its development reports.[17]

For example, a September 2010 issue of *Vision 8* challenged the municipal government's provision of public services in the *comuna*, criticizing "the myopic vision of the territory of some municipal planning representatives," and arguing that "many sectors are not recognized as neighborhoods and are [therefore] discriminated against in [not receiving] public funding/investment; the territorial divisions recognized by the municipal administration do not coincide with the territorial, social, and cultural realities of its inhabitants" (Ciudad Comuna, 2010b, 6). This was just one example of Ciudad Comuna's unfettered critique of the very government branch that funded it.

At the time, the municipal government was actively promoting the autonomy of Medellín's community media, which developed significantly under the Fajardo and Salazar administrations. This growth was largely due to an increase in funding possibilities through the participatory budgeting process; several *comunas* decided to allocate resources to such projects (De La Uribe, 2011). As of 2011, there were an estimated 70–100 community media projects in Medellín. To help promote their independence, the municipal government required that at least 30% of the funding for a community media project come from sources other than the Mayor's Office.

Ciudad Comuna's commitment to openness and autonomy extended beyond their production processes and content to their communication systems and management; they advocated the use of free and open-source software[18] and open publishing and website administration. In 2010, Ciudad Comuna launched a website to serve as a virtual community message board and mobilization tool, using open-source software designed to facilitate participatory publishing. Any local group (whether a formal organization or informal collective) with an interest in Comuna 8 could register to be an administrator and post content. There was no editing or censoring of contributions, aside from hate speech or other harmful content. Using the site was relatively simple, with low barriers to use. Ciudad Comuna offered technical and training assistance, but did not aim to control the content posted by community groups, instead encouraging them to publish autonomously. The site received approximately 119,000 visits in a year.

While the site was built by a software engineer, it was managed through what Ciudad Comuna described as "*administración participativa*" (participatory site administration); at the time of this research, approximately 20 local groups or organizations had registered as administrators.[19] Ciudad Comuna saw the decision to use this open platform as a method of promoting local participation and autonomous content publishing, and of cultivating a sense of ownership over the site among community groups.[20]

Evidence that the site achieved these impacts was limited by its relatively short lifespan. In August 2012, it was hacked and taken offline, drawing attention to some of the potential challenges of using open-source platforms—namely security and stability. A member of Ciudad Comuna expressed concern that the attack appeared to correspond with similar attacks on other community organization websites related to human rights issues, but this was not verifiable. By the time of this writing, the site remained down, and interest among other entities in the *comuna* seemed to have waned. Ciudad Comuna shifted the focus of their online efforts to their own multimedia website, where they continued to publish information from and about local community groups.[21]

In these various ways—ranging from content to communicative architecture to management—Ciudad Comuna is an example of efforts to create relatively horizontal, open, and autonomous spaces of communication, with the aim of provoking critical dialogue and community engagement in local development. While not every decision regarding each production was made horizontally and not all of its attempts at openness proved sustainable, the collective established structures and an organizational culture the fostered citizen participation. The following case of Medellín Digital illustrates some of the implications of choosing more closed communicative spaces.

MEDELLÍN DIGITAL

In 2007, Fajardo's administration launched Medellín Digital with the support of various public and private institutions.[22] The primary aim of the initiative was to reduce the digital divide by increasing free access to communication technologies and especially the Internet in public locations, as a means to promote greater social inclusion and foster digital entrepreneurship and innovation. Despite having the second-largest economy in Colombia (after Bogotá), Medellín was only the sixth highest in terms of computer use (DANE, 2008). It was ranked second-to-last out of 25 major Latin America cities in terms of having "digital citizens," conceived in this case as a populace with easy access to, and making regular use of, digital communication technologies (Convergencia Research, 2009).

Medellín Digital was promoted as part of Fajardo's overall platform of creating a more governable, inclusive, educated, democratic, participatory, and internationally competitive city (Tabares Higuita, 2009). The initial focus was on infrastructure for citywide, public access to broadband connectivity. In contracts with Microsoft and later Hewlett-Packard, Medellín Digital created computer labs with high-speed Internet connections and Microsoft software called "*aulas abiertas*" (open classrooms) in public schools that were meant to be open to both students and the broader community, and added computers with free Internet access in public libraries, business centers, and government locations. Medellín Digital then began to offer digital literacy trainings, which typically included Internet browsing skills, as well as the use of Microsoft programs such as Word, Powerpoint, and Excel. In some instances, publishing and multimedia skills, such as blogging or basic digital animation, were also taught, primarily to students.

A 2013 global study of public access to ICTs in low- and middle-income countries found that over 50% of survey respondents' use of computers would decrease if public access venues were not available (Sey et al., 2013). For many, free public access venues provided their first opportunities to use computers and the Internet; for the majority, those that offered technical support and trainings were the most important places for developing their computer and Internet skills (Ibid.). There is evidence, then, that Medellín Digital's public access and training initiatives would help to cultivate more "digital citizens."

Medellín Digital also developed online platforms and content, including what they described as "a Network of Portals that supports the construction of the digital city through the development of content and services for citizens, and active participation through the use of ICTs."[23] The portals included one for education; entrepreneurship; culture; the network of public libraries; and one with municipal government content including e-government services, such as online tax payment forms and procedures, and vehicle registration. Online "tools for the digital citizen" included services such as filing for passports and citizenship cards; digital literacy trainings that included a guide to browsing and searching the Internet, and opening an email account; and information on employment and educational opportunities.

In 2011, Medellín Digital reported approximately 140,000 registered users for all of their portals, though it is unclear how many were repeat users and what uses they made of the sites.[24] The portals received a total of approximately 9 million cumulative visits per year.[25] In an interview in early 2012, a senior Medellín Digital staff person expressed that they were still hoping for a greater number of unique users, and that interactivity in the portals was not yet as significant as hoped, which the staff person associated in part with technical glitches in the portals, as well as with a lack of a culture of *produsage*[26] in Medellín—the latter justification being debatable, given the significant number of citizens' media projects.

To familiarize citizens with the portals, and also to reach some of the outermost neighborhoods of the city that still lacked broadband access, Medellín Digital launched a mobile computer lab on a bus (El Circular Digital) that traveled what they called a "route of inclusion." Through basic training in computer and Internet use, as well as e-government and other services available through the portals, the Circular Digital aimed to help "convert [users] into digital citizens" (Medellín Digital, n.d.).

What is meant by "inclusion" and "digital citizens"? In 2008, Medellín was awarded first place for "e-Inclusion" by the fifth Premio Iberoamericano de Ciudades Digitales (Latin American Digital Cities Prize). Organizers explained that, "for a Digital City to have meaning, citizens should have digital access, that is, they should have rapid, easy access to diverse telecommunications networks" (XII Encuentro Iberoamericano de Ciudades Digitales, n.d.). This vision of a digital city reflects a technocentric view, conflating access to digital technologies with development and citizenship, and ignoring the cultures and practices of use that imbue such technologies with their meanings and ultimately determine their impacts.

Similarly, in a 2009 study of 25 Latin American cities by Motorola and Convergencia Research, Medellín was ranked first (tied with five other cities) for "participation," defined as "actors using technology to voice opinions, participate, propose ideas and express themselves. In the case of public administration websites, it consists of citizen participation, whether entering into discussion forums or surveys about municipal services, or the existence of blogs and other Web 2.0 tools" (Convergencia Research, 2009, 21). The specific modalities of citizen participation, and the goals or ideal outcomes of this participation remained undefined; the existence of Web 2.0 tools was taken to represent participation—with little detail of the purpose, actual use, and outcomes of using those tools. As is frequently the case in technocentric (and particularly in commercially driven) discourses of new media and democratization, participation was simplistically conflated with technology use and hollowed of its political and cultural significance, as well as of any discussion of the communicative practices and the structures of power within which it is embedded.

Medellín Digital was not a participatory communication initiative, but citizen participation in public life was one of its stated aims—specifically, "the active participation [of citizens] through the use of ICTs."[27] Carpentier (2011) articulates an important structural distinction, however, "between participation on the one hand, and access and interaction on the other," which is, "the key role that is attributed to power, and to equal(ized) power relations in decision-making processes. . . . For instance, in the case of digital divide discourse, the focus is placed on the access to media technologies (and more specifically ICTs)" (Ibid., 131, 129), rather than on cultivating modes of participation that help to equalize relations of power. An analysis of Medellín Digital through the lenses of horizontality, dialogue, openness, and autonomy thus reveals a very different conception of participation than that illustrated by Ciudad Comuna.

Medellín Digital's Communicative Practices and Spaces

According to Medellín Digital, the culture portal provided "participatory tools" and a "space for dialogue."[28] Yet Medellín Digital's portals tended to function most often as channels for mass communication, with far fewer opportunities for horizontal, mass self-communication or dialogue. Medellín Digital's e-government services offered citizens information and provided instruments to ease bureaucratic processes within existing government services, but they did not enable two-way or multidirectional communication with government representatives, nor did they create spaces of citizen participation in government. For example, the portals offered very little in the way of digital tools to support Medellín's participatory budgeting beyond providing limited, basic information about the process. This is surprising, since participatory budgeting was one of the city's flagship examples of citizen participation, and elsewhere such practices have been enhanced with digital communication technologies (see, e.g., Peixoto, 2009).

By 2012, the existing data on Medellín Digital's programs showed that the initiative had the greatest impact in the realm of education, but indicators remained limited to the number of people trained and the skill levels attained, rather than how these have affected the use of the technologies and participation in civic/political life. While the *aulas abiertas* were meant to be a resource for the broader communities in which they were located, the vast majority (approximately 91%) of users and trainees were students. Independent researchers found that the broader community was the least impacted by Medellín Digital's programs, and only about 9% of the *aulas abiertas* users were non-student community members ("Proyecto Medellín," 2011; Grupo de Trabajo de Medellín Digital, 2008, 2; Marulanda, 2012). Further, approximately only one-third of teachers who had benefitted from trainings reported being able to independently use the e-government services intended by Medellín Digital to enhance "digital citizenship."[29]

The extent to which Medellín Digital succeeded in converting Medellín into a "digital city" with more "digital citizens" is thus debatable, further complicated by the lack of clear and consistent measures of what either concept entails. The initiative increased free connectivity to the Internet, primarily for small business people, teachers, and students using the *aulas abiertas* at public educational institutions (Patiño Lemos and Vallejo Gómez, 2011). Between 2007 and 2011, Medellín Digital helped ensure that nearly all formal public educational institutions began providing free Internet access points, as did dozens of government and other public locations, and all public libraries. In 2011, 54.3% of people aged five and older in Medellín used the Internet, up from 31.6% in 2007 (though, of course, this rise in use cannot be exclusively attributed to Medellín Digital). However, by 2011 Medellín's ratio of Internet users per capita nearly exactly paralleled that of the Colombian population as a whole, signaling that Medellín did not yet stand out in Colombia for having an exceptionally "digital" populace (DANE, 2011; Internet World Stats, 2012). Further, these measures say very little about the actual digital capacities and activities of the population, and nothing about the extent to which this enhanced participation in public life.

Through various trainings on the use of online tools for creating UGC and digital animation, Medellín Digital encouraged content production by users, rather than just the consumption of information, though not to the extent of citizens' media projects such as Ciudad Comuna. A study of Medellín Digital's impact by Patiño Lemos and Vallejo Gómez (2011) found that 26% of student respondents were able to independently publish their own content online using social media platforms; the ability to use such Microsoft software as Word and PowerPoint was much higher, at 74%. Only 12% of small business people surveyed reported having skills to produce and publish content online (Patiño Lemos and Vallejo Gómez, 2011).

Carpentier (2011) distinguishes between content-related participation and structural participation, which includes participation in the management and policy of communication/media organizations. Structural participation entails greater participant control over the system (not just the content), and as such, it can be considered a more "maximalist" version of participation (Carpentier, 2011, 131). The ways in which decisions were made about Medellín Digital's platforms were relatively vertical and closed. Structural participation in the development and/or management of Medellín Digital's platforms would have allowed participants some degree of control over the very terms and spaces of their participation, in addition to its content. Medellín Digital, however, allowed far less content-related *and* structural participation than did Ciudad Comuna. The *aulas abiertas* were managed to an extent by local committees at each site, rather than centrally by Medellín Digital. Yet overall, decisions about Medellín Digital's content and strategies remained centralized, with limited participation by members of the communities they served. Despite using social media tools that enabled multidirectional communication, Medellín Digital's platforms and content tended toward unilateral, vertical communication and the delivery of information, rather than horizontal communication, openness, and dialogue.

Medellín Digital's portals and trainings were not being used extensively, if at all, by local community groups like Ciudad Comuna, partly because of Medellín Digital's use of proprietary Microsoft platforms, which conflicted with the groups' goals, philosophies, and practical needs.[30] However, these groups did use other proprietary, closed platforms like Facebook and YouTube, due to the size of the social networks that could be reached. Facebook (the most popular social networking platform in Medellín and in Colombia in general at the time of this research) was used as ubiquitously as mobile phones for connecting with friends and family, but was also used in a very public way to share information; to publicize services, concerts and other events; and to mobilize people to participate in public events. The community groups' online content production and distribution took place on these more mainstream commercial platforms, or on their own blogs or websites. It is important to note that neither Facebook nor YouTube are open source or open administration sites, and therefore not participatory in the "maximalist" sense described by Carpentier (2011)—i.e., participants do not have control over the architecture, administration, and management of the media platform. Yet the widespread use of these platforms by groups that practice participatory cultures illustrates that the two are not necessarily mutually exclusive; limited structural participation in communication platforms may constrain participatory cultures in some important ways, but it does not preclude their existence, nor does it necessarily bar their content-related participation. Further, structural participation is not always of interest to (or feasible for) all users.

We might compare Medellín Digital's approach to government policies in Brazil, where the concepts of digital inclusion, free culture, and social networking "dominate attitudes and values around new media technologies" (Horst, 2011, 437). In addition to civil society and activist efforts, much of this has been driven by the relatively unique stance taken by the Brazilian government to promote the widespread use of minimal-configuration computers with free/open source software, as well as more open copyright policies to promote remix culture. The resulting policies have prioritized the use of nonproprietary technologies, such that Brazil is now:

> a heavy user and promoter of open source software, predominantly Linux, which is used by governments, universities, telecenters, and even supermarkets. . . . The direct support of open source platforms throughout the government institutions was structured as an explicit route to counteracting the hegemony of global corporate actors such as Microsoft and, in turn, to saving money which could be used in other areas of need in Brazil. (Horst, 2011, 447, 454)

These policies and the broader cultures surrounding new media use in Brazil helped to position the country at the avant-garde of digital culture in Latin America. Horst's assessment points to the ways in which this challenged certain relations of power (particularly between global corporate actors and the government of a developing country) and promoted a more participatory digital culture.

Through this lens, Medellín Digital's partnership with Microsoft sustained the dependency of the government on a global corporation, and by basing digital literacy trainings on Microsoft platforms, it mitigated the development of a more open digital culture. Opting for proprietary software was often perceived to be a "safer" and more secure choice (depending on how safety and security was conceived), but it had significant implications for participatory culture and (digital) citizenship.

Pradip Thomas (2012) similarly sees public sector use of open-source software in India as having the potential to bolster public participation in development. At the same time, this may not always be practical or attainable, and Thomas notes that, in much of the world, there is a growing trend toward using mixed software systems in e-governance (Ibid., 80). Of course, there are limits to what technology choices and government policies alone can accomplish toward promoting more participatory public cultures. It is the *combination* of technology choices, policies (and their political economic outcomes), user practices, and cultures of production that together may or may not challenge existing inequalities. But the choice to use free/open source software carries with it broader political and cultural meaning; and Medellín Digital's choice of proprietary software signaled limitations to their commitment to participation.

DIGITAL CITIZENSHIP AND DIGITAL LITERACIES

Medellín Digital's "digital citizens" were seemingly those who could use Microsoft platforms and the Internet for business and basic e-government services, while Ciudad Comuna aimed to cultivate critical, vocal, and engaged citizens using a variety of media. Medellín Digital's web portals used proprietary software, and its content was primarily produced centrally, while Ciudad Comuna used some open-source software and published content produced by local community members. Medellín Digital's initiatives tended to prioritize information delivery and technological access while maintaining the status quo in terms of relationships of power and communication between the government and citizens. Ciudad Comuna's initiatives tended to prioritize the participation of community members in determining content priorities, and saw citizen participation as central to the process of knowledge production for community development (both problem and solution identification), challenging knowledge produced centrally by the government. Ciudad Comuna did not use the rhetoric of digital citizenship as explicitly as Medellín Digital, in part because they were less technologically driven and chose their communication formats (not all of which were digital) based on their broader social goals. However, their commitment to open software, user-generated content, and horizontal/participatory administration implied a more proactive sense of digital citizenship.

The technological architecture of the communicative spaces in which digital citizenship is enacted structure its possibilities. Chris Kelty argues that, "We proliferate ways of governing ourselves and others as we proliferate these tools, technologies, platforms or networks—and in the process change what it means (and meant) to interact, vote, and protest" (2013, 29). Kelty points out that, too often, the agency to participate in social media platforms is collapsed with the agency to participate in governing; the case of Medellín Digital offers an example of this conflation.

Taking the critique further, Christian Fuchs argues that the notion of digital citizenship is part of a misplaced Web 2.0 ideology, in which too much agency is ascribed to users of UGC platforms while ignoring structural limitations to participation, such as proprietary software and corporate control over the architecture of participation; he refers to content-related participation in such sites as "pseudo-participation" (2011, 279). However, being entirely dismissive of the potential for content-related participation in such spaces to yield meaningful social change outcomes denies users any agency and overlooks the significance of storytelling and "challenging codes" (Melucci, 1996) to forge a more participatory and just society; in their own way, then, such critiques also run the risk of technological determinism. Yet drawing critical attention to the limits to structural participation in digital spaces is crucial.

The differences in how (digital) citizenship and participation are conceived result in distinct approaches to educating and training citizens *to participate*; this is apparent in the different kinds of trainings offered by Ciudad Comuna and Medellín Digital. Patiño Lemos and Vallejo Gómez (2011) found that Medellín Digital's digital literacy trainings of teachers increased their uses of some computer and Internet technologies, but the researchers remained skeptical that the trainings enabled the teachers to move beyond basic, instrumental adoption of these technologies in the classroom, to thinking about new pedagogical approaches that digital technologies may enable—e.g., those proposed by Jenkins et al. in *Confronting the Challenges of Participatory Culture: Media Education for the 21st Century* (2006). The authors of this white paper, funded by the MacArthur Foundation's Digital Media and

Learning initiative, argue that the *participation gap*—"the unequal access to the opportunities, experiences, skills, and knowledge that will prepare youth for full participation in the world of tomorrow" (Ibid., 3)—is as much a concern as the digital divide. To close this gap, "new media literacies" training for digital participation should include the capacities of networking; play and problem-solving; performance and improvisation; simulation; remixing content; multitasking; working with distributed cognition and collective intelligence models; the ability to evaluate the sources and credibility of content; transmedia navigation (the ability to follow content across multiple platforms); and the ability to respect and negotiate with differing perspectives and norms (Ibid.).

In Patiño Lemos and Vallejo Gómez's (2011) study, less than one-third of the students surveyed reported doing collaborative work in class using digital communication technologies. Further, they found that despite the goal of having the *aulas abiertas* run for (and with oversight by) the local community, "very few of them have a program that goes beyond offering [basic] training in computer use, to respond to a need in the community."[31] Other critiques of Medellín Digital's approach to digital literacy included the curriculum being largely limited to the use of proprietary Microsoft software, and the lack of development of critical literacy of the technology, including an understanding of its architecture. This would have enabled users to question and understand "how code constitutes the conditions of possibility for different norms, models, and practices of online citizenship, along with the capacity to resist and reshape—to *hack*, if you will—the prevailing terms and conditions of cybercitizenship if they no longer serve our needs" (Longford, 2005, 69, emphasis in original). In this view, digital literacy must include the ability to critically evaluate (and change) the architecture of online communicative spaces, something that neither Medellín Digital nor Ciudad Comuna had yet accomplished among their users/participants. This is a high standard for digital citizenship, and one that, to date, unfortunately remains accessible only to a small number of techno-elite worldwide. Yet these critiques help to maintain an analytical focus on the fundamental questions of power, participation, and influence that are at stake.

Critical questioning of the dynamics that structure relations of power within society in the digital age has often been secondary, if not lost in the widespread rhetoric of participation among techno-utopianists globally. Critical new media literacy would enable the user to do more than just use or consume software and computer technologies at the surface level, but also to analyze them in context, make proactive choices about their use, have the capacity to produce and circulate his or her own content to express voice and engage in dialogue, and, ideally, to shape the very design of these communicative spaces.

THE LIMITS OF (DIGITAL) PARTICIPATION

Along the dimensions of participation discussed above, Ciudad Comuna's approach was clearly stronger, but it functioned on a much smaller scale than Medellín Digital, struggled with resource constraints, and had much more precarious sustainability (as is common in grassroots communication projects). Without the basic infrastructure and access that initiatives like Medellín Digital set out to provide, the project's ability to promote citizen participation and social change would be constrained. Ciudad Comuna should benefit from citywide Internet access provided by Medellín Digital, as more people are more easily able to access and contribute to their content online. Ultimately, participatory public culture may require *both* institutional and grassroots efforts to promote it, ideally working synergistically (Brough, 2014)—more so than had been the case in Medellín at the time of this research. While the Fajardo and Salazar administrations could have done much more to develop this synergy, their pro-community media stance was an important first step in this direction.

Numerous scholars have pointed out that social disparities and segregations offline are frequently mirrored online (e.g., Boyd, 2012; Everett, 2008; Hargittai, 2007; Hargittai and Walejko, 2008; Jenkins, 2009; Jenkins et al., 2006; Seiter, 2008; Watkins, 2009). This, of course, has significant implications for the possibilities of digital citizenship. The online explosion of user-generated content in sites like Twitter, YouTube, and Facebook does not necessarily mean that more, and different, voices are being *heard*—and when they are heard, the question is by whom and with what effect. A flood of expressions can amount to noise that ultimately drowns out voices—particularly minority voices—rather than amplifying them (Juhasz, 2009; Couldry, 2010).

Neither of the cases examined here were particularly successful at promoting public dialogue in online spaces. Ciudad Comuna's communicative architectures were more explicitly designed to promote both online and offline dialogue. Yet a year after launching the community website, the majority of user posts were simple information-sharing, rather than dialogue. One interviewee from Ciudad Comuna attributed this to digital culture being still relatively limited in their community in terms of online interactivity and dialogue, and thought that more training might be necessary to encourage it. Catalyzing online dialogue around issues of public concern remains a challenge in many instances.

Participatory communication practices like Ciudad Comuna have been critiqued for being too localist in focus and unscalable (Carpentier, 2011; Cooke and Kothari, 2001; Gumucio Dagron, 2009; Heeks, 1999; Hickey and Mohan, 2001; Huesca, 2002; Tufte and Mefalopulos, 2009). Institutionally driven programs like Medellín Digital typically aim to find efficient, scalable models of communication for development, but participatory communication is often a locally specific practice, one that needs to be flexible and responsive to contextual particularities (Gumucio Dagron, 2001; see also Quarry and Ramirez, 2009; Servaes and Malikhao, 2005; Tufte and Mefalopulos, 2009). Alfonso Gumucio Dagron has argued that, "In a more reasonable framework for development, scale would have to do with linking communities with similar issues of concern and facilitating exchanges, instead of multiplying rigid models" (2001, 11). This suggests a network mentality—for example, supporting local digital communication projects like Ciudad Comuna to network with others across the city—as opposed to a mentality of replicability and scalability.

A particular challenge for Medellín's locally based communication practices like those of Ciudad Comuna, however, is posed by the complex dynamics of armed violence that surround them, and how these play out across different relations of power in the *comuna*. While a detailed analysis of this is beyond the scope of this chapter, participatory communication in the context of armed violence raises a particular set of challenges, as participants strive to maintain their autonomy amidst sometimes dangerous power struggles (see Rodríguez, 2011).

CONCLUSION

Digital communication technologies are widely seen as affording more citizen participation than traditional media, and yet this is clearly not guaranteed; it depends greatly on the surrounding cultures and practices of use (including digital literacies), as well as on the structure and management of the communication platforms. In this chapter, I considered two very different approaches to promoting citizen participation through digital communication technologies, during a time in Medellín when such participation was an explicit goal of both state and community-level actors. Contrasting these approaches along the dimensions of horizontality, dialogic communication, autonomy, and openness helped avoid technological determinism, and clarified the ways in which participation and digital citizenship were constructed and enacted by each, with very different social implications.

Of course, Medellín Digital's and Ciudad Comuna's approaches were not necessarily contradictory or mutually exclusive; effective public participation requires access to information, and therefore, media/communication systems that promote connectivity and serve the function of information delivery remain important. Ideally, top-down/transmission approaches and bottom-up/participatory approaches would work in synergy, rather than as a binary and in relative isolation. Further, the question of what types of communicative architecture and degrees of openness best support participatory cultures has no simple, single answer—as exemplified by the fact that groups like Ciudad Comuna advocate free/open source software, but also make use of proprietary platforms like Facebook and YouTube.

In practice, it is much easier to capitalize on rhetoric of citizen participation through digital inclusion than it is to actually accomplish it. Efforts to expand civic and political participation through the digital sphere must be linked to communicative practices and spaces characterized by horizontality, dialogue, openness, and autonomy. They must value and promote diverse voices, and connect these to political or other structures for influencing public decision-making. They also necessitate the development of critical new media literacies. Access to digital technologies is clearly only one of several important factors in communication for social change in the contemporary moment.

This analysis began to bridge the literatures on participatory communication, digital media studies, and participatory culture. Much more could be done on this front, both to "update" theorization of participatory communication for the new media context, and to ground the relatively young scholarship on digital media in a longer history of thinking about participation.

In September 2013, Medellín Digital was re-launched as Medellín Ciudad Inteligente (Medellín, a Smart City). Led by a new director, the initiative appeared to be taking steps toward promoting more citizen participation, including adding a strategic focus on *gobierno abierto* (open government) and "effective communication between citizens and authorities" (Portal Educativo de Medellín, 2013). Efforts were also made to develop the synergy between its institutional approach and grassroots communication efforts, through activities such as offering trainings for community radio (Tibaduiza, 2013). As digital citizenship initiatives such as this continue to evolve, applying the dimensions of analysis advocated here would help to mitigate against the technological determinism that ultimately undermines the power of citizen participation.

NOTES

1. Participatory budgeting is a process through which citizens can take part in decision making over part of a public budget through a series of meetings between residents and government authorities.

2. Compromiso Ciudadano translates as both "Citizen Engagement" and "Citizens' Commitment."

3. The concept of a "digital city" was popularized in the mid-1990s, becoming a buzz phrase for certain commercial ICT systems and municipal e-government

initiatives. More recently, rhetoric has shifted toward the less technologically deterministic concept of "smart," rather than digital cities.

4. This and all other translations in this chapter are mine, unless noted.

5. Field research on these cases was carried out in 2011. For an in-depth discussion of the reasons behind this public agenda of citizen participation, see Brough (2014).

6. For a useful overview, see Carpentier, 2011, chapter one.

7. For historical examples such as the Bolivian miners' network of radio stations founded in 1949, see Gumucio Dagron (2001) and Huesca (1995).

8. *Autonomous* here refers to being independent of censorship by government, corporate, or other traditional power holders, or by other funding sources.

9. For a history of the concept of openness in the development of the Internet, see Kelty (2008).

10. Of course, these same technologies can be used to facilitate practices of censorship and surveillance (Andrejevic, 2009; Morozov, 2011); the social practices around these technologies shape their impact. For further discussion of the differences and similarities between the concepts of participatory communication and participatory culture, see Brough (2014).

11. Popular education is a pedagogical philosophy, advanced by Paulo Freire among others, to empower traditionally marginalized groups. It emphasizes horizontal processes of learning and decision making, dialogical communication, and the valuing of local knowledge.

12. Group interview with members of Ciudad Comuna, Medellín, May 13, 2011.

13. At the time of this writing, *Vision 8* was available online at http://issuu.com/ciudadcomuna and http://www.ciudadcomuna.org/presentacion-vision-8.html (accessed February 2014).

14. These numbers were as of March 2014.

15. Local leaders included representatives of the Juntas Administradoras Locales (Local Administrative Boards, or JAL) and the Junta de Acción Comunal (Comuna Action Boards, or JAC).

16. Ciudad Comuna also contracted their production services to other entities to raise funds for their operations.

17. Group interview with members of Ciudad Comuna, Medellín, May 13, 2011

18. According to Graham Longford, the open source coding movement is defined by collaborative and inclusive design; openness and transparency of source code for modification; universal access to software; and nonrestrictive licensing, all of which are in direct response to monopolistic software firms, particularly Microsoft (2005, 89).

19. In 2011, groups posting to the website included several neighborhood organizations, a youth network, and committees of the Local Development Plan, among others.

20. Carpentier cites the Indymedia network as another example of open publishing "to allow for these more intense forms of participation in the media. As Mamadouh (2004:486) explains, 'The main characteristics of open publishing is [sic] that volunteers maintain the software and the public act as publishers'" (Carpentier, 2011, 122).

21. www.ciudadcomuna.org

22. This included the foundation of the public utility company Empresas Públicas de Medellín (EPM), the public, regional telecommunications company UNE EPM, the national Ministries of ICTs and Education, and a private university. To my knowledge, community-based organizations were not significantly involved, a fact that some of my interviewees criticized.

23. www.Medellíndigital.gov.co, accessed August 4, 2012.

24. Interview with senior staff person of Medellín Digital, Medellín, January 12, 2011. Since simply accessing the portals required potential users to register, there were likely fewer regular/repeat users.

25. Figures provided by Medellín Digital in September 2012. Medellín Digital's portal for government services accounted for only approximately 240,000 (about 3%) of those visits; the education portal accounted for the vast majority of visits (over 7 million, cumulatively). However, study data from 2009–2011 showed that less than half of the students surveyed knew of Medellín Digital's education portal (Patiño Lemos and Vallejo Gómez 2011). Further, these figures tell us little about the actual usage of the portals.

26. Axel Bruns (2008) coined this term to refer to the increasingly interconnected practices of user-led production and consumption of online content.

27. www.Medellíndigital.gov.co/ciudadanodigital/Paginas/default.aspx, accessed August 7, 2012.

28. www.Medellíndigital.gov.co/nuestraestrategia/Paginas/contenidos.aspx, accessed August 7, 2012.

29. Interview with researchers in the Educación en Ambientes Virtuales (Education in Virtual Environments, EAV) research group at the Universidad Pontificia Bolivariana, Medellín, January, 25, 2011; Patiño Lemos and Vallejo Gómez (2011).

30. Various interviews with members of Ciudad Comuna and other communication collectives, Medellín, 2011.

31. Group interview with EAV researchers, Universidad Pontificia Bolivariana, Medellín, January, 25, 2011.

REFERENCES

Agencia EFE. 2013, July 10. "Medellín to build LatAm's largest technology district." *GlobalPost*. Accessed November 11, 2013. http://www.globalpost.com/dispatch/news/agencia-efe/130710/Medellín-build-latams-largest-technology-district

Andrejevic, M. 2009. *iSpy: Surveillance and power in the interactive era*. Kansas: University of Kansas Lawrence.

Barranquero, A. 2009. "Latinoamérica: La arquitectura participativa de la Comunicación para el cambio." *Dialogos de la Comunicación* 78: 1–14.

Benkler, Y. 2006. *The wealth of networks: How social production transforms markets and freedom*. New Haven: Yale University Press.

Bennett, W. L. 2003. "New media power: The internet and global activism." In *Contesting media power: Alternative media in a networked world*, edited by N. Couldry, and J. Curran, 17–37. Lanham: Rowman & Littlefield.

Bollier, D. 2008. *Viral spiral: How the commoners built a digital republic of their own.* New York: The New Press.

Bordenave, J. D. 1994. "Participative communication as a part of building the participative society." In *Participatory communication: Working for change and development,* edited by S. A. White, K. S. Nair, and J. R. Ascroft, 35–48. Thousand Oaks: Sage Publications.

Boyd, D. 2012. "White flight in networked publics?: How race and class shaped American teen engagement with MySpace and Facebook." In *Race After the Internet*, edited by L. Nakamura, and P. Chow-White, 203–222. New York: Routledge.

Brough, M. (2014). "Participatory public culture and youth citizenship in the digital age." PhD diss., University of Southern California.

Brough, M., and Z. Li. 2013. "Media Systems Dependency, symbolic power, and human rights online video: Learning from Burma's 'Saffron Revolution' and WITNESS's Hub." *International Journal of Communication 7*: 281–304.

Bruns, A. 2008. *Blogs, Wikipedia, Second Life, and beyond: From production to produsage.* New York: Peter Lang.

Byrne, D. N. 2008. "The future of (the) 'race': Identity, discourse, and the rise of computer-mediated public spheres." In *Learning Race and Ethnicity: Youth and Digital Media,* edited by A. Everett, 15–38. The John D. and Catherine T. MacArthur Foundation Series on Digital Media and Learning. Cambridge, MA: The MIT Press.

Carpentier, N. 2011. *Media and participation: A site of ideological-democratic struggle.* Chicago: Intellect, The University of Chicago Press.

Castells, M. 2009. *Communication power.* Oxford: Oxford University Press.

Chadwick, A. 2006. *Internet politics: States, citizens, and new communication technologies.* Oxford: Oxford University Press.

Ciudad Comuna. 2010a. "Proyectos comunicativos Comuna 8." Medellín: Author.

Ciudad Comuna. 2010b, September. "Un nuevo mapa para la comuna." Memoria y territorio [special edition]. *Visión 8*, 6–7.

Ciudad Comuna. n.d. "Impacto Social." Accessed September 2013. www.ciudadcomuna.org/impacto-social.html

Convergencia Research. 2009. *Ranking Motorola de Ciudades Digitales.* Buenos Aires: Convergencia Research & Motorola.

Cooke, B., and U. Kothari, eds. (2001). *Participation: The new tyranny?* London: Zed Books.

Couldry, N. 2010. *Why voice matters: Culture and politics after neoliberalism.* Thousand Oaks: Sage Publications.

DANE. 2008, September 2. "Indicadores básicos de tecnologías de información y comunicaciones (TIC)." Accessed October, 2011. www.dane.gov.co.

DANE. 2011. "Indicadores básicos de tecnologías de información y comunicación–TIC para Colombia." Accessed October, 2013. http://www.dane.gov.co/files/investigaciones/boletines/tic/bol_tic_2011.pdf.

De La Uribe. 2011, December 4. "La experiencia de los medios comunitarios: Entrevista con Nectalí Cano." *De La Uribe.* Accessed September, 2013. http://delaurbe.udea.edu.co/delaurbe/index.php/ciudad/2414-la-experiencia-de-los-medios-comunitarios

Escobar Arango, D. 2008. "Medellín digital, una ciudad para la inclusión y el salto a la sociedad del conocimiento." *Revista Colombiana de Telecomunicaciones* 15(47): 32–36.

Everett, A. 2008. "Introduction." In Learning race and ethnicity: Youth and digital media, edited by A. Everett. The John D. and Catherine T. MacArthur Foundation series on digital media and learning. Cambridge, MA: MIT Press/MacArthur Foundation.

Freire, P. 2003. *Pedagogy of the oppressed* (M. Bergman Ramos, Trans.). New York: Continuum. (Original work published in 1970).

Fuchs, C. 2011. *Foundations of critical media and information studies*. New York: Routledge.

Grupo de Trabajo de Medellín Digital. 2008. *Medellín Digital: Documento de inversión para el año 2008*. Medellín: Alcaldía de Medellín.

Gumucio Dagron, A. G. 2001. *Making waves: Stories of participatory communication for social change*. New York: The Rockefeller Foundation.

Gumucio Dagron, A. G. 2009. "Playing with fire: power, participation, and communication for development." *Development in Practice* 19(4–5): 453–465.

Ha, A. 2012, October 11. "Co-working space Espacio launches to turn Medellín into a startup hub." *TechCrunch*, Accessed September 30, 2013. http://techcrunch.com/2012/10/11/espacio-medellin/

Hargittai, E. 2007. "Whose space? Differences among users and non-users of social network sites." *Journal of Computer-Mediated Communication* 13(1): 276–297.

Hargittai, E. and G. Walejko. 2008. "The Participation Divide: Content creation and sharing in the digital age." *Information, Community and Society* 11(2): 239–256.

Heeks, R. 1999. "The tyranny of participation in information systems Learning from development projects." Development Informatics Working Paper Series. Manchester: Institute for Development Policy and Management, University of Manchester.

Hickey, S., and G. Mohan. 2001. *Participation: From tyranny to transformation? Exploring new approaches to participation in development*. New York: Zed Books.

Horst, H. 2011. "Free, social, and inclusive: Appropriation and resistance of new media technologies in Brazil." *International Journal of Communication* 5(6): 437–462.

Huesca, R. 1995. "A procedural view of participatory communication: Lessons from Bolivian tin miners' radio." *Media, Culture & Society* 17(1): 101–119.

Huesca, R. 2002. "Tracing the history of participatory communication approaches to development: A critical appraisal." In *Approaches to development communication*, edited by J. Servaes. Paris: UNESCO.

Internet World Stats. 2012. "Colombia: Internet usage indicators and country profile." Accessed October, 2013. http://www.internetworldstats.com/sa/co.htm

Jenkins, H. 2006. *Fans, bloggers, and gamers: Exploring participatory culture*. New York: New York University Press.

Jenkins, H., K. Clinton, R. Purushotma, A. Robison, and M. Weigel. 2006. *Confronting the challenges of participatory culture: Media education for the 21st century*. Chicago: MacArthur Foundation.

Juhasz, A. 2009. "Learning the five lessons of YouTube: After trying to teach there, I don't believe the hype." *Cinema Journal* 48(2): 145–150.

Kahne, J., E. Middaugh, and D. Allen. In review. "Youth, new media, and the rise of participatory politics." In *From voice to influence: Understanding citizenship in the digital age*, edited by D. Allen and J. Light.

Kelty, C. M. 2013. "From participation to power." In *The participatory cultures handbook*, edited by A. Delwiche and J. J. Henderson, 22–31. New York: Routledge.

Kelty, C. M. 2008. *Two bits: The cultural significance of free software*. Durham: Duke University Press.

Longford, G. 2005. "Pedagogies of digital citizenship and the politics of code." *Techné* 9(1): 68–96.

Marulanda, G. 2012. "Medellín crece en conectividad." *TeleMedellín*. Accessed September 5, 2012. http://www.youtube.com/watch?v=yZTRxLU57x0.

Medellín Digital. (n.d.). "El Circular Digital recorre Medellín." Accessed August 4, 2012. www.scribd.com/doc/36067722/circulardigitalMD.

Melucci, A. 1996. *Challenging codes: Collective action in the information age*. New York: Cambridge University Press.

Morozov, E. 2011. *The net delusion: The dark side of Internet freedom*. New York, NY: Public Affairs.

Mossberger, K., C. J. Tolbert, and R. S. McNeal. 2008. *Digital citizenship*. Cambridge: MIT Press.

Patiño Lemos, M.R., and M. Vallejo Gómez. 2011. "Caracterización de prácticas con TIC por actores diferenciados en cuatro comunas de la ciudad de Medellín: Un abordaje para el reconocimiento de la apropiación tecnológica." *Revista Q*, 5(10): 31.

Peixoto, T. 2009. "Beyond theory: e-Participatory Budgeting and its promises for eParticipation." *European Journal of ePractice* 7: 55–63.

Portal Educativo de Medellín. 2013. "Medellín evoluciona de Ciudad Digital a Ciudad Inteligente." Accessed September, 2013. http://www.medellin.edu.co/sites/Educativo/Paginas/Medellin_evoluciona_a_Ciudad_Inteligente.aspx

"Proyecto Medellín Digital va viento en popa." 2011, March 21. *El Tiempo.com*. Accessed April, 2012. http://www.eltiempo.com/colombia/medellin/proyecto-medellin-digital-va-viento-en-popa-con-la-promocion-de-las-_9048660-4

Quarry, W., and R. Ramirez. 2009. *Communication for another development: Listening before telling*. London: Zed Books.

Ribble, M. S., G. D. Bailey, and T. W. Ross. 2004. "Digital citizenship." *Learning & Leading with Technology* 32(1): 6.

Rodríguez, C. 2001. *Fissures in the mediascape: An international study of citizens' media*. Cresskill, NJ: Hampton Press.

Rodríguez, C. 2011. *Citizens' media against armed conflict: Disrupting violence in Colombia*. Minneapolis: University of Minnesota Press.

Seiter, E. 2008. "Practicing at home: Computers, pianos, and cultural capital." In *Digital youth, innovation and the unexpected*, edited by T. McPherson, 27–52. The John D. and Catherine T. MacArthur Foundation Series on Digital Media and Learning. Cambridge, MA: The MIT Press.

Servaes, J. 1999. *Communication for development: One world, multiple cultures*. Cresskill, NJ: Hampton Press.

Servaes, J., T. L. Jacobson, and S. A. White, eds. 1996. *Participatory communication for social change*. London: Sage.

Servaes, J., and P. Malikhao. 2005. "Participatory communication: The new paradigm?" In *Media and glocal change: Rethinking communication for development*, edited by O. Hemer and T. Tufte, 91–103. Buenos Aires: NORDICOM/CLACSO.

Sey, A., C. Coward, F. Bar, G. Sciadas, C. Rothschild, and L. Koepke. 2013. *Connecting people for development: Why public access ICTs matter*. University of Washington. Accessed November 2013. https://dlib.lib.washington.edu/researchworks/bitstream/handle/1773/22753/Global_Impact_Study_2013.pdf?sequence=2

Shirky, C. 2008. *Here comes everybody: The power of organizing without organizations*. New York: Penguin.

Singhal, A., and K. Devi. 2003. "Visual voices in participatory communication." *Communicator* 37: 1–15.

SISBÉN. 2011, December. "Sistema de identificación y clasificación de potenciales beneficiarios de programas sociales: Perfil socioeconómico Comuna 8 Villa Hermosa." Accessed September 2013. http://www.medellin.gov.co/irj/go/km/docs/wpccontent/Sites/Subportal%20del%20Ciudadano/Planeaci%C3%B3n%20Municipal/Secciones/Indicadores%20y%20Estad%C3%ADsticas/Documentos/Estad%C3%ADsticas%20Sisb%C3%A9n/SISBEN%20III%202011-Diciembre/SISBEN%20Diceimbre%202011-certificada/PerfilSocioeconomicoCertificado_diciembre2011/Perfil%20SISB%C3%89N%20Comuna%2008_Certificada_19122011.pdf

Tabares Higuita, L. X. 2009. "Políticas públicas en tecnologías de información y comunicación en la Alcaldía de Medellín: Período 2001–2007." *Revista Q* 4(7): 26.

Thomas, P. N. 2012. *Digital India: Understanding information, communication and social change*. Thousand Oaks: Sage Publications.

Tibaduiza, I. 2013. "Medellín Ciudad Inteligente le apuesta a la apropiación de la tecnología a través de talleres que propician espacios de aprendizaje y trabajo colectivo." Accessed September 2013. http://estrategia.medellin.co/estrategia/?p=339

Tufte, T., and P. Mefalopulos. 2009. *Participatory communication: A practical guide*. Washington, DC: The World Bank.

Watkins, S. C. 2009. *The young and the digital: What the migration to social network sites, games, and anytime, anywhere media means for our future*. Boston, MA: Beacon Press.

XII Encuentro Iberoamericano de Ciudades Digitales (n.d.). "Qué es." Accessed September 14, 2012. http://www.ciudadesdigitales2011.com/que_es.html

Chapter 13

Video Technologies and Participatory Approaches to Peace

From Technological Determinism to Self-Empowerment and Social Change. An Experience from Kenya

Valentina Baú

INTRODUCTION

Through the application of participatory approaches to video-making in the aftermath of a conflict, a process of appropriation of video technology occurs within communities who want to tell their own story. The possibility of using video to facilitate communication between enemy groups, victims and perpetrators, can open the path toward social change and sustainable peace.

In contrast with theories of technological determinism, rather than lessening human agency, technology here empowers people by offering a channel for dialogue. A participatory approach also allows for flexibility in the use of the medium; this prevents the product from being defined by its own technology's structure. Within this context, technology is neither to be viewed as an external element that is forced upon a group, nor as a value-free asset as it is still being regarded in certain practices, but as a new tool that allows communities affected by conflict to take ownership of their own development through communication.

First, this chapter introduces the concept of participatory communication, explaining how communication theory has evolved from a two-step information flow to the recognition of a process of dialogue that takes place between sender and receiver. This section is then followed by an illustration of the relationship between technology and social change, with a presentation of the main theories that were developed in this area. This is accompanied by a discussion that questions these approaches and puts forward an alternative view of how communication technologies can bring about new communication practices that lead to social change in local communities, particularly through the application of a participatory approach to media productions. Views from some of the main authors on the use of the medium of video in this context are also offered.

This theoretical discussion forms the preamble to the analysis of a participatory video project implemented in the Rift Valley of Kenya in the aftermath of the 2007/2008 Post-Election violence. The inquiry I have conducted through my field research demonstrates how participants in the Valley have actively engaged with the technology, recognizing its potential for communication and its actual impact. It also brings to light the perception that participants had of their video-production activities and subsequent screenings as a contribution toward the achievement of sustainable peace in their communities. Despite the wide scholarly debate surrounding the notions of participation and technology that has made both terms the object of a large number of different frameworks, the presentation of the findings arising from the interviews I carried out with those who were directly involved in the participatory media productions in Kenya show how a more focused approach can be utilized in contexts of peacebuilding and social change at the community level. This chapter demonstrates how the introduction of a new communication technology within a social system that was fractured by violence, can establish a new form of dialogue and lay the foundations for sustainable social change when the participation of its members becomes a key component of the media content creation process.

PARTICIPATORY COMMUNICATION AS DIALOGUE

I adopt the definition of participatory communication provided by Singhal and Devi (2003: 2):

> [. . .] a dynamic, interactional, and transformative process of dialogue between people, groups and institutions that enables people, both individually and collectively, to realise their full potential and be engaged in their own welfare.

This definition emphasizes how the concept of dialogue is central in order to initiate a meaningful and sustainable process of development. As we are frequently reminded in the literature, when we want to engage with participatory communication, the starting point for all activities must be the community. This is where problems are discussed and an interaction between different members occurs (Servaes and Malikhao, 2005). The key component is the process of creation, sharing and contestation of meanings and values. What is integral to the dialogic model is the context in which communication occurs, where economic, social, political and cultural factors play an important part. It is through these factors that meanings are shaped. Figueroa et al. (2002) explain the importance of dialogue as a process in which each party tries to understand not only what others believe, but also their own beliefs. This represents a form of collective action, where both sides undergo the experience of shaping and reshaping their convictions. The authors point out how this very process can be regarded as particularly complex when the parties involved have a long history of conflict.

TECHNOLOGY AND SOCIAL CHANGE

Matthewman (2011) summarizes the main approaches that have shaped the philosophical sphere of technology through the following categories:

- Antihumanist approach (or *technological determinism*): technology is the force that shapes social relationships and leads to social change. Humans are secondary to technology, which exists outside of social relations and structures the social.
- Humanist approach (or *social constructionism*): this is the stance taken in sociological theory, where humans are at the center. Technologies cannot be abstracted from the social context. Who takes ownership and control of them is a key issue. They are about action, knowledge, organization and systems.
- Post-humanist approach: scholars from this school go beyond the technology/society dilemma and address the aspects of collective production and interaction.

According to technological determinism, technological change follows its own imperative and acts autonomously from any social pressure. What society is left to do is to limit that "culture shock" that may be caused by the new technology (Murphie and Potts, 2003).

McLuhan's dictum "the medium is the message" suggests that the key aspect of the media is not their content, but the way they shape our perception of the world. Baudrillard (as quoted in Murphie and Potts, 2003, 15) takes these ideas further taking on a pessimistic stance. With his "theory of simulacra," the media become a model for society's behavior, perception and knowledge of the world. According to Baudrillard, what we see, hear and learn from the media, becomes our reality.

It is the elements that characterize a society that determine how technologies are assimilated, employed and even controlled. Murphie and Potts (2003) look at this as the "social shaping of technology" approach:

> Technological change [. . .] is not, by itself, productive of social change. Instead, the direction of change is a product of the particular alignment between the technological possibilities and the society and culture that exists. (Hill, 1989, as cited in Murphie and Potts, 2003, 21)

The claim that the intrinsic properties of technologies do not determine their use, but rather the way people choose to use them, is a refutation of technological determinism that turns technology into a neutral element (Murphie and Potts, 2003).

A "weaker" approach to determinism, on the other hand, argues that technology is one among other factors that can open the path toward social or behavioral change. Chandler (1996) argues that, particularly at the individual level, technologies cannot be regarded as "content-free" and neutral, as they are always biased toward one type of use. For Bernardelli and Blasi (1995) a correlation exists between technological change and cultural transformation. More specifically, the authors contend that new media technologies can transform the communicative relationships between individuals. This view is useful to understand the processes that can take place when communities are enabled to use their own media to open a dialogue between groups.

At the same time, in order to understand how communication and related technologies can contribute to the achievement of social change, it is useful to provide a definition of "social change" that is relevant to this research's context. Gujit (2007) defines social change as a "collective process of conscious efforts to reduce poverty and oppression by changing underlying unequal power relationships" (p. 4). From this definition we can fathom how dialogue is an essential component of this course of action, whose complexity requires the creation of mechanisms developed in a bottom-up fashion.

Hamelink (1994) maintains that communication can both disempower and empower people. In order for empowerment to occur, communication will contribute to processes such as access to knowledge to facilitate informed decision making, sharing of experiences and dialogue. At the same time, the author recognizes that one of the dangers of empowerment lies in the forms of dependency that are generated between the "empowering subject and the agent-to-be-empowered" (Hamelink, 1994, 141). He argues that what is necessary is to transform the "strategies to give voice to the voiceless [in]to strategies by which people speak for themselves" (Hamelink, 1994, 142). In his view, the goal becomes that of *self-empowerment*, where people define their own space, their identity and the solutions to their problems. From a communication stance, self-empowerment can take place through different approaches, one of which is what Hamelink calls *people's media*. These are defined as media that:

> [. . .] are owned and controlled by the powerless with the intention to empower themselves. They are a direct confrontation of the disempowered with the dominant communication structure. They select different themes and discourses, they tell their own stories and articulate their fears and dreams in the cultural idiom of their own communities. Ordinary people become the active participants in the communication process and actively design their own meaning systems instead of passively consuming the meaning system of the prevailing social order (Hamelink, 1994, 142).

These concepts are similarly stressed by Servaes and Malikhao (2005), who recognize that—in order for development to occur—it is a must that people participate both in the planning and in the production of media content. This is also made clear in the approach promoted by UNESCO, which takes forward the Freirian notion of dialogic communication as a tool for liberation from oppression. While Freire considered group dialogue as an instrument for subjugated people, UNESCO's work emphasizes instead how participation should comprise the public's inclusion in all the stages of communication systems' production.

The concept of "participation" in media technologies has been widely explored by scholars such as Carpentier (2007), who claims that in order to understand the potential that these technologies offer toward the achievement of a participatory culture, we need to consider the political-ideological contexts that accompany them. Participation is a contested notion that cannot yet be located on a stable ground due to the minimalist-maximalist debate that encapsulates it. The minimalist view focuses on more symbolic forms of participation, according to which the (limited) function of the media is that of promoting communality. The form of media participation presented by the maximalist view, on the other hand, has a higher degree of intensity and sees nonprofessionals engaging in processes of content-creation. This view also appears to be in line with Hamelink's idea of *people's media*; both of these concepts provide some of the basis for the discussion that follows on the use of video with conflict-affected communities in Kenya.

PARTICIPATORY VIDEO AND THE KENYAN EXPERIENCE

Differences from medium to medium lay in the ways these amplify or enhance some experiences, while inhibiting or minimizing others. Hence, in the process of media selection it is important to bear in mind what or how much needs to be emphasized (Chandler, 1996). Rahim (1994) recognizes how the local use of small electronic media has a strong potential to engage in dialogic communication, while Gumucio-Dagron (2003) maintains that the use of any technology can be tailored to the enhancement of a community participation process when the element of participatory communication is applied to it.

Also Carpentier (2007), again, recognizes that participation is not guaranteed by a specific technology, but by the way one chooses to adopt it. Each technology is characterized by a number of different power dynamics, and their "participatory potential" derives from their organizational structure.

White and Patel (1994) make the case for video as an effective facilitator for participation at the grassroots level. Video can be an instrument for sharing knowledge and experience, and eventually for bringing about social change. Messages produced by local people in local contexts are more accepted than the ones that arrive externally, and this type of communication nourishes a more "human-centered" development.

In her work on participatory media, Rodriguez (1994) defines a participatory video production

> [. . .] as one in which a communicator and a community engage in video production. That is, the whole process of producing a video message (planning, script writing, shooting, viewing, editing, and showing) is shared by both, the "communication expert" and the community. (Rodriguez, 1994, 150)

One of the main characteristics of these productions is their adoption of a horizontal rather than vertical use of video. This type of activities can initiate an exchange of communication at the individual level, but also from group to group and village to village. The messages created by way of the video camera can be taken from one place to the other and shared directly with the people, without the need of negotiating with a gatekeeper. Hence, direct communication can take place between groups without a third party control over the process.

Another important aspect is that the barriers of traditional literacy do not stand when it comes to video, as it can be easily used and controlled. White and Patel (1994) stress how this technology represents an individual's instrument for learning in the context of group activities. In addition, it can be used as a mirror to look at and understand someone's own actions and behaviour. Braden (1999) emphasizes how video offers a way to record discussions and actions without the need of being verbally literate. This way, poor and marginalized people are given a means to communicate, which does not depend on their ability to read and write. Since participants are both the object and subject of production in participatory video processes, the self-discovery of their "filmic lexicon" allows us to gain insight and knowledge of their worldview (Odutola, 2003).

Wheeler (2011) also coins the expression "extended language" to refer to the ability of video to record not only words, but also people's expressions and emotions. In essence, the viewer is exposed to both verbal and nonverbal communication forms that pass on a greater amount of information.

These considerations are particularly important when we observe the fact that most rural people are highly unlikely to see themselves on a screen throughout their lifetime. Video, however, can make this experience possible. This powerful event can influence a person's perspective to the extent of beginning to reconsider and adjust their frame of reference in accordance to the environment they live in.

RESEARCH IN THE RIFT VALLEY

When the results of the December 2007 presidential elections were announced, widespread civil violence broke out in Kenya. This manifested itself mostly in two different forms: the first one was a general protest against the contentious outcome of the vote across the country, particularly in urban centers, where intense backlashes occurred between the police and the militia groups that had materalized as a demonstration of dissent. The second one was a violence fought along ethnic lines, which took place mainly in the Rift Valley. Here, the Kalenjin youths—supporting the Luo candidate Odinga—turned against the Kikuyus, who endorsed their tribesman Kibaki, reelected as the country's president.

The Rift Valley is a largely rural area inhabited predominantly by the tribes of Kalenjin, Kikuyu, Luo, Luhya, Kisii, Kamba and other smaller groups. The area is characterized by low literacy and high unemployment, particularly among youth, and it was the worst affected during the ethnic violence.

Through the research I conducted there at the end of 2012, I have collected the views of ten young people between the age of 19 and 35 and belonging to different ethnic backgrounds (mostly from the formerly warring tribes of Kalenjin and Kikuyu), who took part in the production of participatory videos, facilitated by *Insightshare*, in the aftermath of what has come to be known as the 2007/2008 Post-Election Violence. The youths had both taken part in and been victims of the violence, and they had been the beneficiaries, among others, of a peacebuilding program implemented by international NGO Mercy Corps. The *Local Empowerment for Peace* (LEAP) program addressed the intertribal tensions that had remained as a legacy of the fighting. It incorporated a Sports for Peace component that took the form of a program in itself (*LEAP Sport*), through which young people in the Valley were given the opportunity to develop conflict management skills and employ them in intertribal sports, cash-for-work and income-generating activities.

In the videos produced,[1] whose idea stem from the purpose of program evaluation, twelve of the program beneficiaries tell their story of the conflict and how they had overcome their anger and hatred toward the (former) warring tribe as a result of their participation in the NGO's initiatives.

Steering away from the original purpose that drove the film-making project, my research concentrated specifically on the contribution that the participation of local communities' members in the production of those videos had in transforming the conflict between tribes; it also looked at the impact that the videos themselves had on the communities through the various screenings that were held. In the context of this chapter, I am analyzing participants' answers primarily from the perspective of their experience with the video technology, and how such experience can be recognized to be as one of the paths to change and peace after a conflict.

While the specific cultural settings play an important role, what is regarded as context here is the one that is formed in the aftermath of mass violence, where the social fabric has been destroyed and victims and perpetrators need to learn how to live together again.

In order to present the role that the video technology played in such environment, I am considering four different dimensions that have connected this medium to the communication process of recounting their own stories of violence, along with their new commitment to peace, that the youth engaged with during the video-making. The ensuing discussion clarifies these dimensions and illustrates participants' views.

VIDEO TECHNOLOGY AND CHANGE IN THE AFTERMATH OF THE 2007/2008 POST-ELECTION VIOLENCE

In order to offer insights on the effects that the use of the video technology has had on the young people who shared their stories of the Post-Election Violence and their new-found commitment to peace (thanks to Mercy Corps work) through the participatory video activities, I developed four qualitative indicators. These are loose fields that I felt surfaced in the respondents' answers, despite the questions not specifically addressing their views on the technology.[2] Table 13.1 offers an illustration of these fields.

Appropriation of Technology

The youths who were involved in Mercy Corps participatory video activities highlighted how the video helped to make their storytelling experience more real. They also indicated how, thanks to the medium, they felt they had gained the ability to bring about change through the message of peace they were

Table 13.1 Connections between Video and People's Communication

Appropriation of technology	Appropriation of video technology for communities who want to tell their own story
Self-empowerment	The use of the video technology gives people a means for self-empowerment, by offering a communication channel that they can tailor and control
Technology as facilitator	Video technology facilitates dialogue between enemy groups
Participatory approach to the use of technology	The participatory process of video-making allows people to take ownership of the communication process without it being determined by the technology itself

spreading. Moreover, by way of the video technology, they got the opportunity to watch and listen to themselves back on screen: this was an important aspect of the process, as it allowed for self-reflection on the actions they had chosen to engage with during the conflict.

> I suggest that if the videos will be used in the future, then it will be a great help in Kenya so that people get to learn what is right and what is wrong. People will get to see their reality of life.
>
> *Thomas, 19yo*

—

> Through my experience and what I underwent, I feel like what I said through the video maybe would turn somebody else. Maybe he or she had a negative attitude the way I was at first. So I wanted to tell them that I was thinking like you, I was doing like you, but after I have seen that participating in some activities that would not benefit me and my society is not good. So I decided to share my views.
>
> [. . .] I was happy because I knew I could just express myself through a camera, and my voice even can be heard.
>
> *Andrew, 18yo*

Another element that created a strong connection with the technology was the fact that they were not only able to finally been seen and heard, but also to speak on behalf of others who wanted peace, including those who were no longer there due to the violence.

> What made me take part in that video, I reflected and thought that maybe I could have been one of the people who died, but fortunately for me I was alive and therefore I thought it was worth to share my story with the others.
>
> *Simon, 26yo*

One more aspect showing appropriation is the suggestion that several participants made to broadcast their videos to a larger audience, in order to prevent any possibilities for violence to reoccur.

> I feel good that the whole world sees me on the video; they can witness what happened, they can see the good and bad and make a decision.
>
> *John, 27yo*

—

> I felt so good watching my own story, and listening to myself talk. It was very good for me to listen to my own story.
>
> *Simon, 26yo*

Self-empowerment

The videos and their production process have greatly contributed toward participants' self-empowerment. Beside the opportunity, again, to be seen and heard by a large number of people in their communities and possibly beyond, the medium provided them with the chance to make a difference. All interviewees indicated how they were finally able, thanks to the technology they were using, to pass on an important message; this gave them a new confidence as they were able to see their role of promoting peace among fellow Kenyans as a powerful one. A sense of empowerment also arose in them from the simple fact of learning to master a technology they had never used before, developing the awareness of having gained a new communication channel in the process.

Within this context, it is interesting to make a reference, again, to the work of Bernardelli and Blasi (1995). The authors explain how those who experience a new communicative technology focus their attention on the different possibilities that this technology carries, such as, for example, an increased diffusion power from the type of communication they had before.

During the screenings, participants also began to look at the videos they had produced as learning tools for the communities, and recognized their responsibility of educating others to peace through their own experience of the conflict, which they were able to share thanks to the medium. They felt that their own knowledge was finally being accepted, and saw the time that was given to them to tell their story as a recognition they were proud of.

> I felt happy because someone has recognized my experience and is ready to share it with someone else. So I am happy because I know I have used my little knowledge and experience, and also I was happy because I was involved directly.
>
> *Andrew, 28yo*

—

> It made me feel good because now it means I'm being watched in a video. That, I can see, is the best way of spreading my message of maintaining peace in the community. So myself being seen in the media, as young as I am now, I can effectively pass the message to the people and it's good for me to be seen in that video. It's the easiest means I can pass that message.
>
> *Thomas, 19yo*

—

> OK, I felt that I'm an important person in the society, because I'm there to talk for those who can't talk for themselves, because that thing totaled many people, and I came out to talk for them. I think it will help people.
>
> [. . .] Because I thought, at least I can go somewhere, I can talk to people, yeah. I can impact something on somebody's mind, yeah.
>
> *Rose, 21yo*

Technology as Facilitator

Mowlana (1997) has stated that "[f]uture intercultural cooperation and conflict will be determined by the relationship of communications (technological development) and communication (traditional classical channels for satisfactory human interaction)" (Mowlana, 1997, 240). The scholar also pointed out that the effect of a message on images is based on the value systems of communicators. Therefore, what remains of utmost importance is the ability of the communicator and recipient to find a shared meaning in a particular message. At the same time, since communication has the power to also reinforce group consciousness, "the symbolic environment created along the lines of mutually shared images and meanings [. . .] plays a crucial role in coping with possible dysfunctioning processes of communication technology" (Mowlana,1997, 241).

Within this context, another opportunity for dialogue that the technology offered was that people were able to learn about one another's experience from the films that were produced, thus relate to the stories of people from other tribes.

> It was very good and helpful for the community, because as we took the film and continued with it at least they were able to listen to other people from other communities, listen to their stories and relate closely. Therefore, it was very helpful for the community.
>
> *Simon, 26yo*

—

> In most part, I saw that something like in media, people can come together, dissolve together as one community. Because what happened was the past, like

> you have to let the past go and fight for our future. So I saw that being in one community would help us build our society and live like one people and in one country. I enjoyed that, because I learnt that every person has a different personality, different reasoning, and different principles, so by coming together as a media, help us learn different things, yeah.
>
> *Rose, 21yo*

One of the participants in my study talked about his impressions of the medium as a mediator. He acknowledged the role that the video had in mediating dialogue between the (former) enemy groups that took part in the participatory video activities. His sense was that the video had made it easier for people to sit together and begin to reflect on the violent events that had happened. This would have been difficult if it had not been initiated by the films and had simply taken the form of a face-to-face discussion, as some still feared violence from the rival tribes.

> Different communities, they are conflicting, they could not come, like, in a room and share stories or maybe try to seek common grounds in solving the conflict. So, actually media came in as a common ground. So, actually by coming in and maybe narrating the story on video and then people can hear me or they can watch me talk. It's actually acted as mediator. The media came in as a mediator.
>
> I can narrate my stories and urge the other communities, my feelings, I share my feelings. I talk of what I feel. Maybe I am against what they are doing for I ask them to stop that because I cannot stay with them in one room. [. . .] I was fearing them, just watching them I thought maybe they are still violent. Or maybe, they thought by coming here together I could also retaliate the same and maybe be violent. So actually media . . . I took part in community media to share the story because media was a neutral point to share my stories and also, they themselves can share their stories through the media and maybe all of us could watch and actually have that story help them to understand their problems and also, they themselves can understand our problems through the media.
>
> *Charles, 23yo*

Participatory Approach to the Use of Video Technology

Through participatory video, participants—who were all members of local communities—felt that they were finally given a space to articulate their stories. They were also able to recount episodes of their lives their own way, saying what they felt was important to tell. The process also involved the opportunity for people to come together and work on mixed-tribe activities, while listening to stories from one another. One of the respondents highlighted how preparing the script for his film on the basis of his story, was one of his favourite parts of the workshops. This shows how this process' component of participatory video-making gave him a strong sense of ownership of the media product he had been working on. Another man emphasised the importance of being among people who were coming from the same areas and who had experienced the same difficulties. This contributed to the creation of an environment in which people were able to learn and engage with the use of a new technology, while feeling safe in spite of the difficult context.

> I thought that was very important for these communities who are against each other. To watch the video and then they will learn something from that. That we should come together.
>
> *George, 29yo*

—

> That process was good because the group was trained. Most of the youth who participated they are part of the society who participated in the violence, and were used to explain their experiences that they underwent. So I think that the participatory process was good because we were not a bit anxious because the people that interviewed us are just common people that we are used to and with whom we live together.
>
> *Andrew, 28yo*

—

> [D]uring that time I felt like it was a nice time. It was a nice platform to come out and share my story and use that opportunity that it was initially by somebody else and actually come out and share my feelings and I believed sharing my feelings, or shown my story, was solving half of my problems, solving half of my community's problems because they felt like we offloaded some problems by sharing out to the people through the media.
>
> *Charles, 23yo*

—

> It was a very good experience because it was one of a kind. It was the very first time to see something like that. It was good because each of us—whether you were a perpetrator or a victim—we were able to tell our story without being afraid or being tied to anything. So, it was a very good experience for me and I didn't think, maybe, that it could happen again, after what happened.
>
> After I told my story, the one thing that I liked most was that I was able to build on that; I did a script for the video, and I loved that part.
>
> *Timothy, 35yo*

The set of elements discussed here can be found in the media-centered model of participation that looks at the functioning of community media through an "essentialist" eye. According to McQuail (2007), *essentialism* is one of the pathways to ascertain the consequences that communication technology has brought to human society. In particular, this framework reflects on the impact that the intrinsic attributes of a technology can have on those who are in direct contact with it, such as audience, users and producers. It also considers the difference that is made by that technology by way of its interaction with specific human characteristics, such as the ability of the audio-visual medium (e.g., video) to pass on a message to those who are illiterate. However, as Carpentier (2011) suggests, we should not let the essentialist view lead us to a complete rejection of the idea of neutrality of a technology, as communication technologies (whether used in a participatory way or not) are part of a society's complex system and contribute, as a result, to a social construction.

Again, these views encourage us to reflect on the need of searching for a middle ground in our approach to and adoption of technology. As previously defined, when social change is seen as the process that transforms unequal power relationships in contexts of poverty and oppression (Gujit, 2007), giving people the opportunity to participate in shaping the content and use of a new communication channel offered to them by a media technology can lead to a positive transformation within communities. This is particularly important in contexts where relationships were broken by violence.

CONCLUSION

Theories of technological determinism define the role of technology as one through which control is exerted upon society, with technologies shaping our realities, governing how change occurs and the directions it takes. The answer to this school of thought was provided by the sociological view that inverts this relationship and sees technology as a value-free asset in the hands of society (or groups within it), who decides for its use. An approach that is neither overly deterministic nor neutral, however, allows us to identify emerging communicative practices that arise from the new communication context formed by the medium that has been introduced (Bernardelli and Blasi, 1995). This has been shown through the Kenyan experience of participatory video presented in this chapter.

A deterministic approach can be found in the modernization phase of development, when communication technology was used to pass on information to the local population. The evolution of theory has now placed emphasis on participation and dialogue, whose main component is the creation of shared meanings and values starting from the community. The new model encompasses a participatory type of communication and looks at the ways in which communication technologies can facilitate community-driven processes that bring about social change and sustainable development.

After introducing the views of different authors on the concepts of dialogue, participation and technology, the discussion that arises from the interviews I have conducted in the Rift Valley of Kenya with a group of young people who were involved in a participatory video project in the aftermath of the 2007/2008 Post-Election Violence demonstrates how the participants' experience was shaped by their appropriation of the video production process thanks to the participatory methodology that was applied to it. As one of the respondents stated:

> I think it's good because when you use video to explain something to somebody, a person understands it well because it actually sees it. It's not just like a story because you can see. [. . .] in a video you watch, you listen, and it stays better in your mind, you remember what happened. I feel good I was in that video, because everybody sees it, and the people who are around can understand it.
>
> *John, 27yo*

The youths who took part in the project underwent a process of self-empowerment owed to this new means of communication, which helped them reach a large audience for the important message of peace they were the carriers of. The video was also seen as a mediator between groups in a sensitive context such as the one following a conflict; through the screenings that took place, people found it easier to initiate a dialogue that led to the recognition of the similarities in their experience of violence as well as of the wrongs that were done. The participatory approach allowed for flexibility, as people could use the technology the way they wanted, to tell the story they wanted others to hear. This is another important factor in the aftermath of a conflict, when people fear that what happened to them may never be heard. It also brought together victims and perpetrators to collaborate in the media content production, making video the connecting element.

NOTES

1. The participatory video activities were facilitated by Oxford-based organization *Insightshare* (www.insightshare.org).
2. All the names of the interviewees, whose quotes appear in this chapter, have been changed.

REFERENCES

Bernardelli, A. & Blasi, G. (1995). Introduction. Semiotics and the Effects-of-Media-Change Research Programmes. An Overview of Methodology and Basic Concepts [Special Issue]. *Versus. Quaderni di Studi Semiotici, 72*, 3–28.

Carpentier, N. (2007). Theoretical Framework for Participatory Media (pp. 105–121). In N. Carpentier, P. Pruulmann-Vengerfeldt, K. Nordenstreng, M. Hartmann, P. Vihalemm, B. Cammaerts & H. Nieminen (Eds.) *New Technologies and Democracy in an Enlarged Europe*. Tartu: University of Tartu Press.

Carpentier, N. (2011). *Media and Participation: A Site of Ideological-democratic Struggle*. Intellect: Chicago.

Chandler, D. (1996). Engagement with Media: Shaping and Being Shaped. *Computer-Mediated Communication Magazine, 3*(2).

Gujit, I. (2007). *Assessing and Learning from Social Change: a Discussion Paper.* Brighton: Institute of Development Studies (IDS). Retrieved from: http://www.ids.ac.uk/files/dmfile/ASClowresfinalversion.pdf (12/09/2012).

Gumucio-Dagron, A. (2003). Vertical Minds versus Horizontal Cultures: an Overview of Participatory Processes and Experiences (pp. 68–80). In J. Servaes (Eds.) *Communication for Development and Social Change*. New Delhi: Sage Publications.

Hamelink, C.J. (1994). *Trends in World Communication*. Penang: Southbound.

McQuail, D. (2007). Communication and Technology: beyond Determinism? (pp. 27–40). In N. Carpentier, P. Pruulmann-Vengerfeldt, K. Nordenstreng, M. Hartmann, P. Vihalemm, B. Cammaerts & H. Nieminen (Eds.) *New Technologies and Democracy in an Enlarged Europe*. Tartu: University of Tartu Press.

Miller, S. (January 1997). Futures Work—Recognising the Social Determinants of Change. Social Alternatives, 16(1), 57–58.

Matthewman, S. (2011). *Technology and Social Theory*. Stanford: Stanford University Press.

Mowlana, H. (1997). *Global Information and World Communication*. London: Sage Publications.

Murphie, A. & Potts, J. (2003). Culture and Technology. London: Palgrave.

Potts, J. (2008). Who's Afraid of Technological Determinism? Another Look at Medium Theory. *The Fibreculture Journal*, 12.

Rahim, S.A. (1994). Participatory Development Communication as a Dialogical Process (pp. 117–137). In S.A. White, K.S. Nair & J. Ascroft (Eds.) *Participatory Communication. Working for Change and Development*. New Delhi: Sage Publications.

Servaes, J. & Malikhao, P. (2005). Participatory Communication: the New Paradigm? (pp. 91–103). In O. Hemer & T. Tufte (Eds.) *Media and Glocal Change: Rethinking Communication for Development*. Buenos Aires: Nordicom and CLACSO.

Singhal, A. & Devi. K. (2003). Visual Voices in Participatory Communication. *Communicator*, XXXVIII, 2:1–15.

White, S. A. & Patel, P. K. (1994) Participatory Message Making with Video: Revelations from Studies in India and the USA (pp. 359–386). In S.A. White, K.S. Nair & J. Ascroft (Eds.) *Participatory Communication. Working for Change and Development*. New Delhi: Sage Publications.

Chapter 14

Tiger Gate

A Case Study of Communication Channels and Technologies Used by Activists in a Social Change Action in China

Song Shi

INTRODUCTION

Through a case study of the Tiger Gate event,[1] which is a nationwide grass-roots new media for social change action that questioned the authenticity of a group of wild tiger photos published by government agencies and the according policy to set up a new wild tiger natural reserve in ShanXi province in China, this chapter will explore what communication channels and ICTs were involved in this action, the interactions of the communication channels, and how these communication channels and ICTs were used to mobilize distributed knowledge and skills among different individual activists to construct the collective knowledge that challenges the official claims. Contrary to the traditional technological deterministic understanding that technologies are drivers of social change in modernization paradigm development theories (e.g., Lerner, 1958; Rogers, 1976), I argue that, in this case, technologies are tools that may provide a new potential for combining the information and capacity embedded in computer and Internet system with the knowledge embodied in people (e.g., Servaes and Carpentier, 2006). Moreover, unlike past research on Chinese Internet and social change which primarily focus on new media and the Internet (e.g., Qiu, 2009; Yang, 2009) and contrary to Lévy's understanding of collective intelligence (1999) in which ICTs play a predominant role, I advocate a multichannel perspective to examine and understand new media for social change actions and the collective knowledge construction in these actions.

THEORETICAL FRAMEWORK

Three components constitute the theoretical framework of this case analysis: collective intelligence (Lévy, 1999), convergence culture (Jenkins, 2006), and communication for sustainable social change and development (e.g., Servaes, 1999; Servaes , 2008; Servaes and Carpentier, 2006).

As two well-developed theories in new media studies, the role of convergence culture and collective intelligence in this case is twofold. First, they provide the framework to analyze how ICT connected and empowered individual activists engaged in this grassroots action. Second, they provide a framework to describe the media ecology of this new media for social change action. They will be used to explore how activists with very diverse background and skills including digital photography fans, botanist, wildlife photographers, and common netizens were mobilized to use their knowledge and skills to question the authenticity of the South China Tiger photos and the policymaking regarding the natural reserve.

Convergence culture is a theory proposed by Jenkins (Jenkins, 2006) and has been widely used in new media studies. Three key concepts constitute Jenkins' definition of convergence culture: media convergence, participatory culture, and collective intelligence. By media convergence, Jenkins means two processes, corporate convergence and grassroots convergence, and the two processes coexist together. Corporate convergence is a top-down corporate-driven process in which media corporations "are learning how to accelerate the flow of media content across delivery channels in different media to expand revenue opportunities, broaden markets, and reinforce viewer commitments" (Jenkins, 2006, p. 18). Grassroots convergence is a bottom-up consumer-driven process in which "consumers are learning how to use these different media technologies to bring the flow of media more fully under their control and to interact with other consumers" (Jenkins, 2006, p. 18). By participatory culture, Jenkins means that consumers are active participants rather than passive media spectators in media consumption and production. For collective intelligence, Jenkins borrowed this concept from Lévy's study of collective intelligence in cyberspace. Jenkins believes that the new participatory skills consumers learned through their relation to popular culture and their interaction in fan communities or online communities will eventually develop into the skills that allow new citizens to deliberate together (Jenkins, 2006, p. 259); collectively, consumers will generate strong correctives to the traditional power sources such as state and capitalist corporations.

Lévy defined collective intelligence as "a form of universally distributed intelligence, constantly enhanced, coordinated in real-time, and resulting in the effective mobilization of skills" (Lévy, 1999, p. 13). This definition contains four essential characteristics of collective intelligence. First, collective intelligence or collective knowledge is universally distributed among all individuals. Second, collective intelligence is constantly enhanced. Third, collective intelligence is real-time coordinated. Fourth, collective intelligence implies and results in effective mobilization of skills among communities members (Lévy, 1999). Regarding the role of ICTs in construction of collective intelligence, Lévy believes that computer makes collective intelligence possible. He argues that the Internet and ICTs provide the possibility for community members to realize the real-time coordination of universally distributed intelligence (Lévy, 1999, p. 14). He believes that computer and ICTs make it possible to create a new society that incorporates and enlarges the scope of self-knowledge into a form of group knowledge and collective thought (Lévy, 1999, p. xi).

Regarding the role of ICTs in social change, compared with Lévy, past literature in communication for development and social change studies demonstrates a different approach. Rather than seeing ICTs as drivers of change, ICTs are considered as tools "which may provide a new potential for combining the information embedded in ICT systems with the creative potential and knowledge embodied in people" (Servaes and Carpentier, 2006. p. 5). Past literature also shows that although ICTs are vital tools to achieve development and social change, to create a sustainable social change, other communication channels must also be involved (e.g., Inagaki, 2007; Langley and van den Broek, 2010; Servaes, 2009; Servaes, Carah, Hadlow et al., 2006; Servaes and Carpentier, 2006; Tacchi, 2007). In his study of communication strategies and sustainable development, Servaes argued that communication for sustainable development is a social process to create a consensus for action to achieve sustainability in all fields of society. ICTs are important tools to facilitate this process, but their use is not an end in itself. Interpersonal communication and traditional forms of communication must also play a fundamental role (Servaes, 2009). After reviewing 37 empirical research studies on communication for development published in peer-reviewed journals from 2000 to 2005, Inagaki advocated an integrated communication strategy. She argued that: (1) The effectiveness of communication channels can only be discussed in the context of a specific project. There is no all-purpose communication channel that is most effective in all contexts. (2) Many projects employ more than one communication channel and the interaction among different channels may have significant impacts on the effectiveness of the communication interventions. Servaes, Carah, Hadlow et al. in their study of past ICT for development projects, also argue that ICT for development projects are most successful when deployed as part of an integrated approach to development and social change. They argue that success of ICT project requires not only infrastructure support, but also long-term political backing, skills training, and investment. Thus, interpersonal communication such as in skills training workshops and mass communication such as in advocacy for political support also play an important role. They also found, among past development projects, ICTs and mass media such as radio have widely been used together to enhance the effects of each other. Moreover, Tacchi, among many others (e.g., Dagron, 2001; James, 2004; James, 2005; Girard, 2003), argues that mixed media strategy which combines ICT and other channels benefits the effectiveness of ICT for development projects (Tacchi, 2007).

INTRODUCTION OF THE TIGER GATE EVENT

Before discussing and analyzing the Tiger Gate event, a detailed introduction on the process of the event and key figures involved will be very helpful for the case analysis. In October 12, Forestry Department of Shanxi province held a press conference to announce that on October 3, a peasant named Zhou Zhenglong took a photograph of a supposedly extinct wild South China Tiger in ZhengPing County, ShanXi province; they argued that expert evaluation has shown that the photos, 40 digital and 31 film, are authentic; And the photos showed that wild South China Tiger is not extinct in China. There might thus be a small breeding population of tigers around ZhengPing County. Local media such as HuaShang Daily as well as nationwide media including XinHuaNet.com and China Central Television (CCTV)[2] have reported the news and the photos.

After eight of the photos were uploaded in the Internet on October 12, their authenticity began to be widely questioned by Internet users. Photographers and photography fans at leading photography online forums in Chinese cyberspace, notably, xitek.com, are the first group who began to question the photos online. At the same time, the head botanist at Institute of Botany at Chinese Academy of Science Fu Dezhi, who later became a leading figure in the activists' community, questioned the authenticity of the photos and the news from XinhuaNet.com in his blog and emay.com.cn.[3] Users of major Chinese Internet forums such as Tianya.com, kdnet.com, 163.com as well as local Internet forums in ShanXi such as hsw.cn are all actively involved in the discussion and debate.

As the voice questioning the photos spread rapidly online, on October 17, Shanxi Forestry Department refuted in mainstream media the activists' argument that the photos are faked. Wang WanYun, director of the Office of Wild-Animal Protection of Shanxi Forestry Department, argued that because of Zhou's concern of copyright, they can't publish more photos online to persuade netizens, yet claimed the photos are authentic. And he claimed that Zhou also took photos of footprints of the tiger, which is a solid evidence of the authenticity of the photos. Zhou also responded to the questions by claiming he would guarantee the authenticity of the photos with his life. The official claim and Zhou's respond didn't calm down the increasing questions on the photos from online activists and media. On October 23, Shanxi Forestry Department officials and Zhou went to Beijing to report the finding of wild South China Tiger to the State Forestry Administration. State Forestry Administration had no comments on the authenticity of the photos. October 26, Guan Ke, director of Center of Forestry Information and Propaganda of Shanxi Forestry Department, opened his blog by real name to debate with online activists and later became the leading figure among netizens who supported the official claim of Shanxi Forestry Department. In his second blog post on October 26, he disclosed the parameters of the camera Zhou used to take the photos and the time Zhou took most of the 71 photos: on October 3, 4:38:28 to 5:3:8, 22 minutes in total. The newly disclosed information of the camera and the time Zhou took the photos from Guan's blog are counterproductive for the official claim. They were used by activists to further discredit the photos.

Since November 2, more online activists began to join Fu's rank to question the authenticity of the photos by real name. At the same time, heated debate on the photos continued among anonymous netizens in major forums. Those opposing and those supporting the official claims were called anti-tiger netizens and pro-tiger netizens by media, respectively. Later, deputy director of Shanxi Forestry Department, Zhu Julong and Sun Chengqian debated with the anti-tiger activists and media reports in mainstream media and local media in ShanXi, as well as in their own blogs. International media including Science (Holden, 2007) and Wall Street Journal (Ye, 2008) also reported the event and the online debate. On November 16, the origin of the photos, a New Year picture produced by a small company in ZheJiang province was found and published online by anti-tiger activists. The breaking news was widely reported by mainstream mass media including CCTV, People's Daily, and Southern Metropolitan Daily as well as major portals such as 163.com, sina.com and QQ.com. Pro-tiger figures notably Gu Ke, Zhu Julong, and Zhou Zhenglong as well as some pro-tiger anonymous Internet users continued defending the authenticity of photos and claimed that the New Year picture is faked. On November 26, 163.com published all the 40 digital photos of the South China Tiger taken by Zhou, which provides more materials for grassroots investigation.

On December 9, 2007, the State Forestry Administration required ShanXi Forestry Department to reevaluate the photos but still made no comments on the authenticity of photos. The debate on the photos continued both online and in mass media. On February 4, 2008, ShanXi Forestry Department issued a letter of apology to the public through media for announcing the finding of wild South China Tiger without serious investigation but said the re-evaluation of the photos has not completed and made no comments on the authenticity of the photos.

On February 19, 2008, media reported that officials in ShanXi Forestry Department exposed that the reevaluation of the photos has not begun. The news triggered a new round of outrage among anti-tiger netizens and criticism from media on government inaction. On March 7, the major forum focusing on the Tiger Gate event, South China Tiger forum of hsw.cn was closed by local government in ShanXi. Anti-tiger netizens moved to other major public forums such as www.shunyi.cc/bbs, Tianya.com, and tieba.baidu.com continuing their online actions and debate. With the constant online action and the increasing media pressure, on June 29, ShanXi Forestry Department announced the investigation result of the Tiger Gate event: the photos taken by Zhou Zhenglong are faked. Zhou has been arrested by local police. In the press conference held by government of ShanXi province, police showed the two New Year cameras used to take the South China Tiger photos and the tools to make the tiger footprints used by Zhou. Thirteen governmental officials were held accountable for the scandal, seven of whom were removed from their positions. In November 2008, the provincial court at ShanXi province declared that Zhou, the peasant who claimed to have taken the photos, was sentenced to jail for two years and six months.

DATA AND ANALYSIS

The data of this part of analysis are drawn from my one-year online observation of the event in online communities such as the three major forums focusing on the event (emay.com.cn, www.xitek.com's South China Tiger forum, hsw.cn's South China Tiger forum), popular Chinese Internet forums such as Tianya.com, leading activists' blog, and governmental officials' blog concerning the event. The news coverage and investigation of the event from October 12, 2007 to December 27, 2008 are also collected. After carefully reading the online activists' interactions challenging the official claims, I identified 12 methods and 7 perspectives that have been collectively developed by activists with various expertise and knowledge from different parts of China to challenge the authenticity of the South China Tiger photos (see Table 14.1). In Table 14.1, I summarized the methods and perspectives, time that the methods were developed, the main activists who developed the methods, the geographical location of the activists, background of the activists, and the Internet spaces where the major activists posted their methods.

The Importance of New Media and ICTs

A growing literature has shown that new media and technologies play an important role in social change actions (e.g., Inagaki, 2007; McKee, Erma, Chin and Rachel, 2000; Servaes, 1999; Servaes, 2008). Following the technologies as tools to facilitate social change approach (e.g., Servaes and Carpentier, 2006), from two perspectives, I will explore the importance of new technologies in the Tiger Gate event.

ICTs as communication tools to facilitate construction of collective knowledge and advocacy

From the data in Table 14.1 and the introduction of the Tiger Gate event, we can see that ICTs are crucial tools to facilitate the construction and the sharing of collective knowledge questioning the authenticity of the South China Tiger photos. They provided spaces and tools to mobilize the knowledge and skills distributed among individual activists with various social backgrounds and at different parts of the country. They generate the technology possibility and spaces for the emergence of an activists' network that transcends geographical and social barriers. We can see that Internet forums/BBS (Bulletin Board System) such as xitek.com and Tianya.com provided activists from different parts of China including Beijing, Shanghai, Shenzheng, Nianjing, Shaoxin (Zhejiang province), Weifang (Shangdong province), Panzhihua (Sichuan province), and HaErbing (Heilongjiang province) a space to constantly construct the collective knowledge questioning the South China Tiger photos. This accords with Tai's argument on the role of Internet in social change actions that the Internet transforms the arena of social movements by building networks of activists that are dispersed and separated in the physical world (Tai, 2006, p. 255). A map of the locations of the activists in Table 14.1 provides a visualized image of the cross-regional knowledge construction (see Figure 14.1).

Without new ICTs such as BBS/forum and Internet, it is almost impossible for activists located in different parts of China with such huge geographical distance to construct collective knowledge together.

In this case, we also can see the real-time-coordination effect of ICTs in collective intelligence construction that Lévy has argued (1999, p. 13). The instantaneity of new ICTs such as BBS and Internet forums provided activists the possibility to coordinate their knowledge construction and actions in real time. For example, on November 16, anti-tiger activists (notably smallfish) posted the origin of the South China Tiger photos, a New Year picture, in xitek.com's South China Tiger forum. Four hours after smallfish posted the New Year picture,[4] activists at xitek.com already collaboratively found the manufacture of the New Year picture (see Table 14.2).

Table 14.1 Perspective and Methods to Question the South China Tiger Photos

No.	Time	Perspective	Method	Main activist[1]	Geographical location[2]	Activists background[3]	Internet spaces
1	2007-10-14	Botany	The size of the leaves of the plants around the tiger is disproportionately big	Fu Dezhi	Beijing	Botanist at the Institute of Botany at Chinese Academy of Science	emay.com.cn; Fu's blog
2	2007-10-15	Zoology	Research shows wild South China Tiger was already extinct	Lao Liu	HaErbing, Heilongjiang	Professor at Northeastern Forestry University (China)	wildchina.cn[4]
3	2007-10-15 9:15	Digital photography	Animated GIF of the South China Tiger photos shows the photos are faked	FirstImpression	ShaoXing, ZheJiang	Public employee	xitek.com; Tianya.com
4	2007-10-29	3D photography	3D images made from the South China Tiger photos show the tiger is 3D	flower7166	—	—	xitek.com
5	2007-10-30 17:16	Math (linear algebra)	To prove the tiger is flat, we can use linear algebra	Chimpanzee	Beijing	IT engineer at a governmental agency	xitek.com
6	2007-10-31 16:19	Computer vision	Result from Matlab software shows the tiger is flat	Sancho	ShangHai	Engineer at a computer vision company	xitek.com
7	2007-11-1 7:06	Digital photography	Through increasing the color saturation of the South China Tiger photos, we can see the layer of optical brightener on the tiger	LaoBian	ShenZhen, GuangDong	Professional photographer	xitek.com

8	2007-11-14	—	The video of Zhou returning to the site where he took the photos shows the tiger is faked	Ba Mu	—	—	xitek.com
9	2007-11-16	Wildlife photography	Taking several dozen of photos of a big cat animal for 20 minutes is impossible	Bao Kun	Beijing	Professional photographer	His blog
10	2007-11-16 10:30	Human Flesh Search[5]	Origin of the South China Tiger photos has been found	Smallfish,[6] Panzhihua xydz[7]	WeiFang, ShanDong, Panzhihua SiChuan	Public employee at China Custom	Tianya.com; xitek.com
11	2007-11-23	3D photography	3D photo pair created with the South China Tiger photos show the tiger is flat	cubism	—	—	kdnet.net;hsw.cn
12	2012-12-27	Math (geometry)	Geometry method proves that the tiger is flat	WangYang101	NanJing	College math teacher	His blog; hsw.cn

[1] Main Activist is the online ID of the activists. Some activists never disclosed their real identity.

[2] Geographical location is the geographical locations of the activists that were discoursed by new reports or by activists in their online interactions. Some activists never disclosed their geographical locations.

[3] Activists' background is the background of the main activists disclosed by themselves or by news reports.

[4] www.wildchina.cn is a leading wildlife photography forum in China.

[5] Human Flesh Search is a Chinese Internet phenomenon of massive researching, based on massive human collaboration, using online and offline skills and information to find the information interesting to a specific community or group of people. Its goals are very diverse from exposing individuals to public humiliation, entertainment, to anticorruption (Wang et al., 2010). In this case, Human Flesh Search played an important role to find the origin of the tiger photos taken by Zhou.

[6] Smallfish played the key role in the finding of the origin of the South China Tiger photos. He is an anti-tiger activist at Tianya.com, the biggest Chinese online forum. On 16 November, 2007, Panzhihua xydz told Smallfish's that he/she found that a New Year picture in his/her room is almost the same with the South China Tiger photos. Smallfish posted the finding and the photos of the New Year picture he got from Panzhihua xydz on xitek.com's South China Tiger forum, titled "Result from Human Flesh Search, origin of the tiger has been found!" (Smallfish, 2007, November 16).

[7] Panzhihua xydz is the online ID of an activist from Panzhihua (Sichuan province) at Tianya.com. He/she played a significant role in the finding of the origin of the South China Tiger photos (a New Year Picture). His/her identity and background have never been disclosed.

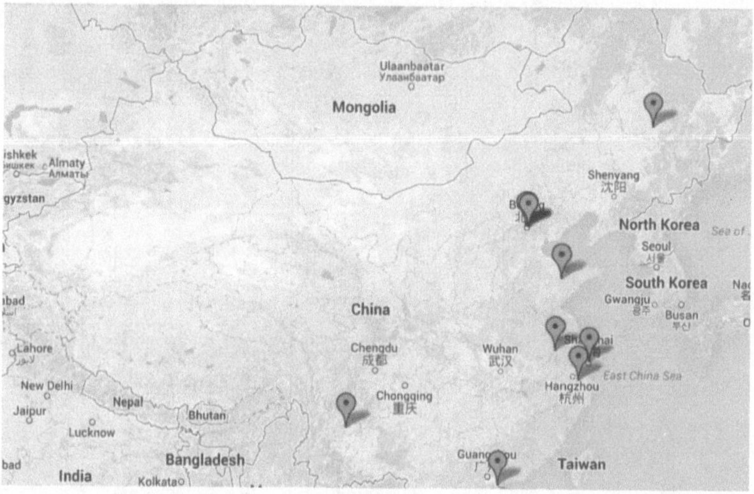

Figure 14.1 Locations of the Activists Questioning the South China Tiger Photos.
Source: Map data ©Google, SK Planet, ZENRIN.

In Table 14.2 from Data No. 1 to Data No. 6, all coordination, interactions, and construction of the knowledge on the South China Tiger photos happened within near 40 minutes. Without the Internet-based ICTs such as forums/BBS, such real-time mobilization and coordination of collective knowledge construction would not be possible.

Also, communication for social change researchers argue that, in advocacy actions, the communication media are critical in creating awareness, generating public interest and demand, and placing the issue on the public agenda and building social support (e.g., Servaes, 2009). In the Tiger Gate event, forums/BBS such as Tianya.com and xitek.com attracted huge amount of visitors reading, discussing, and sharing the information about and opinions on the event, thus largely increased the public awareness of the event in cyberspace and partly sustained the public support and appeal for an investigation of the event to hold those involved in the fraud accountable. For example, after anti-tiger activist FirstImpression started a thread to question the authenticity of the South China Tiger photos with an animated GIF file at Tianya.com (see Table 14.1), from October 17, 2007 to February 2008, 11,000 people have replied to his post in the thread and more than 200,000 have read the thread. In xitek.com, from November 16 2007 to November 26, within ten days, Smallfish's thread, "Result from Human Flesh Search, origin of the tiger has been found!" has been read by 289,000 people.

Blog, as a newly emerged Web2.0 technology in 2007, has also been widely used in the Tiger Gate event both by anti-tiger activists and pro-tiger officials involved in the debate. In Table 14.1, among the anti-tiger activists,

Table 14.2 The Finding of the Manufacture of the New Year Picture

Data No.	*Time*	*User ID*	*Content*
1	2007-11-16 15:12	Danpier	Danpier, a xitek.com activist, started a thread at xitek.ocm titled "the manufacturer of the New Year picture has been found" and posted the website and the logo of the manufacturer in his/her post. . . . (Some reply posts omitted)
2	2007-11-16 15:25	YiShiFengYun	Posted the manufacturer's (Yi Wu WST Printing & Packing Corporation) detailed introduction, address, contact information, and the name of the CEO in the thread. . . .
3	2007-11-16 15:32	OldSaltedFish	Found the product introduction of the New Year picture on Yi Wu WST Printing & Packing Corporation's website, made a screenshot, and posted the screenshot in the thread.
4	2007-11-16 15:35	Corn	Explained why Yi Wu WST Printing & Packing Corporation is the manufacturer of the New Year picture, and proposed advices for further investigation. "1: On the photo of the New Year picture (uploaded by Smallfish), there is a logo on the lower left corner. It is the logo of this manufacturer. 2: If the manufacturer is still there, contact the manufacturer to confirm the production and sales of the New Year picture. . . ." . . .
5	2007-11-16 15:40	cs67	Posted the logo of the New Year picture in the thread. . . .
6	2007-11-16 15:56	xiangboo	Could we "find some friends from Yi Wu[1] to confirm the information about the manufacturer?" . . .

[1]Yi Wu is the location of the manufacture. This information has been exposed in Data No. 2, the address of the manufacture.

Fu Dezhi (botany's perspective), Bao Kun (wildlife photography's perspective), and wangyang101 (mathematics perspective) have used blog as a platform to advocate their methods that question the South China Tiger photos. For example, after emay.com.cn withdrew from the debate over the Tiger Gate event[5] on November 6, Fu used his Sina blog, blog.sina.com.cn/emaycomcn, as his major Internet platform to constantly question the South China Tiger photos and advocate for an investigation to hold those involved in the fraud accountable. Unlike forum/BBS which provides the possibility for real-time collective knowledge construction among large group of activists, the personal diary style blog technology which only allows a single user to create new posts reduced the interaction in blog. It provided Fu the tools to build a personal advocacy platform combining his comments, investigation, actions, as well as news reports, and forwarded other anti-tiger activists' articles. For example, in December 2007, he posted 902 articles in the blog, of which large majority are about Tiger Gate. On December 27, he posted 17 articles about Tiger Gate in which 5 are his comments on the event, 4 are news reports on the event, and 8 are forwarded articles from other anti-tiger activists with his comments. The large amount of forwarded articles from other anti-tiger activists added a certain amount of interactions for the blog.

Other ICTs such as video sharing website and Instant Messenger (IM) have also been used in the actions questioning the South China Tiger photos. For example, on November 14, 2007, activists posted the video that recorded Zhou and an officer of Forestry Office of Zheng Ping County returning to the site where Zhou took the photos on video sharing website. Then after carefully watching the video, activist "Bamu" found a new evidence to show that the South China Tiger photos are faked (see Table 14.1). By watching the video and comparing with the photos, he/she found that the tree branch in Zhou's photo is also in the video. He/she made a screenshot and identified the height of the branch in the photo and video, and posted them at xitek.com. He argued that, from the height of the branch, it is obvious that the tiger photos are faked. For the detailed analysis and the screenshot, please refer to the activists' forum (xitek.com, 2008).

As the most used Internet service in Chinese cyberspace in 2007[6] (CNNIC, 2008), IM has also been used by activists in their action questioning the authenticity of the South China Tiger photos. For example, in the finding of the New Year picture (the origin of the South China Tiger photos), Panzhihua xydz, a netizen at Tianya.com who first found the New Year picture, replied to Smallfish: "Smallfish, add me into your QQ friends. I'll send you some photos. I found a New Year picture in my room is almost the same with the South China Tiger photos except for the background and the ears . . ." (Panzhihua xydz, 2007, November 16). We can see that QQ, the most used IM service in China, is the channel the two chose to communicate the information about the New Year picture. On November 16, 2007, 10:42, Smallfish posed on xitek.com "This New Year picture is found by a friend from Sichuan Pianzhihua in his own room. He tried to send me a high definition photo of the New Year picture through QQ but failed. And now is trying to send it out through other way." We can see QQ is also the first way Panzhihua xydz sent the photos of the New Year picture to Smallfish. Media reports on the Tiger Gate event also show that QQ has been widely used by anti-tiger activists to coordinate their actions. For example, Fu Dezhi, FirstImpression, and Smallfish are all members of an anti-tiger QQ group named "Anti-fake-tiger." On November 16, within five minutes that Smallfish uploaded the New Year picture on xitek.com, the administrator of "Anti-fake-tiger" QQ group already sent the information to all members of the QQ group and asked them to visit xitek.com to study the new findings (Southcn.com, 2007).

Digital technologies as tools in investigation

In last section I have shown that like in most canonical communication for social change actions, technologies were used as communication tools to facilitate the construction of collective intelligence, the real-time collaboration among activists, and advocacy in the Tiger Gate event. In this section, I will focus on a new potential of technologies in new media for social change actions: digital technologies as tools in investigation. I want to show that in the Tiger Gate event, digital technologies have exceeded their conventional role as pure communication tools and demonstrated the potential to be tools for grassroots investigation in social change actions.

Jenkins argued that when common consumers have digital technologies in their own hand, they may become a strong corrective to the traditional power sources such as state and capitalist corporations (Jenkins, 2006). In the specific context of the Tiger Gate event, this on one hand means that activists can use digital technologies as communication tools to mobilize a large group of common people to advocate for a thorough investigation of the event. On the other, it also means that activists can creatively use digital technologies to conduct grassroots investigation to provide knowledge and evidence questioning the official claims. In my previous summary of different perspectives and methods that question the authenticity of the South China Tiger photos in Table 14.1, we can see that digital technologies have been widely used in activists' grassroots investigation on the South China Tiger photos (see Table 14.1).

From the data in the Table 14.1, we can see that among the 12 methods to investigate the South China Tiger photos, five of them, No. 3, No. 4, No. 6, No.7, and No. 11 used digital technologies or knowledge of digital technologies including GIF animation making, digital photography, computer vision, and 3D photography as investigation tools. For example, in No. 3 method, anti-tiger activist FristImpression created an animated GIF (Graphics Interchange Format) file with two of the eight South China Tiger photos published online and one released to the media by ShanXi Forestry Department in its press conference. He placed the three photos on the three frames of the GIF file, respectively, and set up an interval for the three frames to be displayed one by one in GIF animation. He argued that the animated GIF shows that although the three photos were taken from different perspective (more than 15 degree difference) with different focal length at different time, the tiger's face, body, and stripes are completely identical. No living tiger could freeze itself so long for Zhou to take these photos. This means that the photos are not taken of a wild tiger but of a tiger printed out on a cardboard. In No.6 method, Sancho, a 36-year-old engineer working for a computer vision company, developed the computer vision method to question the authenticity of the South China Tiger photos. He used Matlab, a computing language and software widely used in computer vision, developing a short program to perform a test of two photos from the South China Tiger photos taken by Zhou. He concluded that the result of the test shows that the tiger in the photos is flat so it is faked, whereas the grass and branches around the tiger are three dimensional so they are real. With the assistance of new technologies, both FirstImpression and Sancho have revealed the truth of the South China Tiger photos long before the official investigation. These technology-assisted grassroots investigations significantly influenced the public opinion on the Tiger Gate event and strengthened the public support to advocate a thorough investigation. The result from these grassroots investigation have been widely read, discussed, and disseminated by both Internet users and mass media reports. For example, as mentioned previously, from October 17, 2007 to February 2008, FirstImpression's animated GIF file to show that the tiger in the South China Tiger photos is faked, has been read by more than 200,000 people at Tianya.com. It has also been widely reported as one of the most important evidences to question the authenticity of the South China Tiger photos by major mass media and Chinese portals including Southern Metropolitan Daily, CCTV (CCTV, 2008), 163.com, and, Sina.com. Sancho's computer vision method that shows the tiger in the South China Tiger photos is flat, from October 31 to November 23, within one month, has been read by 79,315 people at xitek.com. And it has also been widely considered as the most persuasive one among different methods to question the South China Tiger photos by activists at xitek.com and hsw.cn. Xitek.com user "loveorola" replied in Sancho's thread ". . . this is the most reliable and scientific method!!! Please disseminate it!!!". Xitek.com user "QinLing" replied, "So many days, this is the most persuasive method to test the South China Tiger photos . . ." In the report from Southern Metropolis Weekly titled "Best Anti-Tiger Activists," Sancho was named as the Best Detective for "using Matlab software finding a breakthrough to prove that the tiger in the South China Tiger photos is flat" (Pan, 2007). Sancho's method has widely been reported by mass media and major Chinese portals including Southern Metropolis Weekly, Southern Metropolitan Daily, QQ.com, and 163.com.

Besides the five methods that directly used new technologies, two other methods also involved the use of new digital technologies. For example, anti-tiger activist Wang Yang 101 used Photoshop skills to draw lines and triangles on the Tiger Gate photos to explain his geometry method in a more understandable way for common people (Wang Yang 101, 2007). Fu used Photoshop skills to highlight the disproportionately big leaves around the tiger in the South Tiger photos. Fu later also developed a series of photos with Photoshop skills to further explain his botany method to question the tiger photos (Fu, 2008. January 10).

From what has been discussed above, we can see new technologies have been widely used as communication tools and investigation tools in this new media for social change action. They provided activists the opportunity to combine the information and capacity of the Internet and computer system with the knowledge and skills embodied in people and eventually enhanced activists' capacity to make real social changes.

A New Perspective: Multichannel Perspective to Look at New Media for Social Change

A growing literature in communication for development and social change studies has shown that although ICTs are vital tools to achieve development and social change, to create a sustainable social change, other communication channels such as mass media and interpersonal communication must also be involved (e.g., Inagaki, 2007; Langley and van den Broek, 2010; Servaes, 2009; Servaes, Carah, Hadlow et al., 2006; Servaes and Carpentier, 2006; Tacchi, 2007) Thus, a multichannel perspective to look at new media for social change actions to address ICTs, other communication channels, as well as the interactions among them is crucial in the study of new media for social change.

The Importance of Mass Media

One aspect I disagree with Lévy is the predominant role of ICTs in the construction of collective intelligence in his understanding of collective intelligence (Lévy, 1999). He overlooked that the construction of collective knowledge and collective intelligence happens in a multichannel media environment where mass media and other communication channels also play a significant role. In this section, through analyzing the data from the finding-New-Year-picture episode in the Tiger Gate event, I want to highlight the importance of other communication channels such as mass media in the construction of the collective knowledge questioning the authenticity of the South China Tiger photos. For the specific data, please refer to Table 14.3.

In Table 14.3, from the media column, we can see that mass media such as the Southern Metropolitan Daily, Beijing News, and CCTV, as well as major portals such as 163.com, have been deeply involved in this episode. Specifically, we can see after Data No. 4 (Southern Metropolitan Daily reported the breaking news of the finding of the New Year picture and the methods

Table 14.3 The Finding of the Original Picture of the Tiger Photos

Data No.	*Time*	*Content*	*Media*
1	2007-11-16 1:10am	Smallfish started a thread to advocate the urgent need to find the original picture of the South China Tiger photos. 'Panzhihua xydz' replied in the thread: "I found a New Year picture in my room is almost the same with the South China Tiger photos except for the background and the ears . . ." and sent four photos of the New Year picture to Smallfish.	Tianya.com
2	2007-11-16 10:30 am	Smallfish uploaded the four photos he got from Panzhihua xydz on xitek.com's South China Tiger forum in his thread titled "Result from Human Flesh Search, origin of the tiger has been found!" and asked xitek.com users help to compare the New Year picture and the South China Tiger photos.	xitek.com
3	2007-11-16 10:30 am to 12:00 pm	Activists at xitek.com have developed various methods to show that the New Year picture and the South China Tiger photos are identical. They also argued, to ensure the reliability of the New Year picture, there is an urgent need, first, to continue searching for the same New Year pictures around the country; second, to find the manufacturer of the New Year picture.	xitek.com
4	2007-11-16 12:30	Southern Metropolitan Daily and its online platform, nddaiy.com, reported the breaking news of the finding of the New Year picture and the methods xitek.com's activists developed to show that the New Year picture and the South China Tiger photos are identical.	Southern Metropolitan Daily
5	2007-11-16 12:30	The leading pro-tiger governmental officer, GuanKe, told Southern Metropolitan Daily reporter, the New Year picture is badly PSed.[1]	Southern Metropolitan Daily
6	2007-11-16 12:54:33	Major portals such as 163.com, sohu.com, and xinhuanet.com and major media began to report the breaking news of the finding of the New Year picture from Southern Metropolitan Daily.	163.com, sohu.com, xinhuanet.com ,etc
7	2007-11-16 15:12	Danpier, a xitek.com activist, started a thread titled "the manufacturer of the New Year picture has been found." Through analyzing the logo of the New Year picture, he/she found that the manufacturer of the New Year picture is Yi Wu WST Printing & Packing Corporation at YiWu, Zhejiang province. He/she also found the website of Yi Wu WST Printing & Packing Corporation and post the link in the thread.	xitek.com

8	2007-11-16 15:23:02	163.com reported that they have consulted Hu Huijian, zoologist at South China Institute of Endangered Animals. Hu said that, from the stripes of the tiger, he is sure that the tiger in the New Year picture and the tiger in Zhou's photos is the same.	163.com
9	2007-11-16 16:28:00	Southern Metropolitan Daily reported that the activists at xitek.com have found the manufacturer of the New Year picture and its official website. Tuo Guanglin, CEO of the manufacturer, told Southern Metropolitan Daily's reporter that his company has produced the New Year picture.	Southern Metropolitan Daily
10	2007-11-17	163.com, Southern Metropolitan Daily, and other media including Beijing News continue searching for the same New Year pictures around the country.	163.com, Southern Metropolitan Daily, Beijing News
11	2007-11-17 09:20	163.com's user, Mr. Huang at DongWan GuangDong province, sent text message to 163.com saying he has a New Year picture exactly the same with the one found by xitek.com's activists at home. On 12:50, 163.com's staffs visited Mr. Huang and took photos for the New Year picture he bought two years ago in his apartment.	163.com (a major portal)
12	2007-11-17 afternoon	Mr. Xu, manager of XingLong Picture Store at GuangZhou read Southern Metropolitan Daily's report on the New Year picture and called Southern Metropolitan Daily saying his store has large amount of the same New Year pictures.	Southern Metropolitan Daily
13	2007-11-17 afternoon	Kena, a xitek.com's activist at Beijing, with reporters from CCTV found large amount of the same New Year pictures at a retail store at Beijing. He then, posted the information at xitek.com.	xitek.com CCTV
14	2007-11-17 14:14	User of hsw.cn at Jinan, Shandong province, found the same New Year picture at a book store.	Hsw.cn

[1]PS, among digital technology fans, originally refers to Adobe Photoshop which is a main digital graphic editing tool. Later, it refers to all digital photography technologies, skills, and tools. It can be also used as a verb. For example, the photo is PSed.

activists developed to show that the New Year picture and the South China Tiger photos are identical), this episode has exceeded the boundary of cyberspace. With the active involvement of mass media and major portals such as Southern Metropolitan Daily (in Data No. 4 and Data. 5 for example), Beijing News (in Data No. 10), CCTV (in Data No. 13), and 163.com (in Data No. 6 for example), the discovery of the New Year picture has become a public event in convergence culture, a public event that transcended the boundaries of different media channels. Also, in Data No. 3, the activists community proposed two urgent questions regarding the reliability of the New Year picture: first, to continue searching for the same New Year pictures around the country; second, to find the manufacturer of the New Year picture. The first is solved by the involvement of mass media in Data No. 12 and Data No.13 and portals in Data No. 11. The second was solved by the online activists community (Data No. 7) and was confirmed by mass media (Data No. 9). A more vivid demonstration of the episode can be seen in Figure 14.2.

From the data in Table 14.3 we can also see that mass media played a significant role in the real-time coordination of the construction of the collective knowledge questioning the South China Tiger photos. For example, in Data

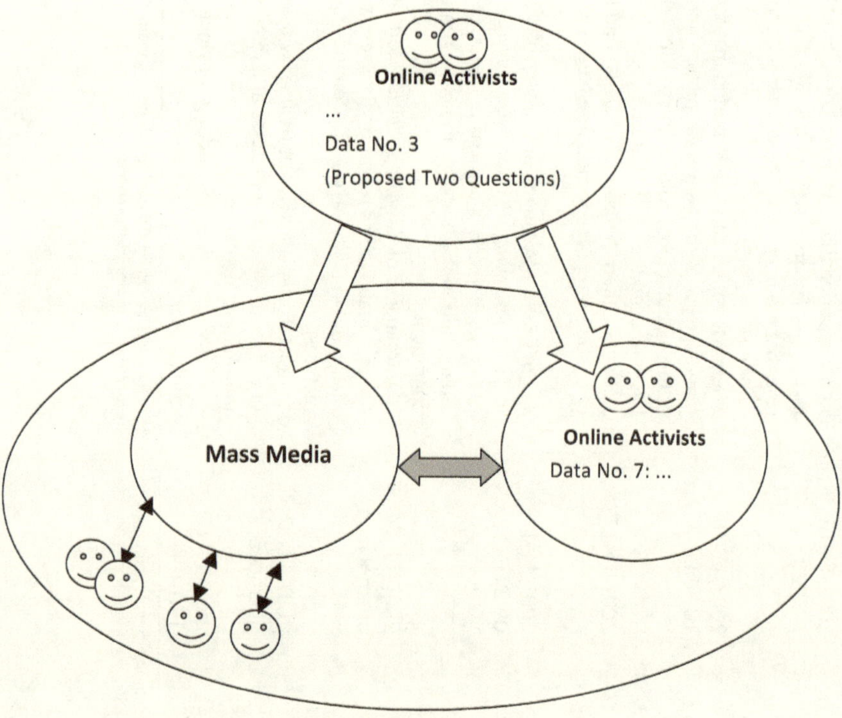

Figure 14.2 Finding-the-New-Year-Picture Episode.

No. 4, one and a half hour after the New Year picture was uploaded on xitek.com, half an hour after the activists proved that the New Year picture and the South China Tiger photos are identical, Southern Metropolitan Daily already reported both the finding of the New Year picture and the methods to show that the New Year picture and the South China Tiger photos are identical. It disseminated the information in real-time to the public and facilitated the real-time coordination to further construct the collective knowledge questioning the South China Tiger photos. Also, in Data No. 5, a half hour after the activists showed that the New Year picture and the South China Tiger photos are identical, Southern Metropolitan Daily already got and reported the response of leading pro-tiger officials to the new finding, which provided information for activists' community to create new knowledge to further question the denial from the pro-tiger figures.

From the data in Table 14.3 and Figure 14.2, we can see that mass media such as Southern Metropolitan Daily also played an important role in the effective mobilization of skills. In Figure 14.2, we can see that mass media largely increased the number of the people who may engage in the event. It thus largely extended the community whose skills and knowledge may be mobilized to construct the collective knowledge questioning the South China Tiger photos. For example, in Table 14.3, Data No. 12, after reading the report from Southern Metropolitan Daily, Mr. Xu told the reporters of Southern Metropolitan Daily that his store has the same New Year pictures. In Data No. 11, Mr. Huang after reading the news reports on the New Year picture, sent text message to 163.com saying he has a New Year picture exactly the same with the one found by xitek.com's activists. From the above Data in Table 14.3, we can see that mass media's reports contributed to the mobilization of common people like Mr. Xu and Mr. Huang to engage in the collective knowledge construction questioning the South China Tiger photos. Also, from Data 9 to Data 13, because of mass media's involvement, journalists in Southern Metropolitan Daily, 163.com, and CCTV are also mobilized to investigate the New Year picture and the South China Tiger photos.

From Table 14.3, we can also see that mass media promoted the constant enhancement of the collective knowledge questioning the South China Tiger photos. For example, from Data No. 10 to Data 13, with the involvement of mass media and the mobilization of journalists and common people at different parts of the country, within one day, a large number of the same New Year picture have been found in three cities in China. An urgent question to ensure the reliability of the New Year picture has been solved. Also, in Data 8, 163.com's interview of Hu Huijian, zoologist at South China Institute of Endangered Animals, enhanced the knowledge on the New Year pictures and the South China Tiger photos from a zoology perspective.

Interactions between New Media and Mass Media

In the media environment of convergence culture where media contents flow across multiple channels and platforms ever more quickly (Jenkins, 2006), in social change actions, the relation between ICT channels (Internet-based communication channels) and other communication channels such as mass media is not an either/or relation. They interact with each and reinforce each other. In this section, through analyzing the data from the finding-New-Year-picture episode in the Tiger Gate event, I want to explore the interactions and reinforcement between ICT channels and mass media channels in new media for social change actions.

In Table 14.3, we can see that mass media and the ICT-based new media are not isolated from each other. They interact with each other, and, more importantly, they reinforce each other in the construction of the constantly enhanced collective knowledge questioning the South China Tiger photos. A more detailed depiction that focuses on the interaction and the reinforcement process can be seen from the data in Table 14.4.

In Data No. 3 and Data No. 5 in Table 14.4, we can see that the collective knowledge constructed by online activists community in Data No. 1, No. 2, and Data No. 4 reinforced the news reports of Southern Metropolitan Daily (a mass media). In Data No. 5, new knowledge on the South China Tiger photos (*the tiger looks like a Indochinese Tiger*) based on the finding of the online community (the New Year picture) was collaboratively developed by mass media reporters and Hu (an expert outside the online community at xitek.com) and was integrated into the news report of Southern Metropolitan Daily (identified in italics in Data No. 5). In Data No. 6, the new knowledge in the news report of Southern Metropolitan Daily (identified in italics for emphasis in Data No. 5) fed back into the online activists' community and reinforced the collective knowledge questioning the South China Tiger photos of online community. Furthermore, based on the new knowledge, online activists proposed a new perspective for future investigation. Figure 14.3 demonstrates the interaction and reinforcement process between mass media and new media in a more vivid way.

The example of Southern Metropolitan Daily demonstrated in Table 14.4 is not an isolated case. The process of interaction and reinforcement between mass media and new media showed in Figure 14.3 is a pattern that widely emerged in the Tiger Gate event. From what has been discussed and analyzed in this part, we can see that, although ICTs are important tools and channels in new media for social change actions, a multichannel perspective to look at new media for social change phenomena and the construction of collective knowledge in social change actions are of great importance.

Table 14.4 The Interaction and the Reinforcement of New Media and Mass Media

Data No.	*Time*	*Content*	*Media*
1	2007-11-16 10:30	Smallfish uploaded the four photos of the New Year picture he got from Panzhihua xydz on xitek.com . . .	xitek.com
2	2007-11-16 10:30 to 12:00	Activists at xitek.com developed various methods to show that the tiger in the New Year picture and the tiger in the South China Tiger photos are identical . . .	xitek.com
3	2007-11-16 12:30	Southern Metropolitan Daily reported the breaking news of the finding of the New Year picture and the methods xitek.com's activists developed to show the New Year picture and the South China Tiger photos are identical.	Southern Metropolitan Daily
4	2007-11-16 15:12	Xitek.com's activist Danpier found the manufacturer of the New Year picture and posted his finding and the website of the manufacturer on xitek.com.	xitek.com
5	2007-11-16 16:28	Southern Metropolitan Daily reported: 1) the activists at xitek.com has found the manufacturer of the New Year picture . . . ; 2) Hu Huijian, zoologist at South China Institute of Endangered Animals, confirmed that the tiger in the New Year picture and the tiger in Zhou's photos are the same tiger and after carefully studying the New Year picture, he thought *the tiger looks like a Indochinese Tiger not a South China Tiger because its ears are too big* . . .	Southern Metropolitan Daily
6	2007-11-16 17:18	Activist at xitek.com, dcaoyuan, cited Southern Metropolitan Daily's report and Hu's words. Then, argued "In the photos taken by Zhou, the ears of the tiger are deliberately covered under some leaves. This shows there must be someone who knows South China Tiger well involved in the fraud . . ."	xitek.com

CONCLUSIONS

Through a case study of the Tiger Gate event, this chapter shows, first, various communication channels including ICT channels and mass media channels as well as other channels are involved in this new media for social action. Second, various ICTs such as BBS/forum, blog, and IM played a significant

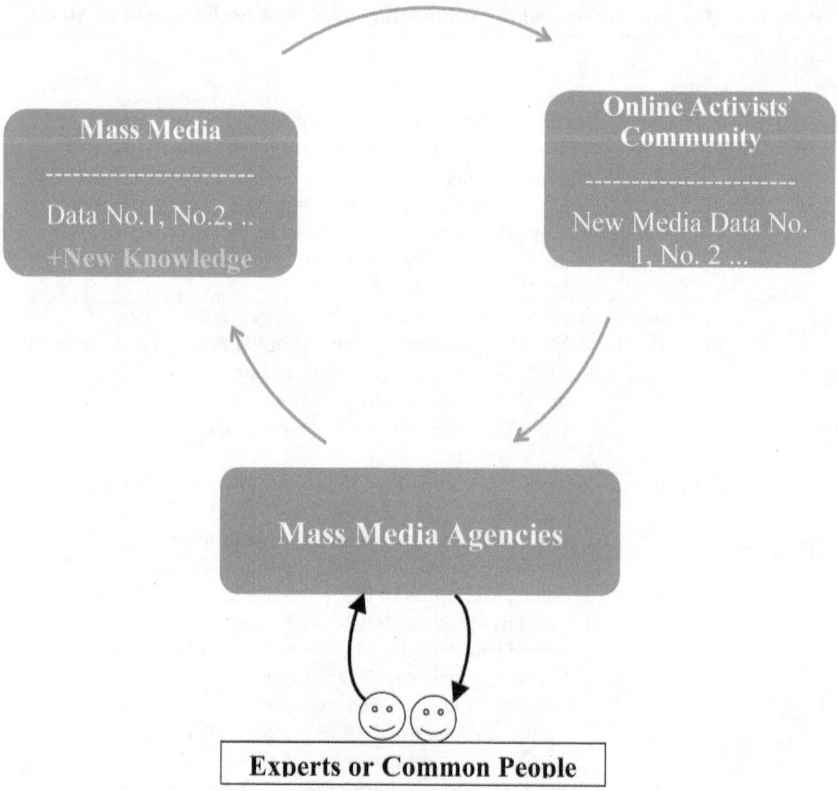

Figure 14.3 Interaction and Reinforcement between Mass Media and New Media.

role to facilitate the cross-region and real-time coordinated construction and the sharing of collective knowledge questioning the official claims. Moreover, in this grassroots social change action, digital technologies have exceeded their conventional role as pure communication tools and demonstrated the potential to be tools for grassroots investigation in social change actions. Third, contrary to Lévy's understanding of collective intelligence (1999) in which ICTs play a predominant role, the case study shows that a multichannel perspective to look at the construction of collective knowledge and new media for social change actions is of great significance for the study of new media for social change in China and for the effectiveness of social change actions. It provides empirical support for the argument in communication for social change studies that, although ICTs are vital tools to achieve development and social change, to create a sustainable social change, other communication channels such as mass media and interpersonal communication must also be involved.

NOTES

1. Tiger Gate event is also known as the South China Tiger event.

2. China Central Television or Chinese Central Television, abbreviated as CCTV, is the predominant state television broadcaster in mainland China.

3. Emay.com.cn is an Internet forum of botany started by Fu and other researchers at Institute of Botany.

4. See Table 14.3 and 14.4 for the time smallfish posted the New Year picture.

5. On November 6, Fu announced that emay.com.cn will withdraw from the debate over the Tiger Gate and will temporarily stop the function to post any new information in the Internet forum. I'll further discuss the withdrawal of emay.com.cn later in the stakeholder relation analysis.

6. CNNIC refers to China Internet Network Information Center that is the administrative agency responsible for Internet affairs in China. According to the report from CNNIC, in 2007, 81.4% of the Chinese Internet users use IM, almost 25% ahead of email (CNNIC, 2008).

REFERENCES

China Central Television (CCTV). (2008). Tiger gate. Retrieved May/29, 2012, from http://news.cctv.com/special/C20076/01/

China Internet Network Information Center (CNNIC). (2008). *Statistical survey report on the internet development in china.* (No. 22). CNNIC. Retrieved from www.cnnic.net.cn/uploadfiles/pdf/2008/7/23/170516.pdf

Dagron, G. (2001). *Making waves: Stories of participatory communication for social change.* New York: Rockefeller Foundation.

Fu, D. (2008, 01–10). Tiger gate event: Demonstrate the disproportion between the size of the tiger and the size of the leaves with figures. Message posted to http://blog.sina.com.cn/s/blog_4ef9035a01008a9o.html

Girard, B. (Ed.). (2003). *The one to watch. radio, new ICTs and interactivity.* Rome: Food and Agriculture Organization.

Holden, C. (2007). Rare-tiger photo flap makes fur fly in china. *Science, 318*(9), 893.

Inagaki, N., & World Bank. (2007). *Communicating the impact of communication for development : Recent trends in empirical research.* Washington, DC: World Bank.

James, J. (2005). Technological blending in the age of the internet: A developing country perspective. *Telecommunications Policy, 29*(4), 285–296. doi:10.1016/j.telpol.2004.11.010

James, J. (2004). *Information technology and development: A new paradigm for delivering the internet to rural areas in developing countries.* London; New York: Routledge.

Jenkins, H. (2006). *Convergence culture: Where old and new media collide.* New York: New York University Press.

Langley, D., & van den Broek, T. (2010). Exploring social media as a driver of sustainable behaviour: Case analysis and policy implications. Retrieved December, 23,

2012, from http://microsites.oii.ox.ac.uk/ipp2010/system/files/IPP2010_Langley_vandenBroek_Paper.pdf

Lerner, D. 1917–1980. (1958). *The passing of traditional society: Modernizing the middle east*. Glencoe, IL: Free Press.

Lévy, P. 1956–. (1999). *Collective intelligence: Mankind's emerging world in cyberspace*. Cambridge, MA: Perseus Books.

McKee, N., Manoncourt, E., Saik Yoon, C., & Carnegie, R. (Eds.). (2000). *Involving people, evolving behaviour*. Penang: Southbound and UNICEF.

Pan, C. (2007, 11/26). Best anti-tiger activists. *Southern Metropolis Weekly, 172*, 19.

Panzhihua xydz. (2007). Reply to Smallfish's post-the origin photo of the south china tiger photos have been found. Retrieved 12/11, 2013, from http://bbs.tianya.cn/post-free-1044130-1.shtml

Qiu, J. L. (2009). *Working-class network society: Communication technology and the information have-less in urban china/*. Cambridge, MA: MIT Press.

Rogers, E. M. (1976). *Communication and development: Critical perspectives*. Beverly Hills, CA: Sage Publications.

Servaes, J., Carah, N., Hadlow, M., Louw, E., & Thomas, P. (2006). Communication for development. making a difference—A WCCD background study. *World congress on communication for development: Lessons, challenges and the way forward* (pp. 209–292). Washington DC: World Bank.

Servaes, J. (2009). Communication policies, good governance and development journalism. *Communicatio: South African Journal for Communication Theory and Research, 35*(1), 50.

Servaes, J. (1999). *Communication for development : One world, multiple cultures/*. Cresskill, NJ: Hampton Press.

Servaes, J. 1952–. (2008). *Communication for development and social change*. New Delhi, India; Thousand Oaks, CA: Sage Publications.

Servaes, J. 1952–, Carpentier, N., & European Consortium for, C. R. (2006). *Towards a sustainable information society: Deconstructing WSIS*. Bristol, UK; Portland, OR: Intellect.

Southcn.com. (2007). Online anti-tiger activists. Retrieved May/12, 2012, from http://news.qq.com/a/20071128/000977_3.htm

Tacchi, J. (2007). Information, communication, poverty and voice. In J. Servaes (Ed.), *Moving targets. mapping the paths between communication, technology and social change in communities* (pp. 125–158). Penang: Southbound.

Tai, Z. (2006). *The internet in china: Cyberspace and civil society*. New York: Routledge.

Wang Yang 101. (2007, 12–26). Mathematic methods could perfectly prove that the south china tiger photos are faked. Message posted to http://blog.sina.com.cn/s/blog_502d921d010083p7.html

xitek.com. (2008). xitek.com's South China Tiger forum, http://forum.xitek.com/forum-96-1.html

Yang, G. (2009). *The power of the internet in china: Citizen activism online*. New York: Columbia University Press.

Part IV

A + A + A

By Way of Conclusion

People Matter

Jan Servaes

Several important trends of the late twentieth century have transformed our world: the Internet and other ICTs, the globalization of our economies and trade relationships, widespread population migrations driven by economic, political or military reasons, the emergence of multiculturalism, and the demise of the nation-state in favor of sub- and supranational systems. Our world has become more interdependent and interconnected, and the nation-state has become more culturally heterogeneous.

In this complex and global context, communication has increasingly become important, if not essential.

By way of conclusion, we wish to elaborate on a few of the issues, which have already been addressed in this book, but perhaps, because of their importance for the future, may need to be highlighted even more: (1) What matters most: Technology or people?; (2) Is "universal access" enough?; and (3) How to assess the impact of technology for social inclusion?.

A distinction between structural and sociocultural participation will allow us to classify the diverse types of policy that promote digital participation for social inclusion. Our argument therefore ends with a plea for new forms of digital literacy.

SOCIAL INCLUSION AND DIGITAL CITIZENSHIP

Social inclusion is an important principle in policy discussions (Atkinson, 2009; Commission of the European Communities, 2005; Fielden and Malcolm, 2009; Kelly, 2010; Talbot, 2004; United Nations, 2005). Social inclusion refers to the equal participation of all social groups in a so-called welfare society. In other words, welfare states claim to be inclusive states. Therefore, no social groups should be left in a marginalized position. This ideal of social inclusion also implies that citizenship in welfare states is more than merely legal citizenship. Citizenship should be a social citizenship, which implies social participation (Ommundsen et al., 2007).

There are many ways to achieve social inclusion, but one that has gained major importance over the last decade is social inclusion that is the result of ICTs possession and use. Citizenship has been called *digital citizenship* (d'Haenens, Koeman and Saeys, 2007; Mertens, 2009; Mossberger, 2010; Mossberger, Tolbert and McNeal, 2008; Servaes, 2003), and contemporary society has been referred to as an *Information/Knowledge society* (Breit and Servaes, 2005; Mansell and When, 1998; Mattelart; 2001; Webster, 1995).

Many international policy documents have promoted the possibilities ICTs could contribute to social inclusion. For instance, in the Bucharest Declaration (2002), a Pan-European policy document on the development of the information society, we read: "The Information Society offers great potential in promoting sustainable development, democracy, transparency, accountability and good governance" (p. 1); and "Access and contribution to knowledge and information broaden the contents of the public domain and foster mutual understanding and respect for diversity" (p. 2). And the "United Nations Global E-government Readiness Report 2005. From E-government to E-inclusion" (UN, 2005, 114) confirms: "The potential [of ICTs] stems from the unique ability to provide access to unlimited information at any time irrespective of the distance and access to each person, and irrespective of location, to participate in proffering what he/she values which should be produced by the society. Information technologies facilitate the dissemination of information and the opportunity of feedback as they promote access to government and are the perfect conduit for citizen-government participation to promote public value, and therefore, inclusion."

This euphoric rhetorical discourse often contrasts with the results one finds in the research literature on the digital divide (for instance, Hargittai and Hinant, 2008; Korupp and Szydlikk, 2005; Ricci, 2013; Van Dijk, 2005, 2012). To bridge the gap between the potentialities of ICTs and the reality of the digital divide a comprehensive ICTs policy is needed. The task of those responsible for ICTs policies is to develop initiatives that make ICTs use and possession more widespread. Indeed, if ICTs have intrinsic capacities to promote social inclusion, it will be necessary to make all social groups participate in the Information/Knowledge society.

THE DIGITAL DIVIDE, ONCE AGAIN

"Many people think the digital divide will be closed as soon as everyone has a computer and a connection to the Internet. In my opinion, the problems of the digital divide and information inequality do not disappear when this happens, as other, deeper divides appear. [. . .] Nevertheless, having material or physical access is a necessary condition for development of the requisite skills and ability to use this technology. The importance of this condition should not be downplayed in stressing other types of access," argues Jan Van Dijk (2005, 45). He and his colleague Hacker (Hacker and van Dijk, 2001), further argue that there are four main hurdles of access to the information society producing these inequalities: (a) lack of basic skills and "computer fear"; (b) no access to computers and networks; (c) insufficient user-friendliness; and (d) insufficient and unevenly distributed usage opportunities (see also Norris, 2002).

These four obstacles should be further analyzed. For instance, *use* is not a homogenous category. Youngsters can use computers and the Internet for entertainment (such as playing games and posting "selfies") or for functional purposes (such as searching information about homework, about politics or about employment). The capacities of digital media use for participation and inclusion are obviously most connected with use for functional purposes. And data show that functional use is determined by socioeconomic status and ethnicity (Heinderyckx, 2003; Jenkins, 2004; Livingstone and Helsper, 2007).

In other words, the impact of new ICTs on civil society, participatory democracy and citizenship is increasingly of immense contemporary concern (Shirazi, Ngwenyama and Morawcynzski, 2010). This impact is usually associated with the demand of universal access. But universal access/service alone does not suffice. The way Stephen Coleman puts it: "If citizenship requires universal access, democracy needs trustworthy channels of information and deliberation if it is to prosper" (2001, 124). In other words, modern citizenship needs the demand for and provision of information in order to develop the proper rights and responsibilities in the conditions and complexities of a Information/Knowledge Society.

BEYOND "ACCESS": THE NEED FOR STRUCTURAL PARTICIPATION

As argued by most authors in this book, the mere fact of providing "access" to new technologies—be it in theory as a stated formal right or law, or in practice—is not enough. In Mertens and Servaes (2011) we tried to study this in the context of Brussels, the capital of Belgium and seat of the European Union and many international/global organizations. This multicultural and multilingual community of more than 1 million people is one of the most international cities in the world.

The dynamic counterpart of "inclusion" is "participation." Inclusion is the consequence of participation activities. Following earlier attempts to classify participation processes (d'Haenens, Van Summeren, Saeys and Koeman, 2004), and using a multi-methodological approach of both quantitative and qualitative data collection, we distinguished between two types of participation, and consequently also between two types of ICTs policies: one enhancing *structural participation* (policies on digital education for youngsters and digital adult education) and the other enhancing *sociocultural participation* (wireless networks in Brussels, and e-governance). In a working document of the European Commission (2005, 19) we read: "Digital and social participation clearly appear as closely intertwined in a society which becomes progressively 'technical', where technology increasingly serves communication and networking purposes, and where the boundaries between on- and offline activities are fading away." Our inventory and evaluation of ICTs policies in Brussels was an attempt to explore this relationship between digital and social participation.

The first type of participation is called *structural participation*. Being part of society is to a large degree the consequence of one's economic position, and having a good economic position means participating structurally. Three key aspects of this dimension could be distinguished.

- The first dimension is the distinction between being employed or unemployed; in other words, having *a job*. Unemployment policy is a key dimension of economic policy, and, in as far as ICTs policy is a policy to promote skills for the unemployed, it is a policy that relates to this dimension of structural participation. Without material access, the development of skills becomes futile; material access may not be sufficient, it is most certainly necessary. One of the benefits of the Brussels ICTs policy is that it has heavily invested in providing schools with computers via the so-called "multimedia plans."
- The second dimension that matters for structural participation is the quality and the level of the *labor position* that people hold. It may well be the case that jobs with a higher level of ICTs use are structurally better jobs. This is at least what Karen Mossberger, Caroline Tolbert and Ramona McNeal (2008, 45) found in their American study on the relation between ICTs use at work and economic opportunity: "Those who use the Internet on the job are more likely to have higher weekly earnings and incomes. The frequency of Internet use at work implies differing levels of capabilities on-line and we find that as the frequency of use rises, so does income."
- A third dimension that matters for structural participation is *education*. The education degree one achieves can have a correlation with one's economic position. Because the level of education has a potential relationship with

the level of income, policy activities oriented towards ICTs education can be considered activities that promote structural participation.

Being able to participate structurally may be an important part of *being socially integrated.* However, it is not enough. Since Putnam (2000), the importance of social networks and *social capital* for a vital society has been emphasized very much in social science literature. The degree to which a society is integrated is related to the contacts that people have with each other. This participation in the network of social contacts can be called *socio-cultural participation.*

In addition, three subtypes of sociocultural participation could be distinguished.

- The first and less institutionalized type of sociocultural participation refers to the contacts people have in their *personal networks* with family and friends. These contacts are essential for a sufficient level of sociocultural participation, but sociocultural participation also implies social contacts in organizations.
- *Social and cultural networks* help people to find their place in society and some organizations have also the intention to influence politics.
- *Politics and political organizations* and the degree to which individuals or social and cultural groups can influence them is the third dimension of sociocultural participation (Arts, Evers, Van Dam and Buitenhuis, 2003; Hamelink, 1999; Norris, 2010).

All three dimensions of sociocultural participation can be influenced by ICTs policy and were part of the Brussels government's attempts to promote sociocultural participation.

IT'S THE "CULTURE," STUPID!

Notwithstanding ambitious "multimedia" plans, mainly targeted at schools and public spaces like libraries, the results of our Brussels study (for more details, see Mertens and Servaes, 2011) turned out to be rather disappointing. As education until the age of 18 years is compulsory in Belgium and 95% of the Brussels schools were covered by the multimedia plans, youngsters theoretically had the possibility to access computers at school. However, the drop-out rate among certain socially vulnerable groups, and the school "culture" are determinants for these youngsters to get good school results and use ICTs. This can be seen as another variation on what has already been described extensively in the classic book by Bourdieu and Passeron (1970) on *social reproduction*. The assessment of these authors is that social groups of youngsters, such as for instance some groups of nonnative youngsters, having problems to catch up with current school standards are having these problems not because of the social-economic differences between their families and the middle class families, but because of cultural differences between them and the middle-class families. There is a deep-rooted distrust within these families toward the school and school culture, for instance because of the elitism associated with school culture or the negative experience with the school system of the parents themselves when they had to go to school. As the multimedia plans are essentially school based, this gap between the school culture and the culture of some families has also its implications for this type of digital access policy. Also in welfare states such as Belgium education policy sometimes fails, and this has its consequences regarding the ability to use a computer, which requires reading and writing skills.

In a way, the results were remarkably consistent with the findings of earlier studies on the introduction of computers in schools, or communication for development projects around the world (Inagaki, 2007; Lennie and Tacchi, 2013; Servaes, 2013a,b). For instance, Tang and Ang (2002: 460) observe that "technological innovation has often been implemented with the assumption of a passive or silent audience. Studies showed that teachers were often overlooked while implementing technology in schools [. . .] On the other hand, research found that technology has not changed the way educators do their work; its impact was limited to improvement in hardware equipment [. . .] Many teachers did not use it in the manner prescribed."

TECHNOLOGY ASSESSMENT FOR SOCIAL INCLUSION

How can we assess the impact of new technologies on social change? *Technology Assessment*, according to Joe Coates (1976), "systematically examines the effects on society that may occur when a technology is introduced, extended, or modified. It emphasizes those consequences that are unintended, indirect, or delayed." Determining the "unintended, indirect, or delayed" societal impacts of a future technological innovation poses severe challenges (Sclove, 1995; Smith and Marx, 1994). Coates (1971) himself provided an interesting example of the problematic nature of pinning down indirect effects. He studied the first through sixth order effects of introducing television in communities. The first effect of the entertainment box lead to less mingling in local clubs; that resulted in less neighborly interaction; followed by isolation; then overdependence on spouses to meet one's psychological needs; and finally, divorce rates escalate. Technology Assessment has, over the years, moved away from attempts to predict exact effects and their timing, but is currently rather interested in identifying potentially important vectors of change. A number of variants to the classical, traditional or expert-based Technology Assessment (TA), that was initially developed have now emerged, which are not fundamentally different from each other. These include (IAIA, 2010):

- participative TA (pTA) which emphasizes the social nature of technology, that experts cannot assess social reaction, and the need to involve the community in decision making,
- interactive TA which seeks to influence the innovation process through early interactions between suppliers, users and parties affected using a Delphi-like procedure,
- constructive TA which argues that the social issues need to be addressed by a cocreation process where social issues influence the design of the technology at a very early stage, and
- real-time TA which seeks to integrate natural science and engineering investigations with social science and policy research from the outset.

However, as far as its *legitimation* is concerned, Thomas Saretzki (2012) distinguishes between two basically different approaches to justify participation in technology assessment and technology policy.

- One is *normative* and refers to basic principles of normative theories of democracy (Held, 1987). "If and to the extent that processes of technization and technologization are major forces in a democratic society, then its citizens or their representatives should have a chance to participate in the assessment and the policy-making processes related to the technologies in question as a consequence of the democratic principle of self-government of free and equal citizens. As the technology assessment process is seen as providing the grounds for judgments of possible problems and options to solve them, TA is conceived of as part of the democratic opinion- and will-formation that precedes and predetermines the decision making in the field of technology policy and thus is also subject to claims for democratic participation" (Saretzki, 2012, 14; see also Genus, 2006).
- The other approach to justify participation in technology assessment and technology policy is *functional*: In this view, participation is justified as being instrumental in achieving certain goals and fulfilling certain functions that provide the basis of its legitimation. New technologies should be introduced (and old technologies should be evaluated) with broad public participation, because participation fulfils certain functions more effectively and efficiently than other (nonparticipatory) ways of assessing technologies and policy-making in technology-related fields. Functional

justifications can focus on informational, social, spatial or time-related dimensions of technization processes, of technology assessment or of technology policy.

Richard Sclove (2010), in arguing that we need a twenty-first-century TA model, advocates for a model that uses *citizen participation*, collaboration and expert analysis to inform and improve decision making on issues involving science and technology. Sclove argues that participation by laypeople is essential to redressing a number of the areas in which the former more expert-led TA's approach was deficient.

Among the reasons to include citizen participation within the repertoire of TA methods are (Sclove, 2010: 24–29; see also Joss and Belluci, 2002):

- *A Matter of Democratic Right*: Lay citizens are ethically entitled to direct participation or effective representation in technological decisions as a matter of democratic right.
- *Social Values*: Experience with participatory Technology Assessment (pTA) shows that laypeople tend to excel in articulating ethical concerns and moral and social values, making and explaining value judgments and developing normatively informed insights and analysis.
- *Broader Knowledge Base*: pTA gathers and integrates life experiences and other social knowledge that expert analysis omits or undervalues.
- *Cost Reduction*: By producing an informed citizen perspective on not-yet-realized technological potentials, pTA can inform government, business and other stakeholder decision making, thereby helping avoid or reduce costly social controversy.
- *Expedited Conclusions*: Participatory methods typically take 2–12 months to orchestrate and complete. This can be much more useful to policy-makers than the two years typical of the former TA's reports.

Also the policies and strategies outlined by UNESCO (Grizzle and Torras Calvo, 2013, 17–18) for a new understanding of *Media and Information Literacy* (MIL) argue that increased citizen participation may lead to more active and democratic participation, a higher awareness of ethical responsibilities for global citizenship, and more opportunities for diversity, dialogue and tolerance.

Comparing the United States and European TA models from a historical perspective, Sclove finds that the United States, while ahead in the past, has lost its advantage and cutting edge to Europe. He concludes, "Juxtaposing the dynamism and sweeping influence of contemporary science and technology with the new organizational forms and participatory capabilities made possible in part by the Internet, the accomplishments of European pTA over the past two decades and an executive branch that is committed to advancing citizen engagement in governance, there are compelling reasons to integrate a modern, expert-and-participatory technology assessment capability into our nation's (read: US) civic life" (Sclove, 2010, 41).

THE POTENTIAL OF THE INTERNET

References have been made to the impact the Internet has. Some have argued that the Internet has the potential to change the rules of the game. The Internet makes it possible potentially to reach large numbers of people, for what is, compared to traditional mass media, a marginal investment. This fact raises the number of people able to broadcast a message and communicate across borders. What the Internet does on the costing side is to lower the transactional costs of doing business (and this includes political busy-ness), and of organizing. It may be said that the Internet is, and—according to some—could increasingly become, a major tool for the self-organizing of social groups. Others are more doubtful and, with references to NSA, Snowden and other Chinese firewalls, are ready to declare "the end of privacy" (Greenwald, 2014).

Combine both the hypothesis of cheaper information diffusion and self-organizing and you get the effect of diffusing "information power" and hence *power* itself, to a much larger number of actors. This could rebalance power relationships in the public sphere, but may perhaps lead, over a certain time, to another kind of society a move away from mass society, toward a more localized kind of *networked society*, based on the coexistence of varied subcultures (Hepp et al., 2008; Van Dijk, 2012, 2013).

Another important innovation of the Internet is the collective discussion that it enables. True, most newsgroups and blogs are of a deplorable quality, but their existence is significant. With more experience and better software, these tools may be adjunct to democratic processes, for example, in the information gathering phases. Used intelligently, certain search facilities and computer conferencing software has positive potential. But even used spontaneously as it is now, it is not without effects on the political process. Again the reason is that such uncensored forums do create uncontrolled information flows that cannot be controlled by the mass media and the "powers that be."

Therefore, the Internet could be viewed as a social laboratory for experimenting with new ways of developing self-government for social groups using the Internet, as advocated by the movement for a *creative commons* or the peer-to-peer foundation. Therefore, Michel Bauwens (2006, 153) envisions a new template of human relationships in the future: "Peer to peer is a specific form of a network, which lacks a centralized hierarchy, and in which the various nodes can take up any role depending on its capabilities and needs. Peer to peer is an 'egalitarian' network if you like, a form of 'distributive and cooperative intelligence.' "

A PLEA FOR THE FUTURE: DIGITAL LITERACY

The cases presented in this book and the analyzed policies in the Brussels case (Mertens and Servaes, 2011) indicate that structural participation in general, but especially in the education system and the labor market, should be the common goal of future policies. This should be complemented by socio-cultural participation, driven or coordinated by governmental initiatives.

Notwithstanding all good intentions and attempts "gaps" remain. Therefore, attention should be focused on the *fundamental questions*: What have these gaps in common? Why are some youngsters not catching up with the Information/Knowledge society? Why are there so many adults that don't receive digital adult education? Why is e-governance not as widely accepted, as it should be?

Too often, the answers to these questions are based for a large part on the belief in technology, which was the starting point of this book. The myth that, once technology is introduced everything will be alright, remains deeply entrenched in our worldview. Almost everybody, including those socially excluded, seems to believe that technology is the solution. Almost everyone accepts it and only few "luddites" look for alternatives.

In spite of the recurrent claims of ever more user-friendliness, ICTs' use remains strongly subordinated to a set of specific skills. While the Internet will naturally be a dominant medium for knowledge workers (soon to be half of the working population), it will require a public effort to "digitize" larger sectors of the population. These evolve along with innovation, but tend to grow in importance as the complexity of the technologies as well as the scope of their applications extends (MacKenzie and Wajcman, 1985; Servaes and Heinderyckx, 2002; Storsul and Krumsvik, 2013). These sets of skills go well beyond managing the interfaces needed to operate them. Broadly speaking, new media are increasingly associated with new writing, hence of new reading, not to mention new ways to organize, treat, retrieve and control data, information and knowledge in its broadest sense.

This so-called new literacy will soon lead developed societies into difficulties comparable to that of illiteracy in the 19th century. Like the illiterate of those times, the new illiterate will be, as we can clearly see from the diffusion patterns of new technologies, of lower social status, with an associated lower income and level of education. Medium-term developments may lead to a dichotomized social body made of, on the one hand, wealthier, better educated and new literates having the skills and the means to access and use ICTs and, on the other hand, poorer, less educated and new illiterates kept out of the new tech scene and deprived of most technologies and hence denied access to an increasing amount (some may say: *overload*) of information and culture.

Therefore, the common denominator of all these problems is the *need for education, the need for digital literacy*. In this context, digital literacy should be understood as "the ability to use digital technology, communication tools or networks to locate, evaluate, use and create information. It also refers to the ability to understand and use information in multiple formats from a wide range of sources when presented via computers, or to a person's ability to perform tasks effectively in a digital environment. Digital literacy includes the ability to read and interpret media, reproduce data and images through digital manipulation, and evaluate and apply new knowledge gained from digital environments" (Grizzle and Torras Calvo, 2013, 187; Wilson et al., 2011, 183).

Access to ICTs is a necessity, and the importance of access should not be trivialized, but a plea for digital literacy implies recognizing that access is the beginning of ICTs policies and not the end of it.

The development of digital education should not be limited to those on the margins of society, although this should indeed be a first priority. The inclusion of everybody in the Information/Knowledge society through universal access is only the first step, the use of these ICTs in appropriate ways is more important and often more challenging. Digital literacy requires using the Internet in socially and culturally useful ways. Why not start with a reflection on what kinds of uses are appropriate and what kinds of uses have less socially desired consequences? Technology matters, sure. But, people matter more.

REFERENCES

Arts, F., Evers, V., Van Dam, N. & Buitenhuis, E. (2003). E-government voor de multiculturele samenleving. Amsterdam: Universiteit van Amsterdam.

Atkinson, Anthony B. (2009). The EU and social inclusion: facing the challenges. Bristol: Policy Press.

Bauwens, M. (2006). "Peer to Peer: From Technology to Politics." Servaes, J. & Carpentier. N. (Eds.). Towards a sustainable Information Society. Bristol: Intellect.

Bourdieu, P. & Passeron, J.C. (1970). La reproduction : élements pour une théorie du système d'enseignement. Paris: Les editions de Minuit.

Breit, R. & Servaes, J. (Eds.) (2005). Information Society or Knowledge Societies. Penang: Southbound.

Bucharest Declaration. (2002). Towards an Information Society: Principles, Strategy and Priorities for Action (November 9).

http://www.unesco.org/new/fileadmin/MULTIMEDIA/HQ/CI/CI/pdf/stategy_doc_bucharest_declaration.pdf

Coates, J.F. (1971). "Technology Assessment: The Benefits . . . the Costs . . . the Consequences", The Futurist, Vol. 5 (December), 225–231.

Coates, J.F. (1976). "Technology Assessment—A Tool Kit", Chemtech (June), 372–383.

Genus, A. (2006). "Rethinking constructive technology assessment as democratic, reflective, discourse", Technological Forecasting and Social Change 73(1), 13–26.

Joss, S., & Bellucci, S. (Eds.) (2002). Participatory Technology Assessment: European Perspectives, London: The Athenaeum Press.

Coleman, S. (2001). "The Transformation of Citizenship", Axford, B, and Huggins, R, (Eds.), New Media and Politics, London, Sage.

Commission of the European communities (2005). E-Inclusion revisited: The Local Dimension of the Information Society. Commission staff working document. Commission of the European communities: Brussels.

d'Haenens, L., Koeman, J., & Saeys, F. (2007). "Digital citizenship among ethnic minority youths in the Netherlands and Flanders." New Media and Society, 9(2), 278–299.

d'Haenens, L., Van Summeren, C., Saeys, F. & Koeman, J. (2004). Integratie of identiteit? Mediamenu's van Turkse en Marokkaanse jongeren. Amsterdam: Boom.

Fielden, K. & Malcolm, P. (2009, July). Local E-government in New Zealand: digital strategy, social inclusion and liveability. Paper presented at 53rd Meeting of The International Society for the Systems Sciences, University of Queensland, Brisbane.

Greenwald, Glenn. (2014). No place to hide. Edward Snowden, the NSA and the Surveillance State. London: Hamish Hamilton.

Grizzle, A. & Torras Calvo, M.C. (Eds.) (2013). Media and Information Literacy. Policy and Strategy Guidelines. Paris: UNESCO.

Hacker F. & Van Dijk J. (2001). Digital Democracy: Issues of Theory and Practice. London: Sage.

Hamelink, C. (1999). Digitaal fatsoen. Mensenrechten in cyberspace. Amsterdam: Boom.

Hargittai, E. & Hinant, A. (2008). "Digital inequality. Differences in young adults' use of the internet", Communication Research, 35(5), 602–621.

Heinderyckx, F. (2003). "Issues in measuring Information Society adoption in Europe." Servaes, J. (Ed.), The European Information Society. A Reality Check. Bristol: Intellect.

Held, D. (1987). Models of Democracy. Stanford: Stanford University Press.

Hepp, A., Krotz, F., Moores, S. & Winter C. (Eds.) (2008). Connectivity, Networks and Flows. Conceptualizing Contemporary Communications. Cresskill NJ: Hampton Press.

IAIA. (2010). Technology Assessment. International Association for Impact Assessment. http://www.iaia.org/iaiawiki/techassess.ashx (accessed, 24 May 2014)

Inagaki, N. (2007). Communicating the impact of communication for development. Recent trends in empirical research. Washington DC: The World Bank, Working Paper Series, Development Communication Division.

Jenkins, H. (2004). "The myths of growing up online. Alarmist and polarized rhetoric is distorting important new findings about the risks and benefits of children's use of the Internet." Technology Review. http://www.technologyreview.com/Biotech/13773/ (Retrieved March, 22, 2009).

Kelly, R. (2010). Social Inclusion Policy. Brussels: European Commission.

Korupp, S.E. & Szydlikk, M. (2005). "Causes and trends of the digital divide", European Sociological Review, 21(4), 409–422.

Lennie, June & Tacchi, Jo. (2013). Evaluating Communication for Development. A framework for social change. New York: Routledge.

Livingstone, S. & Helsper, E. (2007). "Gradations in digital inclusion: children, young people and the digital divide", New Media and Society, 9(4), 671–696.

MacKenzie, D. & Wajcman, J. (1985). The Social Shaping of Technology. Milton Keynes: Open University Press.

Mansell, R. & When U. (1998) (Ed.), Knowledge Societies. Information Technology for Sustainable Development. Oxford: Oxford University Press.

Mattelart, A. (2001). Histoire de la sociétè de l'information. Paris: La Decouverte.

Mertens S. & Servaes J. (2011), "ICT policies on structural and socio-cultural participation in Brussels", Adomi E. (Ed.), Handbook of Research on Information Communication Technology: Trends, Issues and Advancements. Vol. 1, IGI Global, Hershey-New York, (pp. 319–336).

Mertens, S. (Ed.) (2009). Burgerschap en democratie in een veranderende wereld. Gent: Academia Press.

Mossberger, K. (2010). "Toward digital citizenship. Addressing inequality in the information age." Chadwick, Andrew & Philip N. Howard (Eds.). Routledge handbook of Internet politics, London: Routledge.

Mossberger, K., Tolbert, C.J. & McNeal, R.S. (2008). Digital Citizenship.The Internet, society and participation. Cambridge/London : MIT Press.

Norris, Pippa. (2002). Digital Divide? Civic Engagement, Information Poverty and the Internet Worldwide. Cambridge: Cambridge University Press.

Norris, Pippa. (Ed.) (2010). Public sentinel. News media & governance reform. Washington DC: World Bank.

Ommundsen, Wenche, Michael Leach & Andrew Vandenberg. (Eds.) (2007). Cultural Citizenship and the Challenges of Globalisation, Cresskill NJ: Hampton.

Putnam, R. (2000). Bowling Alone: The Collapse and Revival of American Community. New York: Simon and Schuster.

Ricci, A. (2013). The Early Political Web (1995–2005): A ten-year observational research seeking evidence of 'eDemocracy' in the information architecture of political parties web sites worldwide. Brussels: PhD Universite Libre de Bruxelles (ULB).

Saretzki, Thomas. (2012). "Legitimation problems of participatory processes in technology assessment and technology policy." Poiesis Praxis, Nov. 2012, 9(1–2): 7–26.

Sclove, Richard. (1995). Democracy and technology. New York: The Guilford Press.

Sclove, Richard. (2010). Reinventing Technology Assessment: A 21st Century Model. Using Citizen Participation, Collaboration and Expertanalysis to Inform and Improve Decision-Making On Issues Involving Science and Technology. Washington, DC: Science and Technology Innovation Program, Woodrow Wilson International Center for Scholars (April 2010).

Servaes J. & Heinderyckx F. (2002). "The 'new' ICTs environment in Europe: Closing or widening the gaps?", Telematics and Informatics, 19 (2): 91–116.

Servaes, J. (2003) (Ed.). The European Information Society: A reality Check, Bristol: Intellect.

Servaes, J. (2003). "Digital citizenship and information inequalities: challenges for the future." Servaes, J. (Ed.), The European Information Society: A reality check (pp. 231–238). Bristol : ECCR Book Series Intellect.

Servaes, J. (2013a) (Ed.). Sustainability, Participation and Culture in Communication. Theory and Praxis. Bristol-Chicago: Intellect-University of Chicago Press.

Servaes, J. (2013b) (Ed.). Sustainable Development and Green Communication. African and Asian Perspectives. London-New York: Palgrave MacMillan.

Shirazi, Farid, Ojelanki Ngwenyama & Olga Morawczynski. (2010). "ICT expansion and the digital divide in democratic freedoms: An analysis of the impact of ICT expansion, education and ICT filtering on democracy," Telematics and Informatics, 27: 21–31.

Smith, M.R. & Marx, L. (1994). Does Technology Drive History? The Dilemma of Technological Determinism. Cambridge MA: The MIT Press.

Storsul, T. & A. Krumsvik, A. (Eds.) (2013). Media Innovations. A Multidisciplinary Study of Change. Goteborg: Nordicom.

Talbot, H. (2004). Social inclusion trough ICT: A local authority study. University of Newcastle upon Tyne, Centre for Rural Economy, Working Paper 72.

Tang, P.S. & Ang, P.H. (2002). "The Diffusion of Information Technology in Singapore Schools: a Process Framework." New Media & Society, 4, 457–478.

United Nations. (2005). UN Global E-government Readiness Report 2005. From E-government to E-inclusion. New York: United Nations.

Van Dijk, J. (2005). The deepening divide. Inequality in the Information Society. London/Thousand Oaks/New Delhi: Sage publications

Van Dijk, J. (2012). The Network Society: Social Aspects of New Media. London: Sage.

Van Dijk, J. (2013). "Digital Democracy: Vision and Reality." I. Snellen & W. van de Donk (Eds.), Public Administration in the Information Age: Revisited', Amsterdam: IOS- Press

Webster F. (1995). Theories of the Information Society. London: Routledge

Wilson, C., Grizzle, A., Tuazon, R., Akyempong, K. & Cheung, C-K. (2011). Media and Information Literacy. Curriculum for Teachers. Paris: UNESCO

Index

About the Contributors

EDITOR

Jan Servaes (PhD, 1987, Catholic University of Louvain, Belgium) is chair professor and head at the Department of Media and Communication, City University of Hong Kong, UNESCO Chair in Communication for Sustainable Social Change at the University of Massachusetts, Amherst; honorary guest professor at the Huazhong University of Science and Technology (HUST), Wuhan, China; researcher at the Brussels Center for Journalism Studies (BCJS), Belgium; editor-in-chief of the Elsevier journal *Telematics and Informatics: An Interdisciplinary Journal on the Social Impacts of New Technologies*; editor of the Southbound Book Series Communication for Development and Social Change; and editor of the Lexington Book Series Communication, Globalization and Cultural Identity.

He has been president of the European Consortium for Communications Research (ECCR, www.ecrea.eu) and vice-president of the International Association of Media and Communication Research (IAMCR, www.iamcr.org), in charge of Academic Publications and Research, from 2000 to 2004. He chaired the Scientific Committee for the World Congress on Communication for Development (Rome, October 25–27, 2006), organized by the World Bank, FAO and the Communication Initiative.

Servaes has taught international communication and development communication in Australia (Brisbane), Belgium (Brussels and Antwerp), the United States (Cornell), the Netherlands (Nijmegen), and Thailand (Thammasat, Bangkok) in addition to several teaching stints at about 100 universities in 43 countries.

He has undertaken research, development, and advisory work around the world and is known as the author of journal articles and books on such topics as international and development communication; ICT & media policies; intercultural communication and language; participation and social change; and human rights and conflict management. Some of his most recent book titles include: (2013) Servaes, Jan (Ed.) *Sustainable Development and Green Communication. African and Asian Perspectives*, London/New York: Palgrave MacMillan; (2012) Servaes, Jan (Ed.) *Sustainability, Participation and Culture in Communication. Theory and Praxis*. Bristol-Chicago: Intellect-University of Chicago Press; (2008) Servaes, J., *Communication for Development and Social Change*, Los Angeles, London, New Delhi, Singapore: Sage; (2007) Servaes, J. & Liu, S. (Eds.), *Moving Targets. Mapping the Paths Between Communication, Technology and Social Change in Communities*, Penang: Southbound; (2006) Thomas, P. & Servaes, J. (Eds.), *Intellectual Property Rights and Communications in Asia*, New Delhi: Sage; (2006) Servaes, J. & Carpentier, N. (Eds.), *Towards a Sustainable European Information Society*, ECCR Book Series, Bristol: Intellect; (2005) Shi-Xu, Kienpointner, M. & Servaes, J. (Eds.), *Read the Cultural Other. Forms of Otherness in the Discourses of Hong Kong's Decolonisation*, Berlin: Mouton De Gruyter; (2003) Servaes, J. (Ed.), *The European Information Society: A Reality Check*, ECCR Book Series, Bristol: Intellect; (2003) Servaes, J. (Ed.), *Approaches to Development. Studies on Communication for Development*, Paris: UNESCO Publishing House; and (2002) Servaes, J., *Communication for Development. One World, Multiple Cultures*, Cresskill: Hampton Press (Foreword by Jan Pronk). Email: 9cssc9@gmail.com

AUTHORS

Valentina Baú is completing her PhD at Macquarie University (Sydney, Australia). Both as a practitioner and as a researcher, her work has been focused on the use of communication in development. In the past, Valentina has collaborated with different international NGOs, the United Nations, and the Italian Development Cooperation, while living in various African countries. At present, her research looks at the use of communication for development in peacebuilding, particularly through the use of participatory media. Valentina's work has been published in *Global Media Journal* (African edition), *Peace Studies Journal, Conflict & Communication Online* and *Community Development Journal*. Email: valentina.bau@students.mq.edu.au

Melissa Brough recently completed her PhD at the University of Southern California. She works in the fields of participatory and civic media, digital media and learning, and communication for social change. She was a 2010

research associate with the MacArthur Foundation's Digital Media and Learning Hub, and has collaborated with a number of youth and community media projects, including Mobile Voices and the Chiapas Media Project. Prior to pursuing her PhD, she worked in development communication as a program officer for FilmAid International, and in documentary filmmaking. She was a 2010–2011 Fulbright grantee to Colombia. Email: mbrough@gmail.com

John Hartley, AM, is professor of cultural science and director of the Centre for Culture & Technology, Curtin University, Western Australia; and professor of journalism, media and cultural studies at Cardiff University, Wales. He was foundation dean of the Creative Industries, faculty, co-founder and research director of the ARC Centre of Excellence for Creative Industries and Innovation, and ARC Federation Fellow at Queensland University of Technology. His research and publications cover TV and media studies, cultural and communication studies, journalism, and the creative economy. His current research interest is in cultural science: using evolutionary and complexity theory to reconceptualize culture and creativity. Currently he is working on an Australian Research Council project designed to create an Indigenous-language wiki encyclopedia in Noongar, the Aboriginal language of South-Western Australia. Hartley's most recent books include: *Cultural Science* (2014, Bloomsbury, with Jason Potts); *Key Concepts in Creative Industries* (2013, Sage, with coauthors); *Digital Futures for Cultural and Media Studies* (2012, Wiley-Blackwell), and *A Companion to New Media Dynamics* (2013, Wiley-Blackwell, edited with Jean Burgess and Axel Bruns). Hartley is a fellow of the Australian Academy of the Humanities, life fellow of the Royal Society of Arts (UK), and fellow of the International Communication Association (USA). He was awarded the Order of Australia (AM) in 2009. Email: john.a.hartley@curtin.edu.au

Ellen Hommel is senior lecturer social and cultural education at Amsterdam University of Professional Education. She has a background in youth work, with a Masters in social pedagogy. After working several years with street youth and directing as well as producing documentaries, she was assigned educator of social and cultural professionals. She consecutively held positions as head of the Department of Social and Cultural Education, and as lead of the research program of Cultural and Social Dynamics. Currently, she is teaching participatory video and writing on issues of media literacy, visual research and visual interventions. Email: e.hommel@hva.nl

Yalong Jiang is a PhD candidate at the Department of Media and Communication, City University of Hong Kong. His research interests are focused on communicative behaviors on the Internet and their social impact, political participation and community involvement, web mining, and social influence on media content.

Rico Lie (PhD 2000, Catholic University of Brussels) is a social anthropologist working at the research group Knowledge, Technology and Innovation, Wageningen University, the Netherlands. He previously worked at the University of Brussels in Belgium and the Universities of Nijmegen and Leiden in the Netherlands. At Wageningen University he is an assistant professor in international communication with an interest in the areas of development communication and intercultural communication. Email: rico.lie@wur.nl

Rich Ling (PhD, University of Colorado, Sociology) is the Shaw Foundation professor of media technology, Wee Kim Wee School of Communication and Information, Nanyang Technological University, Singapore. He works at Telenor Research and has an adjunct position at the University of Michigan. Ling has studied the social consequences of mobile communication for the past two decades. He has written *The Mobile Connection* (Morgan Kaufmann, 2004), *New Tech, New Ties* (MIT, 2008) and most recently *Taken for Grantedness* (MIT, 2012). He is a founding coeditor of *Mobile Media and Communication* (Sage) and the Oxford University Press series *Studies in Mobile Communication.* Email: riseling@gmail.com

Patchanee Malikhao received her PhD in sociology from the University of Queensland in Australia, a Master of Arts degree in mass communication from Thammasat University in Thailand, a Master of Science degree in printing technology from Rochester Institute of Technology (RIT), Rochester, New York, and a Bachelor of Science (Hons.) in photographic science and printing technology from Chulalongkorn University in Thailand. She has worked and received extensive trainings in the fields of communication for social change, imaging technology, social science research methods, and data analyses in Belgium, Australia, and the Unites States. She was a recipient of many scholarships and awards, including the Fulbright Scholarship, the Australian Postgraduate Award, and the Outstanding Teacher Award. She worked from 2008 to July 2011 as a researcher and a lecturer in the School of Public Health at the University of Massachusetts, Amherst, in the Unites States. Since then she joined Thammasat University in Bangkok, Thailand. Email: pmalikhao@gmail.com

David Morley is professor of communications at Goldsmiths College, London University, where he directs the PhD program in the Department of Media and Communications. He has held several international visiting positions including the University of Stockholm, University of Madison–Wisconsin, Annenberg Centre for Communications USC, Northwestern University Chicago, University of Western Sydney, and University of Paris II/Maison des Sciences de L'Homme.

His current work is concerned with the material dimensions of communications and with the articulation of the fields of media studies and transport geography.

His publications include *Television, Audiences and Cultural Studies* (Routledge, 1992), *Home Territories: Media, Mobility and Identity* (Routledge, 2001), and *Media, Modernity and Technology: The Geography of the New* (Routledge, 2006). Email: D.Morley@gold.ac.uk

Christine Ogan is professor emerita from Indiana University's School of Journalism and School of Informatics and Computing. She is currently an adjunct lecturer at Warrington College of Business, University of Florida, where she teaches a course in intercultural communication. She has taught a doctoral course at Indiana University in communication and development for more than twenty years, and has published many articles that relate to development communication. Now she focuses on the Turkish diaspora and issues related to multiculturalism and development, especially in uses of social media in several European countries. Her most recent publications include: "The Rise of Anti-Muslim Prejudice: Islamophobia in Europe and the United States," *International Communication Gazette,* 2014, which she authored with colleagues at Indiana University; "Internet-using Children and Digital Inequality: A Comparison between Majority and Minority Europeans," *Communications: The European Journal of Communication Research*, 2013, written with Leen d'Haenens; and "Round Pegs in Square Holes? Is Mass Communication Theory a Useful Tool in Conducting Internet Research?" (2014). *Blackwell International Handbook of Media and Communication Theory,* John Wiley. Email: ogan@indiana.edu

Emily Polk is a lecturer in the Program in Writing and Rhetoric at Stanford University. She has a PhD in communication from the University of Massachusetts, Amherst and a Masters in human rights from Columbia University. Emily's research is focused on the communication of sustainability and resiliency and the mobilization of global social movements in the digital and public spheres. Her work has been published in the *Global Media Journal* and *Development in Practice*, among others. Prior to getting her doctorate, Emily worked for nearly ten years around the world as a media professional, helping to produce radio documentaries in Burmese refugee camps, and facilitating a human rights-based newspaper in a Liberian refugee camp. She has also worked as a senior editor at CSRwire, a leading global source of corporate social responsibility news and information. Her writing and radio documentaries have appeared in *National Geographic Traveler, the Boston Globe, NPR, The National Radio Project, AlterNet, the Indian Express, Central America Weekly, the Ghanaian Chronicle*, and *Whole Earth Magazine*, among others. Email: empolk@stanford.edu

Yong Jin Park is associate professor at the School of Communications, Howard University. His research interests lie in social and policy concerns of communication technologies as his recent works concern privacy, digital skill disparities, and mobile access and use. He received his PhD from the University of Michigan and MA from the University of Southern California.

Song Shi is a postdoctoral fellow at McGill University. He received his PhD degree in the Department of Communication at the University of Massachusetts, Amherst. His research interests include three interrelated fields: the influences of new media on global social change and development; digital divide, new media, and global ICT development; and communication and new media policy. His recent research focuses on the use of the Internet and the development of the Chinese civil society organizations, Internet policy in China and in the United States, global ICT development, and new media for social change and development in social context. In his research, he employs both qualitative methods and quantitative methods in communication studies. Email: song.shi2@mcgill.ca.

Marko M. Skoric is associate professor at the Department of Media and Communication, City University of the Hong Kong. His teaching and research interests are focused on new media and social change, with particular emphasis on civic and political implications of new communication technologies. He received his PhD from the University of Michigan and his BSc from the University College London, UK. Email: mskoric@cityu.edu.hk

Anneke Smelik is professor of visual culture on the Katrien van Munster chair at the Radboud University Nijmegen (Netherlands). She (co-)edited *Performing Memory in Art and Popular Culture*; *The Scientific Imaginary in Visual Culture*; *Technologies of Memory in the Arts*; and *Bits of Life: Feminism at the Intersections of Media, Bioscience, and Technology*. She is author of *And the Mirror Cracked: Feminist Cinema and Film Theory* and several books in Dutch on issues of visual culture, most recently a book on cyborgs in popular culture (*Ik cyborg*, 2012). She is project leader of two research programs on fashion: "Dutch Fashion in a Globalised World" (book forthcoming in 2015) and "Crafting Wearables; Fashionable Technology" (2013–2018). Email: a.smelik@let.ru.nl

Colin Sparks is professor of media studies in the School of Communication at Hong Kong Baptist University. Prior to taking up this appointment in 2011 he was director of the Communication and Media Research Institute at the University of Westminster. He has written extensively on different aspects of the mass media, including upon development and communication, notably in his book *Development, Globalization and the Mass Media* (Sage Publications, 2007). He was one of the founders of the journal *Media, Culture and Society* and remains an active member of its working Editorial Board. Email: sparksc@hkbu.edu.hk

Jo Tacchi is professor in the School of Media and Communication and director of research in the College of Design and Social Context at RMIT University in Melbourne, Australia. She is a media anthropologist who has, for the past decade, focused on communication and development, largely in South Asia. Her most recent book is *Evaluating Communication for Development: A Framework for Social Change* (Routledge, 2013). Email: jo.tacchi@rmit.edu.au

Karin Gwinn Wilkins (PhD, University of Pennsylvania) is professor of media studies, director, Center for Middle Eastern Studies, and chair, Global Studies Bridging Disciplines Program at the University of Texas at Austin. Wilkins has won numerous awards for her research, service and teaching, and chaired the Intercultural/Development Division of the International Communication Association. Her work addresses scholarship in the fields of development communication, global communication, and political engagement. Selected works include *Handbook of Development Communication and Social Change* (ed., Wiley-Blackwell, 2014), *New Agendas in Global Communication* (ed., Routledge, 2013); *Questioning Numbers: How to Read & Critique Research* (Oxford University Press, 2011); *Home/Land/Security: What we learn about Arab Communities from Action Adventure Film* (Lexington Books, 2008), *Re-Developing Communication for Social Change* (ed., Rowman & Littlefield, 2000) as well as numerous journal publications, venues such as in *Nordicom; Communication Theory*; *Media, Culture & Society*; *Communication, Culture & Critique; Journal of Communication*; *Journal of Middle East Media; International Journal of Communication; Gazette: The International Journal for Communication Studies; Development & Change*; *Perspectives on Global Development and Technology*; *Critical Studies in Media Communication;* and *Global Media Journal.* Email: karin.wilkins@austin.utexas.edu

www.ingramcontent.com/pod-product-compliance
Lightning Source LLC
LaVergne TN
LVHW101444200926
848483LV00001B/48